Principles of Nuclear Rocket Propulsion

Principles of Nuclear Rocket Propulsion

William Emrich, Jr.
Senior Engineer,
NASA/Marshall Space Flight Center,
Huntsville, AL, USA

AMSTERDAM • BOSTON • HEIDELBERG • LONDON
NEW YORK • OXFORD • PARIS • SAN DIEGO
SAN FRANCISCO • SINGAPORE • SYDNEY • TOKYO

Butterworth-Heinemann is an imprint of Elsevier

Butterworth-Heinemann is an imprint of Elsevier
The Boulevard, Langford Lane, Kidlington, Oxford OX5 1GB, United Kingdom
50 Hampshire Street, 5th Floor, Cambridge, MA 02139, United States

Copyright © 2016 Elsevier Inc. All rights reserved.

No part of this publication may be reproduced or transmitted in any form or by any means, electronic or mechanical, including photocopying, recording, or any information storage and retrieval system, without permission in writing from the publisher. Details on how to seek permission, further information about the Publisher's permissions policies and our arrangements with organizations such as the Copyright Clearance Center and the Copyright Licensing Agency, can be found at our website: www.elsevier.com/permissions.

This book and the individual contributions contained in it are protected under copyright by the Publisher (other than as may be noted herein).

Notices
Knowledge and best practice in this field are constantly changing. As new research and experience broaden our understanding, changes in research methods, professional practices, or medical treatment may become necessary.

Practitioners and researchers must always rely on their own experience and knowledge in evaluating and using any information, methods, compounds, or experiments described herein. In using such information or methods they should be mindful of their own safety and the safety of others, including parties for whom they have a professional responsibility.

To the fullest extent of the law, neither the Publisher nor the authors, contributors, or editors, assume any liability for any injury and/or damage to persons or property as a matter of products liability, negligence or otherwise, or from any use or operation of any methods, products, instructions, or ideas contained in the material herein.

Library of Congress Cataloging-in-Publication Data
A catalog record for this book is available from the Library of Congress

British Library Cataloguing-in-Publication Data
A catalogue record for this book is available from the British Library

ISBN: 978-0-12-804474-2

For information on all Butterworth-Heinemann publications
visit our website at https://www.elsevier.com/

Working together
to grow libraries in
developing countries

www.elsevier.com • www.bookaid.org

Cover image: NASA Artwork by Pat Rawlings

Publisher: Joe Hayton
Acquisition Editor: Carrie Bolger
Editorial Project Manager: Carrie Bolger
Production Project Manager: Nicky Carter
Designer: Victoria Pearson Esser

Typeset by TNQ Books and Journals

Contents

Preface ix

CHAPTER 1 Introduction 1
    1. Overview 1
    2. Historical Perspective 2
        2.1 Background 2
        2.2 NERVA 3
        2.3 Particle Bed Reactor 5
        2.4 Russian Nuclear Rockets 8
    References 10

CHAPTER 2 Rocket Engine Fundamentals 11
    1. Concepts and Definitions 11
    2. Nozzle Thermodynamics 15
    Reference 20

CHAPTER 3 Nuclear Rocket Engine Cycles 21
    1. Nuclear Thermal Rocket Thermodynamic Cycles 21
        1.1 Hot Bleed Cycle 21
        1.2 Cold Bleed Cycle 23
        1.3 Expander Cycle 24
    2. Nuclear Electric Thermodynamic Cycles 27
        2.1 Brayton Cycle 27
        2.2 Stirling Cycle 28
    Reference 30

CHAPTER 4 Interplanetary Mission Analysis 31
    1. Summary 31
    2. Basic Mission Analysis Equations 32
    3. Patched Conic Equations 37
    4. Flight Time Equations 43
    Reference 53

CHAPTER 5 Basic Nuclear Structure and Processes 55
    1. Nuclear Structure 55
    2. Nuclear Fission 59
    3. Nuclear Cross Sections 63
        3.1 $1/V$ Region 64

    3.2 Resonance Region 64
    3.3 Unresolved Resonance Region or Fast Region 64
  4. Nuclear Flux and Reaction Rates 67
  5. Doppler Broadening of Cross Sections 69
  6. Interaction of Neutron Beams with Matter 72
  7. Nuclear Fusion 74
  References 80

CHAPTER 6 Neutron Flux Energy Distribution 81
  1. Classical Derivation of Neutron-Scattering Interactions 81
  2. Energy Distribution of Neutrons in the Slowing-Down Range 84
  3. Energy Distribution of Neutrons in the Fission Source Range 86
  4. Energy Distribution of Neutrons in the Thermal Energy Range 86
  5. Summary of the Neutron Energy Distribution Spectrum 88

CHAPTER 7 Neutron Balance Equation and Transport Theory 89
  1. Neutron Balance Equation 89
    1.1 Leakage (L) 89
    1.2 Fission Production Rate (P_f) 90
    1.3 Scattering Production Rate (P_s) 90
    1.4 Absorption Loss Rate (R_a) 91
    1.5 Scattering Loss Rate (R_s) 91
    1.6 Steady-State Neutron Balance Equation 91
  2. Transport Theory 92
  3. Diffusion Theory Approximation 94
  References 96

CHAPTER 8 Multigroup Neutron Diffusion Equations 97
  1. Multigroup Diffusion Theory 97
  2. One Group, One Region Neutron Diffusion Equation 100
  3. One Group, Two Region Neutron Diffusion Equation 106
    3.1 Core 106
    3.2 Reflector 107
    3.3 Core + Reflector 107
  4. Two Group, Two Region Neutron Diffusion Equation 110

CHAPTER 9 Thermal Fluid Aspects of Nuclear Rockets 117
  1. Heat Conduction in Nuclear Reactor Fuel Elements 117
  2. Convection Processes in Nuclear Reactor Fuel Elements 122
  3. Nuclear Reactor Temperature and Pressure Distributions in
      Axial Flow Geometry 128
  4. Nuclear Reactor Fuel Element Temperature Distributions in
      Radial Flow Geometry 138

    5. Radiators 143
    References 148

CHAPTER 10   Turbomachinery 149
    1. Turbopump Overview 149
    2. Pump Characteristics 151
    3. Turbine Characteristics 157
    References 163

CHAPTER 11   Nuclear Reactor Kinetics 165
    1. Derivation of the Point Kinetics Equations 165
    2. Solution of the Point Kinetics Equations 169
    3. Decay Heat Removal Considerations 173
    4. Nuclear Reactor Transient Thermal Response 177
    References 181

CHAPTER 12   Nuclear Rocket Stability 183
    1. Derivation of the Point Kinetics Equations 183
    2. Reactor Stability Model Including Thermal Feedback 185
    3. Thermal Fluid Instabilities 195
    References 201

CHAPTER 13   Fuel Burnup and Transmutation 203
    1. Fission Product Buildup and Transmutation 203
    2. Xenon 135 Poisoning 207
    3. Samarium 149 Poisoning 211
    4. Fuel Burnup Effects on Reactor Operation 213

CHAPTER 14   Radiation Shielding for Nuclear Rockets 217
    1. Derivation of Shielding Formulas 217
        1.1 Neutron Attenuation 219
        1.2 Prompt Fission Gamma Attenuation 220
        1.3 Capture Gamma Attenuation 225
        1.4 Radiation Attenuation in a Multilayer Shield 227
    2. Radiation Protection and Health Physics 228
    References 233

CHAPTER 15   Materials for Nuclear Thermal Rockets 235
    1. Fuels 235
    2. Moderators 239
    3. Control Materials 241
    4. Structural Materials 243
    References 246

CHAPTER 16 Nuclear Rocket Engine Testing 249
1. General Considerations 249
2. Fuel Assembly Testing 251
3. Engine Ground Testing 254
   References 256

CHAPTER 17 Advanced Nuclear Rocket Concepts 259
1. Pulsed Nuclear Rocket (Orion) 259
2. Open Cycle Gas Core Rocket 273
   2.1 Neutronics 273
   2.2 Core Temperature Distribution 277
   2.3 Wall Temperature Calculation 280
   2.4 Uranium Loss Rare Calculations 284
3. Nuclear Light Bulb 289
   3.1 Neutronics 292
   3.2 Fuel Cavity Temperature Distribution 294
   3.3 Heat Absorption in the Neon Buffer Layer 295
   3.4 Heat Absorption in the Containment Vessel 298
   3.5 Heat Absorption in the Hydrogen Propellant 300
   References 303

Problems 305
Appendix 323
Index 327

Preface

This book is based on a one semester course in nuclear rocket propulsion, which I have taught over a number of years at the University of Alabama Huntsville. The content presented here, however, is sufficient to expand the course to a full academic year if desired. The aim of this book is to provide the reader with an understanding of the physical principles underlying the design and operation of nuclear rocket engines. The need for this book was felt because while there are numerous texts available describing rocket engine theory and numerous texts available describing nuclear reactor theory, there are no (recent) books available describing the integration of the two subject areas. While the emphasis in this book is primarily on nuclear thermal rocket engines wherein the energy of a nuclear reactor is used to heat a propellant to high temperatures, which is then expelled through a nozzle to produce thrust, other concepts are also touched upon. For example, there is a section in the book devoted to the nuclear pulse rocket concept wherein the force of externally detonated nuclear explosions is used to accelerate a spacecraft.

The prerequisites for this course are a knowledge of mathematics through advanced calculus and undergraduate courses in thermodynamics, heat transfer, and fluid mechanics. A knowledge of nuclear reactor physics is also helpful, but not required. Nuclear reactor physics is covered in sufficient detail in the book to provide a basic understanding of the mechanisms by which nuclear reactors operate and how these mechanisms might affect the operation of a nuclear rocket engine. The phenomena associated with describing the neutron distribution (and hence power distribution) within a nuclear reactor are presented almost exclusively through the framework of neutron diffusion theory. Neutron transport theory is touched on briefly, but only to provide a rationale for the use of the simpler diffusion theory approximation. Many of the derivations in the electronic version of this book are illustrated through the use of interactive calculators which demonstrate how variations in the constituent parameters affect the physical process being described. It is hoped that this visual presentation of the behavior of the various physical processes occurring within a nuclear rocket engine will provide the reader with a clearer understanding as to which parameters in the derivations are important and which are not. In addition, many of the three-dimensional figures in the book may be scaled, rotated, and so on to better visualize the nature of the object under study.

As with any textbook of finite size, decisions had to be made as to which topics would be covered and which topics could be safely ignored. There are so many diverse engineering fields which would be involved in the development of a nuclear rocket system that, no doubt, many interesting and important topics have been left out which could easily have been included in this book. Nevertheless, it is hoped that a sufficient number of topics are covered in the book so that the reader will have at least a modest appreciation for the diversity of engineering involvement which would be required to design a viable nuclear rocket engine.

I have been aided by many people in the preparation of this book. Over the years, many of my students have provided very helpful suggestions and comments as to how the book could be made clearer or more pertinent to their interests. In addition, a number of my colleagues at the NASA Marshall Space Flight Center have critiqued this work, especially with regard to adding additional material.

Much gratitude also goes to my parents who from an early age instilled in me a love for learning and a desire to do my best. I would also like to acknowledge my dear children, Ethan, Joshua, and Rebekah, each of whom supported me throughout all my endeavors with their unfailing love, understanding, and pride. Finally, I would like to thank my wife, Lady, for putting up with the seemingly endless nights of typing and revising this work and for encouraging me to see it through to fruition.

Solutions, videos, and interactive figures and tables available on companion website http://booksite.elsevier.com/9780128044742.

CHAPTER 1

INTRODUCTION

The application of nuclear energy to space propulsion systems has long been seen as a means to enable missions to outer space which are not achievable by any currently conceivable chemical-based propulsion system. During the latter half of the 20th century, nuclear rocket propulsion programs were initiated by the United States and Russia. To a large extent, these efforts were successful; however, for a variety of reasons these nuclear rocket programs were never carried forward to completion in spite of the fact that the test programs for these engines demonstrated efficiencies twice those of the best chemical rocket engines.

1. OVERVIEW

Future crewed space missions beyond low earth orbit will almost certainly require propulsion systems with performance levels exceeding that of today's best chemical engines. A likely candidate for that propulsion system is the solid-core *nuclear thermal rocket or NTR*. Solid-core NTR engines are expected to have performance levels which significantly exceed those achievable by any currently conceivable chemical engine. Nuclear engines are, generally speaking, quite simple conceptually in that all they do is use a nuclear reactor to heat a gas (generally hydrogen) to high temperatures before expelling it through a nozzle to produce thrust. The devil, as they say, is in the engineering details of the design that includes not only the thermal, fluid, and mechanical aspects always present in chemical rocket engine development, but also nuclear interactions and some unique materials restrictions. The purpose of this book is to provide an introduction to some of these engineering challenges which must be addressed during the design of a nuclear rocket engine.

Before beginning, a small bit of terminology description is in order. In chemical engines, fuel ignites to form a gas which is subsequently discharged through a nozzle to produce thrust. The fuel in this case is also the propellant; that is, the substance which is used to generate the thrust. When speaking about chemical engines, therefore, the terms *fuel* and *propellant* are often used interchangeably since the fuel and the propellant are one and the same. In nuclear engines, however, the propellant is simply the working fluid being heated by the nuclear reactor to produce the thrust. The fuel in this case is actually the fissioning uranium in the nuclear reactor. When speaking about nuclear engines; therefore, the term *fuel* is used to describe the fissioning uranium in the nuclear reactor, and the term *propellant* is used to designate the working fluid being expelled through the nozzle.

Solid-core NTR engines are expected to have at least two to three times the efficiency of the best chemical liquid oxygen/hydrogen engines. As is seen later, the efficiency of rocket engines depends upon, among other things, the temperature of the engine propellant exhaust gases (eg, the higher the temperature, the higher the rocket efficiency). In chemical engines, the temperature of the exhaust gases is limited by the amount of energy which may be extracted from the fuel as it reacts. Thus, *chemical engines are said to be energy limited* in their efficiency.

An NTR engine, as mentioned earlier, operates by using nuclear fission processes to heat the propellant to high temperatures. Because the energy released from the fissioning of nuclear fuel is extremely high as compared to that available from chemical combustion processes, the propellant in an NTR can potentially be heated to temperatures far in excess of that possible in chemical engines. The main limitation of these engines results from restrictions on the rate at which this energy can be extracted from the nuclear fuel and transferred to the propellant. This rate of energy transfer is limited by the maximum temperature the nuclear fuel can withstand, and it is this limitation which puts an upper limit on the maximum efficiency attainable by these engines. As such, *NTR engines are said to be power density limited* in their efficiency.

2. HISTORICAL PERSPECTIVE
2.1 BACKGROUND

There have been several programs in the past which have sought to develop solid-core nuclear rocket engines. In the late 1950s, an NTR program called Nuclear Engine for Rocket Vehicle Applications or NERVA [1] was instituted, which resulted in the construction of a number of prototypical nuclear engines. The nuclear reactors in these rocket engines used prismatic fuel elements through which holes were drilled axially to accommodate the flow of the hydrogen propellant.

Since the last of the NERVA tests of the 1970s, NTR development work has continued off and on at modest levels to the present day. In particular, the former Soviet Union (Lutch) and various national laboratories in the United States have worked on new fuel forms which have the potential of performing considerably better than the earlier fuel designs. These fuels generally fall into two groups comprised of uranium carbides and cermets (ceramic metals). In the carbide fuels, the uranium is either in a composite form where it is heterogeneously distributed in a graphite matrix or in more advanced fuel designs, it is formed into solid solutions consisting of a compound of uranium combined with zirconium, tantalum, etc. and carbon. In cermet fuels, uranium oxide (ceramic) is combined with a high melting point material such as tungsten (metal).

The US Air Force also briefly worked on an innovative NTR engine concept called a Particle Bed Reactor or PBR [2] in which the hydrogen in the nuclear fuel element flowed radially through a packed bed of fuel particles. This engine had a very high thrust-to-weight ratio and was to be used in a ballistic missile interceptor in a top secret program called Timberwind. This program, however, was cancelled in the early 1990s after the fall of the former Soviet Union.

The former Soviet Union also sought to develop a nuclear rocket engine [3] as a response to the work being done in the United States on the NERVA program. This nuclear rocket program, which lasted from 1965 through the 1980s, eventually developed the RD-410 nuclear rocket engine which was fairly small as compared to the NERVA engines. The fuel elements in this engine, however, were made of a uranium/tungsten carbide material which allowed them to operate at temperatures somewhat higher than those achievable in NERVA. As a result, the RD-410 was slightly more efficient than the NERVA engines.

2. HISTORICAL PERSPECTIVE

2.2 NERVA

The reactor development portion of the NTR program in the United States (called ROVER) began at the Los Alamos National Laboratory in 1953 with the intent being to design light, high-temperature reactors that could form the basis of a nuclear-powered rocket. This program was conceived as an alternative to the chemical rocket engines then under development which were being designed to lift payloads into orbit.

The reactor portion of the nuclear rocket development effort fell under the auspices of several programs whose purposes were to advance different aspects of the nuclear engine design. The first program element was designated Kiwi. The Kiwi (a flightless bird) program was so named because the engines developed were designed to advance the basic technology of nuclear thermal rockets and not to fly. A follow-on program to Kiwi was called, Phoebus which was constituted to take the Kiwi results and develop engine designs suitable for interplanetary voyages. Toward the end of the program, the Peewee reactor was designed and built to test smaller, more compact reactor designs along with the nuclear furnace development reactor which was designed to test advanced high-temperature fuels in addition to examining concepts for reducing emissions of radioactive material into the atmosphere.

In 1961 the NERVA program began designing and building working rockets based upon the research previously done under the ROVER program. NASA issued a request for proposals and established the Space Nuclear Propulsion Office (SNPO) to manage the NERVA program. The rockets developed under the NERVA program were envisioned for use in human missions to Mars and beyond. NERVA was organized as a joint effort between what was then the Atomic Energy Commission (AEC) and NASA. The AEC had the expertise and the authority to oversee the design of nuclear reactors for civilian use, and NASA had the task of developing the rockets and vehicles that would use the nuclear engines.

The NERVA program achieved the following milestones over the life of the project:

- Nuclear rocket testing occurred between 1959 and 1973.
- A total of 23 reactor tests were performed.
- Highest power achieved was 4500 MW.
- Highest temperature achieved was 4500°F (2750 K).
- Maximum thrust achieved was 250,000 pounds.
- Maximum specific impulse achieved was 850 s.
- Maximum burn time in one test was 90 min.

Fig. 1.1 illustrates a complete NERVA system and its component parts and Fig. 1.2 is a photograph of the Phoebus 2A (250,000 pound thrust) NERVA engine under test.

As was mentioned in the preceding paragraphs, the fuel elements in the NERVA engines were in the shape of hexagonal prisms. These fuel elements were about 55 in. long and about 1 in. flat to flat. The fuel elements in addition contained 19 holes through which flowed the hydrogen propellant. These fuel elements initially were composed of a graphite matrix within which coated uranium fuel particles were embedded. Later fuel forms were fabricated using a uranium/graphite composite. An entire NERVA core contained roughly 1000 of these fuel elements. The fuel elements were held in place in the core by means of support elements. These support elements supported the six adjacent fuel elements in a grouping called a cluster.

4 CHAPTER 1 INTRODUCTION

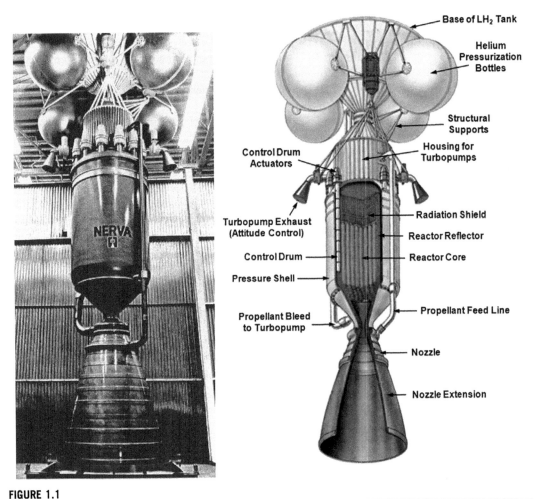

FIGURE 1.1

Nuclear engine for rocket vehicle applications (NERVA).

The fuel and support elements were surrounded by a reflector region composed of beryllium. The reflector was used to reflect back into the core neutrons emanating from the fuel which would normally escape the reactor. This configuration has the effect of conserving neutrons and leads to smaller more compact engine designs. If this region were not present, the reactor would not operate because too many neutrons would escape the core. Just how this works is discussed later.

Control drums embedded in the reflector serve as a control mechanism by which the reactor power can be varied. The control drums operate by varying the number of neutrons which escape the core. The drums are composed of beryllium cylinders with a sheet of material which strongly absorbs neutrons attached to one side. When the absorbing material (usually boron carbide) is close to the core, many neutrons which would be reflected back into the core are instead absorbed thus causing the

2. HISTORICAL PERSPECTIVE

FIGURE 1.2

NERVA test firing (Phoebus 2A). (Video available on companion website http://booksite.elsevier.com/9780128044742).

reactor to decrease in power or shutdown. When the absorbing material is away from the core, the beryllium portion of the control drum reflects the escaping neutrons back into the core where they can be used by the fuel to cause the engine to startup or increase in power.

Fig. 1.3 illustrates a fuel cluster and the manner in which it is integrated into the reactor with the reflector region and control drums.

2.3 PARTICLE BED REACTOR

Active interest in PBR systems dates back to 1982 when discussions of the PBR focused on its potential for supplying high levels of burst-mode electric power. A few years later the Air Force identified the PBR as a potentially attractive candidate for orbital-transfer vehicle applications. Finally, in late 1987 the Strategic Defense Initiative Organization established a highly classified program, code named Timberwind, to evaluate the PBR rocket engine for application in long-range antimissile interceptors.

In early 1991, much of the work under the Timberwind program was declassified, and the technology was evaluated for a wider range of applications, including space launch vehicles and piloted interplanetary missions. The concept promised significant reductions in system mass over solid-core reactors, made possible by the significant increase in the heat-transfer surface area of the particle fuel elements. The PBR's superior heat-removal characteristics result from the 20-fold greater surface area

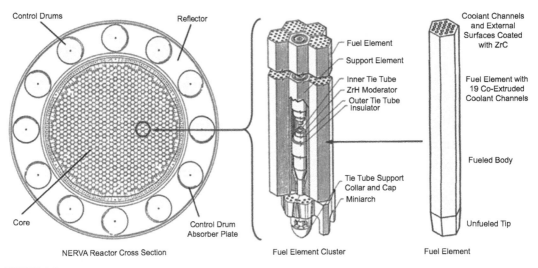

FIGURE 1.3

NERVA core and fuel segment cluster detail.

to volume ratio of the fuel particles over the prismatic fuel used in NERVA. In addition, the PBR has a lower core pressure drop than NERVA-derived systems due to the short flow paths through the particle bed.

The PBR Timberwind program achieved the following milestones over the life of the project:

- Nuclear element testing was performed in 1988 and 1989.
- Conducted zero-power critical experiments to confirm physics benchmarks.
- Conducted two sets of powered fuel element tests in the Annular Core Research Reactor (PIPE-1 and PIPE-2).
- Achieved power densities of 1.5–2.0 MW/L.
- Achieved hydrogen outlet temperatures of 2950°F (1900 K).
- Cold frit flow blockages in PIPE-2 as a result of thermal cycling caused severe core damage.
- Later analyses also indicated that the core was susceptible to thermal instabilities.

Fig. 1.4 shows several component pieces of the PBR fuel element and the fuel particles of which the fuel bed is composed.

In the PBR the flow path through the fuel element is quite different from that found in the NERVA fuel elements. In particular, the PBR fuel element employs a radial flow configuration consisting primarily of two concentric porous pipes (called frits) in between which is supported a bed of tiny fuel particles. Hydrogen propellant flows through the walls of the outer cold frit, through the fuel particle bed where it is heated to high temperatures, and finally exits through the walls of the inner hot frit. The propellant then leaves the fuel element through the central cavity where it is expelled through a nozzle as illustrated in Fig. 1.5.

Because of the high surface-to-volume ratio of the fuel particles in the PBR concept, extremely high heat-transfer rates are possible, potentially resulting in very compact reactor designs having high

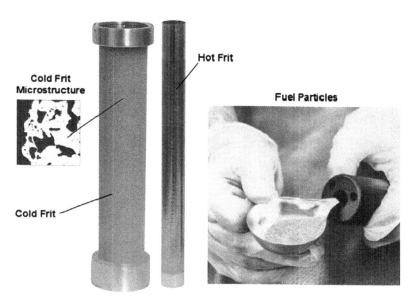

FIGURE 1.4

PBR frits and fuel particles.

thrust-to-weight ratios. Unfortunately, it was discovered during testing that because the particle bed design does not constrain the hydrogen propellant to flow along well-defined flow paths, the fuel element proved to be thermally unstable under certain circumstances. This instability was manifest by the appearance of potentially dangerous local hot spots within the fuel bed. Modifications to the particle bed design have since been proposed which are intended to address the flow instability problem by seeking to constrain the hydrogen flow to well-defined paths within the fuel region. These modifications include using grooved fuel rings instead of fuel particles, employing frits with graded porosity, using perforated foil fuel, and so on.

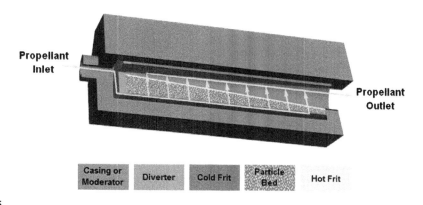

FIGURE 1.5

PBR fuel element detail.

2.4 RUSSIAN NUCLEAR ROCKETS

The development of NTRs in the USSR began in 1955 when it was proposed that a rocket with a nuclear engine be built to enhance the defensive power of the country as a response to the work being done in the United States on the ROVER program. The activities to develop the nuclear engine were distributed among several different research facilities in the Soviet Union with each facility overseeing various aspects of the engine design. The thermal hydraulics of the engine were to be developed at Research Institute-1 (now Research Institute of Thermal Processes, *RITP*), the work on the neutronics of the reactor was to be performed at the Obninsk Physical Energy Institute, *PEI*, and at the Kurchatov' Atomic Energy Institute, *AEI*, and the fuel element design work was to be performed at Research Institute-9 (now A.A. Bochvar All-Union Research Institute of Inorganic Materials, *ARIIM*). Later in 1962 another research institute was formed in the Soviet Union to provide an experimental facility which would allow for the rapid development and production of new types of nuclear fuel. This institute was called the Research Institute of Heat Releasing Elements, *RIHRE* (now "NPO Luch" Research and Production Association, *RPA*).

Initially, two separate nuclear engine designs were to be pursued. In the first engine design ("A" scheme), a simple solid-core reactor was to be developed using refractory materials. In the second, more advanced design ("B" scheme), a gas-core reactor was to be developed that confined and controlled a fissioning uranium plasma. Because of the severe heating problems which were associated with the "B" scheme, it was later decided to pursue only the "A" scheme design with the "B" scheme design continuing on as a research study.

The Soviet scientists developing the fuel for the reactors took a somewhat different path than their American counterparts. Rather than using graphite-based fuel element which has good thermal strength but suffers from the fact that it reacts vigorously with hot hydrogen, the Soviet scientists chose to investigate uranium carbide compounds. The uranium carbide compounds are much more stable in a hot hydrogen environment, but suffer from the fact that they are quite brittle and prone to crack and break. The reasoning of the Soviet scientists was that protecting the graphite from interaction with the hydrogen through the use of coatings was a more difficult problem than designing fuel elements which could successfully resist breaking due to high thermal stresses. Eventually, a fuel element was developed wherein numerous small "twisted ribbon" fuel pieces were bound together in a tube. The hydrogen propellant would flow through the tube and spiral down the fuel pieces. Fig. 1.6 illustrates the design of these fuel elements.

During the time period from 1971 to 1978, NPO Luch began tests of prototype "Kosberg" nuclear engines at a facility southwest of Semipalatinsk-21 (later Kurchatov, Kazakhstan). Also during the time period from 1970 to 1988, a more advanced facility was constructed south of Semipalatinsk-21 to test another type of prototypical nuclear engine called the Baikal-1. In all, 30 tests were performed on these nuclear engines without a failure. From this work a "minimum" engine called the RD-410 was eventually developed. Fig. 1.7 illustrates the RD-410 nuclear rocket engine which was finally constructed.

The RD-410 rocket engine achieved thrust levels of about 7700 pounds and was able to fire continuously for up to an hour. It could also be restarted up to 10 times. Because of its advanced fuel element design, the engine was able to achieve efficiencies about 7% greater than those attained in the NERVA engines. The collapse of the Soviet Union effectively ended all work on nuclear propulsion.

2. HISTORICAL PERSPECTIVE

FIGURE 1.6

"Twisted ribbon" fuel pieces and the RD-410 fuel element.

FIGURE 1.7

RD-410 nuclear rocket engine.

REFERENCES

[1] J.L. Finseth, Overview of Rover Engine Tests — Final Report, NASA George C. Marshall Space Flight Center, February 1991. Contract NAS 8—37814, File No. 313-002-91-059.
[2] R.A. Haslett, Space Nuclear Thermal Propulsion Final Report, Phillips Laboratory, May 1995. PL-TR-95—1064.
[3] B. Harvey, Russian Planetary Exploration History, Development, Legacy, and Prospects, Springer-Praxis Books in Space Exploration, 2007. ISBN 10: 0—387-46343-7.

CHAPTER 2

ROCKET ENGINE FUNDAMENTALS

Rocket engines operate by expelling a high-temperature gas through a nozzle to produce thrust. This thrust acts to accelerate a spacecraft in the direction opposite to that of the expelled gas through the application of Isaac Newton's third law of motion: "For every action, there is an equal and opposite reaction." In chemical rocket engines, the hot gas is created in a combustion chamber where the propellants are ignited and burned. Nuclear thermal rocket engines, on the other hand, use nuclear reactors to supply the heat needed to heat the propellant to high temperatures. The high-temperature gas exiting the rocket engines is introduced into nozzle assemblies where the thermal energy of the hot propellant gas is converted to kinetic energy in the form of a directed high-speed exhaust flow.

1. CONCEPTS AND DEFINITIONS

The primary purpose of rocket engines, be they chemical or nuclear, is to apply a propulsive force or *thrust* to a spacecraft so as to accelerate the vehicle to high speeds. In nuclear rocket engines, thrust is produced as a result of a hot gaseous propellant exiting from a nuclear reactor being discharged through a nozzle. The purpose of the nozzle is to convert the thermal energy in the hot propellant to kinetic energy in the form of a directed high-speed exhaust flow parallel to the line of flight, but in the opposite direction. Applying the principle of *conservation of momentum* as posited by Isaac Newton in his *third law of motion*: "For every action, there is an equal and opposite reaction," this high-velocity propellant exhaust flow has the effect of forcing the spacecraft forward as illustrated in Fig. 2.1.

The thrust is defined to be the force produced by the rocket engine as a result of the time rate of change of momentum of the exhaust gas which is accelerated through the rocket engine nozzle. If it is assumed that nuclear rockets will operate only in space, then there will be no external forces

FIGURE 2.1

Rocket thrusting and the conservation of momentum.

acting on the spacecraft due to atmospheric drag, and so on. Since momentum is conserved, the total time rate of change of momentum of the rocket plus the exhaust is then equal to zero and may be written as:

$$F_{ext} = 0 = \frac{1}{g_c}\frac{\Delta P}{\Delta t} = \frac{(P + \Delta P) - P}{(t + \Delta t) - t} = \frac{1}{g_c}\frac{(m + \Delta m)(V + \Delta V) + \Delta m U - mV}{\Delta t} \quad (2.1)$$

where F_{ext} = external forces on the spacecraft; P = momentum = mass × velocity; m = spacecraft mass; V = velocity vector of the spacecraft; U = velocity vector of the propellant exhaust; t = time; and g_c = conversion factor.

Expanding Eq. (2.1) and rearranging terms then yields an expression of the form:

$$0 = \frac{1}{g_c}\frac{mV + V\Delta m + m\Delta V + \Delta m \Delta V + \Delta m U - mV}{\Delta t} = \frac{m}{g_c}\frac{\Delta V}{\Delta t} + \frac{1}{g_c}\underbrace{(U + V)}_{-v}\frac{\Delta m}{\Delta t} \quad (2.2)$$

where v = velocity of the propellant exhaust with respect to the spacecraft (usually constant).

Note that the grayed out terms all either cancel out or are assumed to be negligible. Taking the limit of Eq. (2.2) as time goes toward zero and applying Newton's *second law of motion* which states that "The force on an object is equal to its mass multiplied by its acceleration" then yields:

$$0 = \frac{1}{g_c}\lim_{t \to 0}\left(m\frac{\Delta V}{\Delta t} - v\frac{\Delta m}{\Delta t}\right) = \frac{m}{g_c}\underbrace{\frac{dV}{dt}}_{a} - \frac{1}{g_c}v\underbrace{\frac{dm}{dt}}_{\dot{m}} \Rightarrow F_{mom} = \frac{ma}{g_c} = \frac{\dot{m}v}{g_c} \quad (2.3)$$

where F_{mom} = thrust due to momentum transfer from propellant exhaust; a = spacecraft acceleration; and $\dot{m}$ = propellant mass flow rate.

In addition to the force resulting from propellant momentum transfer, there is also a force which results from the pressure of the exhaust gases at the nozzle exit as illustrated in Fig. 2.2.

This pressure force may be described by:

$$F_{pres} = (P_e - P_a)A_e \quad (2.4)$$

where F_{pres} = force due to propellant exhaust pressure; P_e = pressure at the nozzle exit due to propellant gases; P_a = external pressure (in space, this term is zero); and A_e = nozzle cross-sectional area at the nozzle exit.

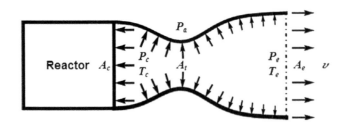

FIGURE 2.2

Rocket nozzle characteristics.

The total thrust of the engine is equal to the sum of the forces due to the propellant momentum transfer and the propellant exhaust pressure. Therefore, summing Eqs. (2.3) and (2.4) and assuming that the engine is operating in space yields an expression for the total engine thrust such that:

$$F = F_{\text{mom}} + F_{\text{pres}} = \frac{\dot{m}v}{g_c} + P_e A_e = \dot{m}\left(\frac{v}{g_c} + \frac{P_e A_e}{\dot{m}}\right) = \frac{\dot{m}v_e}{g_c} \quad (2.5)$$

where F = total engine thrust and v_e = effective propellant exhaust velocity.

Note that Eq. (2.5) can be rearranged so as to define a new term called *specific impulse* where:

$$I_{\text{sp}} = \frac{v_e}{g_c} = \frac{F}{g_c \dot{m}} \quad (2.6)$$

in which I_{sp} = specific impulse.

It turns out that the specific impulse is a very useful parameter in determining the efficiency of a rocket engine. Generally, specific impulse has units of seconds and physically represents the length of time one pound of propellant can produce one pound of thrust (or produce 1 N of thrust from 1 kg of propellant). Specific impulse is analogous to miles per gallon for an automobile.

Rewriting the differential portion of Eq. (2.3) using the definition for specific impulse from Eq. (2.6) yields:

$$0 = m\frac{dV}{dt} + g_c I_{\text{sp}}\frac{dm}{dt} \Rightarrow -g_c I_{\text{sp}}\int_{m_0}^{m_f}\frac{dm}{m} = \int_0^{V_f} dV \quad (2.7)$$

Performing the integrations presented in Eq. (2.7) then yields an expression for the maximum velocity increment attainable for a given vehicle mass fraction and engine-specific impulse such that:

$$V_f = -g_c I_{\text{sp}} \ln\left(\frac{m_f}{m_0}\right) = -g_c I_{\text{sp}} \ln(f_m) \quad (2.8)$$

where V_f = total velocity increment available from the vehicle; m_f = vehicle dry system mass; m_0 = fully fueled vehicle mass; and f_m = vehicle mass fraction = $\frac{m_f}{m_0}$.

Eq. (2.8) is known as the *rocket equation*, and its solution yields the maximum velocity attainable by a space vehicle for a given engine-specific impulse and vehicle mass fraction. To determine a value for the specific impulse in terms of the propellant flow thermodynamic properties, it is necessary to use the *first law of thermodynamics* to relate the propellant temperature to propellant velocity in addition to the definition of I_{sp} from Eq. (2.6) such that:

$$Q = \dot{m}(h_c - h_e) = \dot{m}c_p(T_c - T_e) = \frac{1}{2}\dot{m}v_e^2 = \frac{1}{2}\dot{m}g_c^2 I_{\text{sp}}^2 \quad (2.9)$$

where Q = thrust power; c_p = specific heat of propellant at constant pressure; h_c = enthalpy of propellant in reactor chamber after leaving core; h_e = enthalpy of propellant after leaving nozzle; T_c = temperature of propellant in reactor chamber after leaving core; and T_e = temperature of propellant after leaving nozzle.

Rearranging terms from Eq. (2.9) then yields for the specific impulse:

$$I_{\text{sp}} = \frac{1}{g_c}\sqrt{2(h_c - h_e)} = \frac{1}{g_c}\sqrt{2c_p T_c\left(1 - \frac{T_e}{T_c}\right)} \quad (2.10)$$

CHAPTER 2 ROCKET ENGINE FUNDAMENTALS

Noting the following specific heat relationships for ideal gases:

$$R = \frac{R_u}{mw} = c_p - c_v; \quad \gamma = \frac{c_p}{c_v} \tag{2.11}$$

where R = propellant gas constant; R_u = universal gas constant; c_v = specific heat of propellant at constant volume; γ = ratio of the specific heats; and mw = propellant molecular weight. From Eq. (2.11) it is found that:

$$c_p = \frac{\gamma}{\gamma - 1} R = \frac{\gamma}{\gamma - 1} \frac{R_u}{mw} \tag{2.12}$$

Substituting Eq. (2.12) into Eq. (2.10) then yields:

$$I_{sp} = \frac{1}{g_c} \sqrt{\frac{2\gamma}{\gamma - 1} \frac{R_u}{mw} T_c \left(1 - \frac{T_e}{T_c}\right)} \tag{2.13}$$

In order to determine the nozzle outlet temperature, the assumption is made that the propellant flow is isentropic. Then by applying the *Gibbs equation* for simple systems as derived from the *second law of thermodynamics* it is possible to obtain:

$$0 = T\,ds = dh = du + P\,dv = c_v\,dT + P\,dv \tag{2.14}$$

where u = propellant internal energy; P = propellant pressure; s = propellant entropy; and v = propellant-specific volume.

Recalling that ideal gas law may be written as:

$$P = \rho R T = \frac{RT}{v} \tag{2.15}$$

Expressing the ideal gas law in Eq. (2.15) in differential form then yields:

$$\frac{dP}{P} = \frac{dT}{T} - \frac{dv}{v} \tag{2.16}$$

Substituting Eqs. (2.11) and (2.16) into Eq. (2.14) and rearranging terms then gives:

$$\frac{dP}{P} = \frac{\gamma}{\gamma - 1} \frac{dT}{T} \tag{2.17}$$

Integrating both sides of Eq. (2.17) between the engine chamber and the nozzle exit plane yields an expression of the form:

$$\int_{P_c}^{P_e} \frac{dP}{P} dP = \frac{\gamma}{\gamma - 1} \int_{T_c}^{T_e} \frac{dT}{T} \Rightarrow \frac{T_e}{T_c} = \left(\frac{P_e}{P_c}\right)^{\frac{\gamma-1}{\gamma}} \tag{2.18}$$

Substituting Eq. (2.18) into Eq. (2.13) then yields an equation for the specific impulse which consists solely of the thermodynamic properties of the propellant:

$$I_{sp} = \frac{1}{g_c} \sqrt{\frac{2\gamma}{\gamma - 1} \frac{R_u}{mw} T_c \left[1 - \left(\frac{P_e}{P_c}\right)^{\frac{\gamma-1}{\gamma}}\right]} \tag{2.19}$$

If the propellant exit pressure approaches zero (implying the rocket engine operates in space with an infinite nozzle area ratio), Eq. (2.19) reduces to:

$$I_{sp} = \frac{1}{g_c}\sqrt{\frac{2\gamma}{\gamma-1}\frac{R_u}{mw}T_c} \qquad (2.20)$$

Eq. (2.20) is often used to provide quick estimates of engine I_{sp} as a function of temperature and propellant molecular weight; however, it overestimates the specific impulse since for any finite-sized nozzle the exit pressure will always be greater than zero.

2. NOZZLE THERMODYNAMICS

To determine the exit pressure for finite-sized nozzles a simple compressible flow analysis (see, eg, Ref. [1]) assuming that the flow in the nozzle is isentropic is undertaken. In performing this analysis, the first law of thermodynamics is first used to calculate the change in enthalpy (or equivalently, temperature) resulting from stopping a propellant stream having some given flow velocity such that:

$$\dot{m}(h_0 - h) = \dot{m}c_p(T_0 - T) = \frac{1}{2}\dot{m}V^2 \Rightarrow T_0 = T + \frac{V^2}{2c_p} \qquad (2.21)$$

where T = static temperature (fluid temperature as seen by an observer moving with the fluid) and T_0 = stagnation temperature (fluid temperature after its velocity has been reduced to zero).

From thermodynamic considerations [1], the speed of sound (c) in a fluid may be given by:

$$c = \sqrt{\gamma RT} \qquad (2.22)$$

Noting that the *Mach number* (M) is defined as the ratio of the fluid velocity to the speed of sound in the fluid, it is possible to obtain using Eq. (2.22):

$$M = \frac{V}{c} = \frac{V}{\sqrt{\gamma RT}} \Rightarrow V = Mc = M\sqrt{\gamma RT} \qquad (2.23)$$

When the Mach number is less than 1, the fluid flow is traveling at a velocity less than the speed of sound and the flow is said to be *subsonic*. Likewise, when the Mach number is greater than 1, the fluid flow is traveling at a velocity greater than the speed of sound and the flow is said to be *supersonic*. Substituting Eq. (2.23) into Eq. (2.21) and using the specific heat definition from Eq. (2.12) then yields:

$$T_0 = T + \frac{V^2}{2c_p} = T + \frac{\gamma RT}{2c_p}M^2 = T\left(1 + \frac{\gamma R}{2c_p}M^2\right) = T\left(1 + \frac{\gamma-1}{\gamma R}\frac{\gamma R}{2}M^2\right) = T\left(1 + \frac{\gamma-1}{2}M^2\right) \qquad (2.24)$$

Eq. (2.24) may be understood to represent the temperature change occurring in a flowing compressible fluid when it is stopped or stagnated isotropically. To determine the pressure change resulting from isotropically stagnating a flowing compressible fluid, Eq. (2.18) may be incorporated with a suitable variable change into Eq. (2.24) to obtain:

$$\left(\frac{P_0}{P}\right)^{\frac{\gamma-1}{\gamma}} = \frac{T_0}{T} = 1 + \frac{\gamma-1}{2}M^2 \Rightarrow \frac{P_0}{P} = \left(1 + \frac{\gamma-1}{2}M^2\right)^{\frac{\gamma}{\gamma-1}} \qquad (2.25)$$

CHAPTER 2 ROCKET ENGINE FUNDAMENTALS

Noting that the propellant mass flow rate may be determined from the continuity equation such that:

$$\dot{m} = \rho V A \tag{2.26}$$

it is possible now to use the propellant mass flow rate from Eq. (2.26) along with the definition for the Mach number from Eq. (2.23) and the ideal gas law relationship from Eq. (2.15) to obtain a new expression for the propellant mass flow rate of the form:

$$\dot{m} = \rho A V = \rho A M \sqrt{\gamma R T} = \frac{P}{RT} A M \sqrt{\gamma R T} = \frac{P}{R} A M \sqrt{\frac{\gamma R}{T}} \tag{2.27}$$

Rearranging Eq. (2.27) and using the temperature and pressure relationships expressed in Eq. (2.25), it is possible to obtain another expression for the propellant mass flow rate such that:

$$\dot{m} = P\frac{AM}{R}\sqrt{\frac{\gamma R}{T}} = P_0 \frac{P}{P_0} \frac{AM}{R} \sqrt{\gamma R} \sqrt{\frac{T_0}{T}} \sqrt{\frac{1}{T_0}} = \frac{P_0}{\sqrt{T_0}} AM \sqrt{\frac{\gamma}{R}} \left(1+\frac{\gamma-1}{2}M^2\right)^{\frac{-\gamma}{\gamma-1}} \sqrt{1+\frac{\gamma-1}{2}M^2}$$

$$= \frac{AM\sqrt{\frac{\gamma}{RT_0}}P_0}{\sqrt{\left(1+\frac{\gamma-1}{2}M^2\right)^{\frac{\gamma+1}{\gamma-1}}}}$$

$$\tag{2.28}$$

Since the propellant mass flow rate at any axial point in the nozzle must remain constant regardless of the cross-sectional area at that location, it is possible to relate the conditions at the sonic point (eg, where $M = 1$) to the conditions at any other point in the nozzle. Therefore, using Eq. (2.28) it can be shown that:

$$\dot{m} = \frac{AM\sqrt{\frac{\gamma}{RT_0}}P_0}{\sqrt{\left(1+\frac{\gamma-1}{2}M^2\right)^{\frac{\gamma+1}{\gamma-1}}}} = \frac{A^*(1)\sqrt{\frac{\gamma}{RT_0}}P_0}{\sqrt{\left[1+\frac{\gamma-1}{2}(1)^2\right]^{\frac{\gamma+1}{\gamma-1}}}} \Rightarrow \frac{A}{A^*} = \frac{1}{M}\sqrt{\left[\frac{2\left(1+\frac{\gamma-1}{2}M^2\right)}{\gamma+1}\right]^{\frac{\gamma+1}{\gamma-1}}} \tag{2.29}$$

where X^* = quantity "X" evaluated at the nozzle throat where $M = 1$.

Eq. (2.29) is known as the Mach–area relationship. This relationship allows the area ratio at any point in the nozzle as referenced to the area at the nozzle sonic point to be determined. The pressure ratio corresponding to the area ratio presented above may be determined from Eq. (2.25) by noting that:

$$\frac{P_0/P^*}{P_0/P} = \frac{P}{P^*} = \frac{\left[1+\frac{\gamma-1}{2}(1)^2\right]^{\frac{\gamma}{\gamma-1}}}{\left(1+\frac{\gamma-1}{2}M^2\right)^{\frac{\gamma}{\gamma-1}}} = \left[\frac{\gamma+1}{2\left(1+\frac{\gamma-1}{2}M^2\right)}\right]^{\frac{\gamma}{\gamma-1}} \tag{2.30}$$

Likewise, the temperature ratio corresponding to the area ratio as evaluated from the Mach–area relationship given by Eq. (2.29) may be evaluated from Eq. (2.25) by noting that:

$$\frac{T_0/T^*}{T_0/T} = \frac{T}{T^*} = \frac{1+\frac{\gamma-1}{2}(1)^2}{1+\frac{\gamma-1}{2}M^2} = \frac{\gamma+1}{2\left(1+\frac{\gamma-1}{2}M^2\right)} \tag{2.31}$$

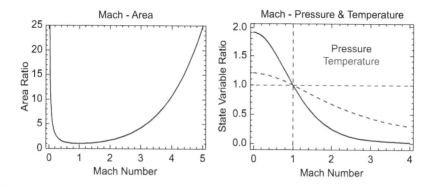

FIGURE 2.3

Compressible flow relationships.

The plots in Fig. 2.3 illustrate the Mach—area and Mach—pressure and temperature relationships where $\gamma = 1.4$. Note in the Mach—area plot that in order to increase the velocity of the exhaust propellant to supersonic speeds, the cross-sectional flow area in the subsonic regime must initially decrease in order to raise the propellant flow velocity. As sonic conditions are reached and the propellant flow transitions into the supersonic regime, the flow area must begin increasing to further raise the propellant flow velocity. The exact converging/diverging area profile chosen for a particular nozzle is dependent on a number of factors which are normally associated with assuring that the flow is as close to isentropic as possible. Note also that the Mach—area relationship becomes asymptotic at some finite maximum Mach number which represents an infinite area ratio. This infinite area Mach number yields the maximum possible engine-specific impulse.

Using the relationships as shown in Fig. 2.3, it is now possible to calculate the pressure ratio and temperature ratio between the engine exhaust plenum and the nozzle exit given that the nozzle area ratios between the nozzle inlet and throat, and nozzle exit and throat are known. Using these area ratios, therefore, the nozzle exit temperature and pressure ratios may be determined by using the following procedure:

1. Calculate the core exit Mach number (subsonic value) by employing Eq. (2.29) and knowing the area ratio between the engine exhaust plenum and the nozzle throat $\left(\frac{A_c}{A^*}\right)$.
2. Calculate the pressure ratio $\left(\frac{P_c}{P^*}\right)$ and temperature ratio $\left(\frac{T_c}{T^*}\right)$ between the engine exhaust plenum and the nozzle throat by using Eqs. (2.30) and (2.31) and the Mach number found in Step 1.
3. Calculate the nozzle exit Mach number (supersonic value) by employing Eq. (2.29) and knowing the area ratio between the nozzle exit and the nozzle throat $\left(\frac{A_c}{A^*}\right)$.
4. Calculate the pressure ratio $\left(\frac{P_c}{P^*}\right)$ and temperature ratio $\left(\frac{T_c}{T^*}\right)$ between the nozzle exit and the nozzle throat by using Eqs. (2.30) and (2.31) and the Mach number found in Step 3.

Table 2.1 Isentropic Nozzle Calculations

Input Parameters	Mach Number	Pressure Ratio (P/P*)	Temperature Ratio (T/T*)	
Area ratio = 2	0.3059	1.77398	1.17795	Subsonic
$\gamma = 1.4$	2.1972	0.17781	0.61052	Supersonic

5. Calculate overall engine pressure ratio and temperature ratio by using the subsonic and supersonic pressure ratios found in Steps 2 and 4:

$$\frac{P_e}{P_c} = \left(\frac{P_e}{P^*}\right)\left(\frac{P^*}{P_c}\right) \quad \text{and} \quad \frac{T_e}{T_c} = \left(\frac{T_e}{T^*}\right)\left(\frac{T^*}{T_c}\right)$$

The various Mach number, pressure, and temperature ratios described in the preceding paragraphs may be easily determined through the use of Table 2.1.

The effects of propellant molecular weight and the various area ratios and their effect on engine-specific impulse as described by Eq. (2.19) can be examined by adjusting those parameters in Fig. 2.4.

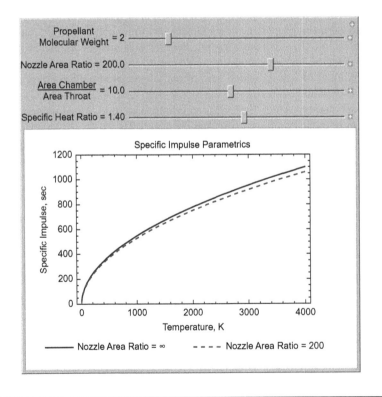

FIGURE 2.4

Specific impulse parametrics.

2. NOZZLE THERMODYNAMICS

In the interactive version of Fig. 2.4, it can be seen that for the space shuttle main engine (SSME), the maximum specific impulse of which the engine is capable is about 450 s assuming a mixture ratio of 6 (equivalent molecular weight ≈ 10), a nozzle area ratio of 77, a specific heat ratio of 1.33, and a chamber temperature of 3500 K. An NTR, on the other hand, using hydrogen propellant with a molecular weight of 2, a nozzle ratio of 77, a specific heat ratio of 1.41, and a chamber temperature of 3000 K yields a specific impulse of around 900 s. What should be noted from this example is that NTR engines achieve their specific impulse advantages not from greater chamber temperatures, which are actually somewhat lower than those produced in space shuttle main engines (SSMEs), but rather from the lower molecular weight which is characteristic of the hydrogen propellant.

EXAMPLE
The expander cycle nuclear rocket engine described in the previous section introduces the hydrogen propellant into the reactor core where it is heated to 3000 K. Given that the area ratio between the reactor exhaust plenum and the nozzle throat is 10 and the area ratio between the nozzle throat and the nozzle exit is 200, determine the nuclear rocket engine's specific impulse. Assume that the specific heat ratio for hydrogen is 1.4.

Solution
The first step in the calculation involves determining the subsonic Mach number of the hydrogen propellant in the core exit plenum. To accomplish this task, it is necessary to implicitly solve the Mach–area relationship expressed by Eq. (2.29). Note that this calculation and the ones that follow can be performed easily using the calculator found in Table 2.1:

$$\frac{A_c}{A^*} = \frac{1}{M}\sqrt{\left[\frac{2\left(1+\frac{\gamma-1}{2}M^2\right)}{\gamma+1}\right]^{\frac{\gamma+1}{\gamma-1}}} = 10 = \frac{1}{M}\sqrt{\left[\frac{2\left(1+\frac{1.4-1}{2}M^2\right)}{1.4+1}\right]^{\frac{1.4+1}{1.4-1}}} \Rightarrow M = 0.0580 \quad (1)$$

With the Mach number known from Eq. (1), the temperature and pressure ratios between the core exit plenum and the nozzle throat may be determined from Eqs. (2.30) and (2.31):

$$\frac{P_c}{P^*} = \left[\frac{\gamma+1}{2\left(1+\frac{\gamma-1}{2}M^2\right)}\right]^{\frac{\gamma}{\gamma-1}} = \left[\frac{1.4+1}{2\left(1+\frac{1.4-1}{2}0.0580^2\right)}\right]^{\frac{1.4}{1.4-1}} = 1.88848 \quad (2)$$

$$\frac{T_c}{T^*} = \frac{\gamma+1}{2\left(1+\frac{\gamma-1}{2}M^2\right)} = \frac{1.4+1}{2\left(1+\frac{1.4-1}{2}0.0580^2\right)} = 1.19919 \quad (3)$$

At this point, it is necessary to perform calculations similar to those just performed to determine the supersonic Mach number of the hydrogen propellant as it leaves the nozzle assembly. In this case, the Mach–area relationship from Eq. (2.29) must be solved using the area ratio between the nozzle throat and the nozzle exit:

$$\frac{A_e}{A^*} = \frac{1}{M}\sqrt{\left[\frac{2\left(1+\frac{\gamma-1}{2}M^2\right)}{\gamma+1}\right]^{\frac{\gamma+1}{\gamma-1}}} = 200 = \frac{1}{M}\sqrt{\left[\frac{2\left(1+\frac{1.4-1}{2}M^2\right)}{1.4+1}\right]^{\frac{1.4+1}{1.4-1}}} \Rightarrow M = 8.0893 \quad (4)$$

With the Mach number known from Eq. (4), the temperature and pressure ratios between the nozzle exit and the nozzle throat may again be determined from Eqs. (2.30) and (2.31):

$$\frac{P_c}{P^*} = \left[\frac{\gamma+1}{2\left(1+\frac{\gamma-1}{2}M^2\right)}\right]^{\frac{\gamma}{\gamma-1}} = \left[\frac{1.4+1}{2(1+\frac{1.4-1}{2}8.0893^2)}\right]^{\frac{1.4}{1.4-1}} = 0.00018 \qquad (5)$$

$$\frac{T_c}{T^*} = \frac{\gamma+1}{2\left(1+\frac{\gamma-1}{2}M^2\right)} = \frac{1.4+1}{2(1+\frac{1.4-1}{2}8.0893^2)} = 0.08518 \qquad (6)$$

From Eqs. (3) and (5), it is now possible to calculate the pressure ratio between the core exit plenum and the nozzle exit such that:

$$\frac{P_e}{P_c} = \left(\frac{P_e}{P^*}\right)\left(\frac{P^*}{P_c}\right) = 0.00018 \times \frac{1}{1.88848} = 9.5315 \times 10^{-5} \qquad (7)$$

Using the pressure ratio from Eq. (7) in conjunction with Eq. (2.19), the specific impulse for this nuclear rocket engine may be evaluated to yield:

$$I_{sp} = \frac{1}{g_c}\sqrt{\frac{2\gamma}{\gamma-1}\frac{R_u}{mw}T_c\left[1-\left(\frac{P_e}{P_c}\right)^{\frac{\gamma-1}{\gamma}}\right]}$$

$$= \frac{1}{9.8\frac{m}{s^2}}\sqrt{\frac{2 \times 1.4}{1.4-1} \times \frac{8314.5\frac{g\,m^2}{K\,mol\,s^2}}{2\frac{g}{mol}} \times 3000\,K\left[1-(9.5315 \times 10^{-5})^{\frac{1.4-1}{1.4}}\right]} = 919\,s \qquad (8)$$

The specific impulse value in Eq. (8) may be compared to the plot presented in Fig. 2.4 to verify these calculations and illustrate how the specific impulse would be affected by changes to the various parameters.

REFERENCE

[1] A.H. Shapiro, The Dynamics and Thermodynamics of Compressible Fluid Flow vol. 1, Ronald Press, 1953, ISBN 978-0-471-06691-0.

CHAPTER 3

NUCLEAR ROCKET ENGINE CYCLES

Nuclear rockets operate using one of the several types of thermodynamic cycles which vary in complexity and efficiency. For nuclear thermal rockets, these thermodynamic cycles are "open," in that during operation, the working fluid is discharged through the nozzle to produce thrust after circulating only once through the engine system. These engines typically use a turbopump to highly pressurize the propellant prior to being introduced into the reactor where the propellant is heated to high temperatures before being discharged to the nozzle. The pump is normally driven by an integrated turbine system which is powered by propellant that has been warmed somewhat using waste heat from the reactor. Nuclear systems, which are designed to produce electricity for some type of electric or ion rocket, generally incorporate "closed" thermodynamic cycles in which the working fluid always remains within the system and is circulated continuously during operation. Nuclear electric systems require, in addition to the turbopump assembly, radiators to reject the waste heat that results from the thermal-to-electric conversion process.

1. NUCLEAR THERMAL ROCKET THERMODYNAMIC CYCLES
1.1 HOT BLEED CYCLE

The hot bleed cycle was the rocket engine cycle of choice for both the NERVA and the Timberwind programs. While the Timberwind program never reached the stage of development, where the cycle could be implemented into the engine system, it nevertheless was the reference cycle design for that program. The NERVA program, on the other hand, did reach the stage of development where the hot bleed cycle could be implemented into the engine system, and in fact the cycle proved quite successful during the various engine tests.

The hot bleed cycle's main advantages are the high cycle efficiencies resulting from the low bleed flow required to drive the turbopump and the relative simplicity of the engine. The main disadvantage of the cycle is that the portion of the bleed flow which is diverted from the core exit plenum will be quite hot and hard on any valves and piping in contact with it prior to its being mixed with the bleed flow shunted off before it would have entered the reactor core. A schematic of the hot bleed cycle along with its thermodynamic characteristics is presented in Fig. 3.1.

The hot bleed cycle characteristics are as follows:

1–2 Liquid propellant from the tank is raised to the operating pressure after passing through the pump portion of the turbopump.

2–3 After passing through the turbopump, the propellant circulates through the nozzle, support elements, chamber walls, and so on, gasifying the propellant.

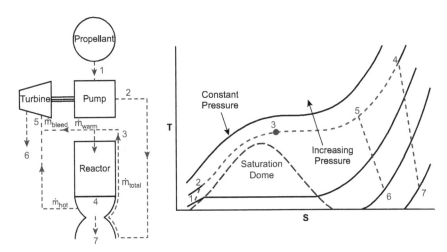

FIGURE 3.1

Hot bleed cycle.

3–4 The gaseous propellant flow splits, with the majority of the flow being directed into the reactor core where it is heated to several thousand degrees before exiting the core into the engine exhaust plenum.

3–5 The rest of the gaseous propellant flow mixes with hot propellant bled from the reactor exhaust plenum and enters into the turbine portion of the turbopump.

5–6 The mixed propellant flow, which is now at a temperature consistent with the maximum acceptable turbine blade material limits, passes through the turbine portion of the turbopump where it gives up some of its energy to drive the pump portion of the turbopump. After passing through the turbopump, the propellant flow is discharged through a small nozzle.

4–7 The remainder of the hot gaseous propellant in the engine exhaust plenum is directed through the main nozzle where the heat energy is changed to directed kinetic energy producing thrust.

$$\dot{m}_{total}(h_2 - h_1) = \dot{m}_{bleed}(h_5 - h_6) \tag{3.1}$$

where

$$\dot{m}_{bleed} h_5 = (\dot{m}_{warm} + \dot{m}_{hot}) h_5 = \dot{m}_{warm} h_3 + \dot{m}_{hot} h_4 \Rightarrow h_5 = \frac{\dot{m}_{warm} h_3 + \dot{m}_{hot} h_4}{\dot{m}_{warm} + \dot{m}_{hot}}$$

with $\dot{m}_{bleed}$ = total propellant mass flow rate diverted to drive the turbopump turbine; $\dot{m}_{warm}$ = propellant mass flow rate diverted from the reactor inlet; $\dot{m}_{hot}$ = propellant mass flow rate diverted from the reactor outlet; $\dot{m}_{total}$ = total propellant mass flow rate; and h_n = enthalpy at position "n" in the cycle.

1.2 COLD BLEED CYCLE

The cold bleed cycle is a possible rocket engine cycle which could be implemented as an alternative cycle to the hot bleed cycle used in NERVA and Timberwind. However, it has never been implemented in any rocket engines of any kind to date.

The cold bleed cycle's main advantages are the high turbopump reliability that results from the low turbine inlet temperatures and the relative simplicity of the engine. The main disadvantages of the cycle are that the chamber pressures tend to be low because of the limited amounts of power available to the turbopump from the nozzle and chamber regenerative cooling flow and the relative inefficiency of the cycle due to the waste of significant amounts of propellant from the bleed dump discharge from the turbopump. A schematic of the cold bleed cycle along with its thermodynamic characteristics is presented in Fig. 3.2.

The cold bleed cycle characteristics are as follows:

1–2 Liquid propellant from the tank is raised to the operating pressure after passing through the pump portion of the turbopump.
2–3 After passing through the turbopump, the propellant circulates through the nozzle, support elements, chamber walls, and so on, gasifying and warming the propellant.
3–4 The warm gaseous propellant flow splits and part of the flow (the bleed flow) is directed into the turbine portion of the turbopump where the pressure and temperature drop as the propellant releases some of its energy to drive the pump portion of the turbopump. After passing through the turbopump, the bleed flow is discharged to the environment through a small nozzle.

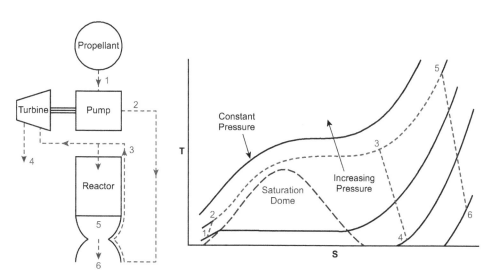

FIGURE 3.2

Cold bleed cycle.

3–5 The remainder of the propellant flow is directed into the reactor core where it is heated to several thousand degrees after which it is directed into the engine exhaust plenum.
5–6 The hot gaseous propellant in the engine exhaust plenum is directed through the main nozzle where the heat energy is changed to directed kinetic energy producing thrust.

The bleed flow and the heating rate to the propellant from the nozzle, support elements, chamber walls, and so on required to drive the turbopump to the extent necessary to achieve a desired propellant pressure at the reactor inlet can be determined from a system energy balance given by:

$$\dot{m}_{total}(h_2 - h_1) = \dot{m}_{bleed}(h_3 - h_4) \tag{3.2}$$

1.3 EXPANDER CYCLE

The expander cycle is a very efficient rocket engine cycle, which has been most notably implemented in the RL-10 chemical rocket engine by Pratt & Whitney. However, it has never been considered thus far for use in NTR engines.

The expander cycle's main advantages are the high turbopump reliability that results from the low turbine inlet temperatures and the efficient use of propellant resulting from the fact that no propellant bleed dump is required as in the bleed cycles. The main disadvantage of the cycle is that the chamber pressures tend to be low because of the limited amounts of power available to the turbopump from the nozzle and chamber regenerative cooling flow. Another disadvantage of the cycle is the relative complexity of the engine itself resulting from the additional flow paths required. A schematic of the expander cycle along with its thermodynamic characteristics is presented in Fig. 3.3.

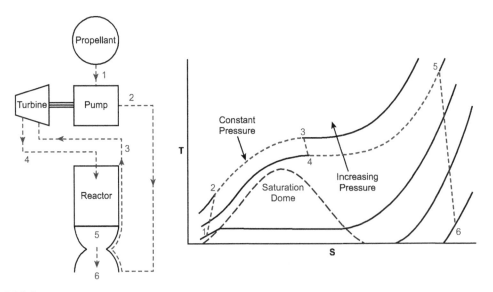

FIGURE 3.3

Expander cycle.

1. NUCLEAR THERMAL ROCKET THERMODYNAMIC CYCLES

The expander cycle characteristics are as follows:

1–2 Liquid propellant from the tank is raised to the operating pressure after passing through the pump portion of the turbopump.
2–3 After passing through the turbopump, the propellant circulates through the nozzle, support elements, chamber walls, and so on, gasifying and warming the propellant.
3–4 The warm gaseous propellant is directed into the turbine portion of the turbopump where the pressure and temperature drop as the propellant releases some of its energy to drive the pump portion of the turbopump.
4–5 After leaving the turbopump, the propellant is directed into the reactor core where it is heated to several thousand degrees after which it is directed into the engine exhaust plenum.
5–6 The hot gaseous propellant in the engine exhaust plenum is directed through the main nozzle where the heat energy is changed to directed kinetic energy producing thrust.

The heating rate to the propellant from the nozzle, support elements, chamber walls, and so on required to drive the turbopump to the degree necessary to achieve a desired propellant pressure at the reactor inlet can be again determined from a system energy balance which is of the form:

$$\dot{m}_{total}(h_2 - h_1) = \dot{m}_{total}(h_3 - h_4) \Rightarrow h_2 - h_1 = h_3 - h_4 \Rightarrow h_3 - h_2 = h_4 - h_1 \qquad (3.3)$$

EXAMPLE
A nuclear rocket engine operates on an expander cycle. The engine uses hydrogen as the propellant which is introduced into the pump portion of the engine turbopump as a saturated fluid at 20K. The pump isentropically increases the pressure on the hydrogen to circulate it through the hot reactor structure where the hydrogen is gasified and increased in temperature to 100K. If the hydrogen is to be introduced into the reactor at 7 MPa, to what pressure must the hydrogen be increased prior to entering the turbine portion of the turbopump? Assume that the turbine portion of the turbopump also operates isentropically.

Solution
Beginning at the entrance to the pump portion of the turbopump, the thermodynamic conditions of the hydrogen may be found to be:

$$T_1 = 20 \text{ K (saturated)} \Rightarrow P_1 = 0.09072 \text{ MPa } \& \ h_1 = -3.6672 \ \frac{\text{kJ}}{\text{kg}} \ \& \ s_1 = -0.17429 \ \frac{\text{kJ}}{\text{kg K}} \qquad (1)$$

Guessing that the pressure drop required across the turbine portion of the turbopump is negligible, assume an inlet pressure to the turbine of 7 MPa. Also recall from the problem statement that the turbine is isentropic. Therefore:

$$s_1 = s_2 = -0.17429 \ \frac{\text{kJ}}{\text{kg K}} \ \& \ P_2 = 7 \text{ MPa} \Rightarrow T_2 = 23 \text{ K } \& \ h_2 = 90.179 \ \frac{\text{kJ}}{\text{kg}} \qquad (2)$$

Making the assumption that no pressure drop occurs in the hydrogen, as it circulates through the hot reactor structure, it is found that:

$$T_3 = 100 \text{ K } \& \ P_2 = P_3 = 7 \text{ MPa} \Rightarrow h_3 = 1221.56 \ \frac{\text{kJ}}{\text{kg}} \ \& \ s_3 = 20.9 \ \frac{\text{kJ}}{\text{kg K}} \qquad (3)$$

Recalling that the turbine is isentropic and using Eq. (3.3) to determine the turbine exit enthalpy, it is found that:

$$h_4 = h_3 - h_2 + h_1 = 1135.05 \frac{kJ}{kg} \text{ \& } s_3 = s_4 = 20.9 \frac{kJ}{kg\,K} \Rightarrow T_4 = 92\text{ K \& } P_4 = 5.64\text{ MPa} \qquad (4)$$

Since the reactor inlet pressure from Eq. (4) is found to be 5.64 MPa, it is obvious that the initial assumption of negligible pressure drop across the turbine is incorrect. For the next iteration, guess a higher value for the pump outlet pressure. For this iteration, assume a pump outlet pressure of 10 MPa. Since the pump inlet conditions have not changed, it is found that:

$$s_1 = s_2 = -0.17429 \frac{kJ}{kg\,K} \text{ \& } P_2 = 10\text{ MPa} \Rightarrow T_2 = 24\text{ K \& } h_2 = 129.36 \frac{kJ}{kg} \qquad (5)$$

Again making the assumption that no pressure drop occurs to the hydrogen as it circulates through the hot reactor structure, it is found that:

$$T_3 = 100\text{ K \& } P_2 = P_3 = 10\text{ MPa} \Rightarrow h_3 = 1201.2 \frac{kJ}{kg} \text{ \& } s_3 = 19.17 \frac{kJ}{kg\,K} \qquad (6)$$

Recalling that the turbine is isentropic and using Eq. (3.3) to determine the turbine exit enthalpy, it is found that:

$$h_4 = h_3 - h_2 + h_1 = 1068.17 \frac{kJ}{kg} \text{ \& } s_3 = s_4 = 19.17 \frac{kJ}{kg\,K} \Rightarrow T_4 = 89\text{ K \& } P_4 = 7.17\text{ MPa} \qquad (7)$$

In this case the reactor inlet pressure from Eq. (7) is found to be slightly high at 7.17 MPa. For the next iteration, perform a linear interpolation on the pump outlet pressure using the results from Eqs. (4) and (7). This interpolation yields a pump outlet pressure of 9.67 MPa. Again, since the pump inlet conditions have not changed, it is found that:

$$s_1 = s_2 = -0.17429 \frac{kJ}{kg\,K} \text{ \& } P_2 = 9.67\text{ MPa} \Rightarrow T_2 = 24\text{ K \& } h_2 = 125.09 \frac{kJ}{kg} \qquad (8)$$

As before, assuming no pressure drop in the hydrogen as it circulates through the hot reactor structure, it is found that:

$$T_3 = 100\text{ K \& } P_2 = P_3 = 9.67\text{ MPa} \Rightarrow h_3 = 1202.98 \frac{kJ}{kg} \text{ \& } s_3 = 19.34 \frac{kJ}{kg\,K} \qquad (9)$$

Once again recalling that the turbine operates isentropically and using Eq. (3.3) to determine the turbine exit enthalpy, it is found that:

$$h_4 = h_3 - h_2 + h_1 = 1074.24 \frac{kJ}{kg} \text{ \& } s_3 = s_4 = 19.34 \frac{kJ}{kg\,K} \Rightarrow T_4 = 89\text{ K \& } P_4 = 7\text{ MPa} \qquad (10)$$

In this case the reactor inlet pressure is found from Eq. (10) to be almost exactly the value desired (eg, 7 MPa) requiring a pump outlet pressure of 9.67 MPa.

2. NUCLEAR ELECTRIC THERMODYNAMIC CYCLES
2.1 BRAYTON CYCLE

The Brayton cycle is a fairly old power cycle, being first proposed by George Brayton in the 1870s for use in reciprocating oil-burning engines. Today the Brayton cycle is widely used to supply power to aircraft, ships, and stationary power plants. Coupled with a nuclear reactor and space radiators, the Brayton cycle is also appropriate for use in space power applications such as for a power supply for electric propulsion systems.

A schematic of a nuclear electric ion propulsion system based on a Brayton cycle is presented in Fig. 3.4 along with its thermodynamic state point characteristics.

The Brayton cycle characteristics are as follows:

1—2 The gaseous working fluid passes through the compressor portion of the turbopump where it is adiabatically (ideally isentropically) raised to a high pressure.

2—3 After passing through the compressor, the working fluid enters the nuclear reactor where at constant pressure, it is heated to high temperatures.

3—4 Once the hot working fluid leaves the reactor, it enters the turbine portion of the turbopump where the enthalpy of the working fluid is adiabatically (and ideally isentropically) converted to mechanical energy. A portion of the energy from the turbine is used to drive the compressor portion of the turbopump, and the remainder of the energy is used to drive an electric generator.

4—1 Upon leaving the turbine portion of the turbopump, the working fluid, which is now at a low pressure, enters a space radiator where the temperature of the working fluid is reduced until it reaches the inlet state conditions of the compressor portion of the turbopump.

The Brayton cycle power balance (neglecting inefficiencies in the generator and power conversion system) is then given by:

$$W = \dot{m}(h_3 - h_4) - \dot{m}(h_2 - h_1) = \dot{m}(h_3 - h_4 - h_2 + h_1) \quad (3.4)$$

where W = net work performed by the cycle; $\dot{m}$ = mass flow rate of the working fluid; and h_n = enthalpy at position "n" in the cycle.

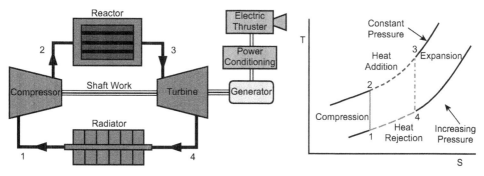

FIGURE 3.4

Brayton cycle.

The inlet temperature to the turbine from the nuclear reactor is limited by the maximum temperature that the turbine blades can withstand. This temperature also limits the maximum pressure ratio which can be used in the cycle. A high pressure ratio is desirable since it results in a more efficient and thus more compact system. To increase the efficiency of a Brayton cycle system without requiring ever higher pressure ratios, there are several techniques which may be employed to good effect albeit at the expense of increased system complexity.

In one technique called recuperation, the working fluid leaving the turbine, which is still quite hot, is used to heat the working fluid leaving the compressor. This technique reduces the amount of heat which must be dissipated by the heat rejection portion of the system and thus reduces the size and weight of the radiator. Another efficiency-raising method often employed in Brayton cycle systems is called intercooling. In this technique, a multistage compressor is configured so as to cool the working fluid between stages. By using several stages of intercooling, the compression process may be made nearly isothermal thus significantly reducing the amount of work required to compress the working fluid.

2.2 STIRLING CYCLE

A Stirling cycle engine is a closed cycle regenerative heat engine that operates by cyclically compressing and expanding a gaseous working fluid at different temperatures such that there is a net conversion of heat energy to mechanical work. Similar to Brayton cycle, the Stirling cycle is a fairly old power cycle, being first proposed by Robert Stirling in 1816. Originally, the Stirling cycle was conceived as an alternative cycle to that used steam engines because of its high efficiency and ability to use almost any heat source. Today the Stirling cycle has found limited use in certain niche applications, mainly because of the requirement that a high-temperature differential across the device is necessary for efficient operation. This high-temperature differential requirement results in material and fabrication issues which can be quite demanding. For space application, where the use of exotic materials and their associated cost is less of an issue, the Stirling cycle engine may well be the power cycle of choice [1].

Several configurations of a Stirling engine are possible. In an alpha configuration, two power pistons are contained within separate hot and cold cylinders with the working fluid being driven between the two cylinders by the pistons. In a beta configuration, there is only a single cylinder which contains a power piston and a displacer piston whose purpose is to drive the working fluid between the hot and cold parts of the engine. In a third configuration called a gamma configuration, there are again two cylinders with one cylinder containing the power piston and the other cylinder containing the displacer piston. There is also generally a regenerator heat exchanger between the hot and cold portions of the cavity containing the piston mechanisms. This heat exchanger, which may be attached to the displacer piston, serves the dual purpose of acting as a thermal barrier between the hot and cold cavities and also as a thermal storage medium to preheat the cold working fluid as it is transferred from the cold cavity to the hot cavity. Often this heat exchanger consists of little more than a porous material through which the working fluid may flow. A schematic of a nuclear electric propulsion system based on a Stirling engine in the beta configuration is presented in Fig. 3.5 along with its thermodynamic state point characteristics.

The Stirling cycle characteristics are as follows:

1−2 The power piston isothermally (constant temperature) compresses the working fluid at the cold end temperature. Because the working fluid is cold, relatively little work is required to compress the working fluid. The displacer piston also moves so as to begin transferring the working fluid to the hot end of the engine.

1−3 The displacer piston continues moving the working fluid to the hot end of the engine where it is heated isochorically (constant volume).

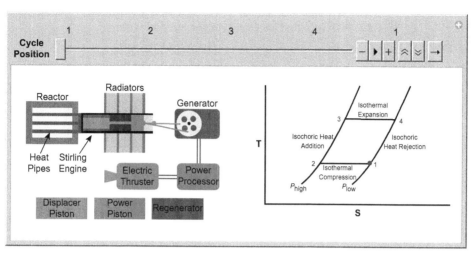

FIGURE 3.5

Stirling cycle (Beta configuration).

1–4 The heated working fluid increases in pressure and expands isothermally (constant temperature) so as to drive the power piston forward to its farthest stroke. The energy released through the power piston motion is greater than that required for subsequent working fluid compression.

4–1 The displacer piston moves such that the working fluid is transferred isochorically (constant volume) back to the cold end of the engine where the heat in the working fluid is rejected by the radiators.

The net power balance for a Stirling engine can be calculated by considering the integral of cyclic $P\,dV$ work performed during the movement of the power piston. This integral may be represented by:

$$W = \oint P\,dV \tag{3.5}$$

where: P = working fluid pressure and V = working fluid volume.

The integral in Eq. (3.5) only needs to be evaluated over the isothermal heat rejection (eg, stroke 1–2) and heat addition (eg, stroke 3–4) portions of the cycle since no work is done over the isochoric portions of the cycle, therefore:

$$W = \int_3^4 P\,dV + \int_1^2 P\,dV \tag{3.6}$$

If it is now assumed that the ideal gas law holds for the working fluid being used, it is found that:

$$PV = mRT \Rightarrow P = \frac{mRT}{V} \tag{3.7}$$

where T = working fluid temperature; m = mass of working fluid in the engine; and R = gas constant of the working fluid.

Table 3.1 Specific Gas Constant for Several Gases

Working Fluid	Gas Constant, R (J/kg/K)
Air	319.3
Ammonia	488.2
Argon	208.0
Carbon dioxide	188.9
Helium	2077.0
Hydrogen	4124.2
Nitrogen	296.8

Recalling that the work is performed isothermally during the power strokes and that the system is closed (eg, temperature and mass are constant), and incorporating Eq. (3.7) into Eq. (3.6), it is found that:

$$W = mRT_{high} \int_3^4 \frac{dV}{V} + mRT_{low} \int_1^2 \frac{dV}{V} = mRT_{high} \ln\left(\frac{V_4}{V_3}\right) + mRT_{low} \ln\left(\frac{V_2}{V_1}\right) \quad (3.8)$$

Noting that working fluid volume is the same at states 1 and 4 and at states 2 and 3, Eq. (3.8) may be rewritten to yield a final equation for the work output of an ideal Stirling cycle to be:

$$W = mR(T_{high} - T_{low}) \ln\left(\frac{V_4}{V_3}\right) \quad (3.9)$$

Eq. (3.9) shows that the work output of a Stirling cycle engine can be enhanced by increasing the temperature difference between the heat source and heat rejection temperatures and by increasing the compression ratio of the working fluid. The equation also shows that the work output of a Stirling cycle engine can be enhanced by increasing the mass of the working fluid in the engine or by increasing the working fluid gas constant. Table 3.1 shows the gas constants for several potential working fluids which could be used in a Stirling engine.

It is interesting to note as an example from Table 3.1 that by using helium as a working fluid rather than argon, it is possible for a Stirling engine to increase its potential power output by an order of magnitude.

REFERENCE

[1] P. McClure, D. Poston, Design and Testing of Small Nuclear Reactors for Defense and Space Applications, Invited Talk to ANS Trinity Section, Santa Fe, LA-UR-13–27054, 2013.

INTERPLANETARY MISSION ANALYSIS

CHAPTER 4

Most interplanetary missions using high-thrust propulsion systems do not apply thrust for the entire flight but rather have a series of thrusting maneuvers near the departure and destination planets with relatively long coast periods between the planets. Normally, at least four major propulsive maneuvers are required for round-trip missions. These main propulsion system burns include the following: (1) a departure acceleration burn from home planet, (2) an arrival deceleration burn at the destination planet, (3) a departure acceleration burn from the destination planet, and (4) an arrival deceleration burn back at the home planet. Closed-form solutions to the orbital mechanic equations for these missions are generally not possible; however, by using what is called the patched conic approximation the mission characteristics of the interplanetary voyage may be determined to a high degree of accuracy.

1. SUMMARY

While employing Eq. (2.19) to calculate specific impulse yields a simple means by which rocket engine efficiencies may be estimated, further analyses which examine the degree to which specific impulse impacts the operational characteristics of various interplanetary missions may provide a more useful means to evaluate the performance different rocket engine concepts. To this end, equations are derived, which relate the specific impulse of a rocket engine to the transit time and fuel requirements necessary to accomplish interplanetary missions. Such mission analyses can be quite complicated; however, the problem can be simplified considerably by using what is called the patched conic approximation [1]. This approximation, which has been found to be fairly accurate in most circumstances, breaks up an analytically unsolvable "N" body problem into several analytically solvable two-body problems that are "patched" together using conic sections of Kepler orbits. The patching occurs at what is called the planet's sphere of influence which is defined as that radius where planetocentric (planet-centered) gravitational effects on the space vehicle end and heliocentric (sun-centered) gravitational effects begin. In the analyses which follow, it will be assumed that from the point of view of the sun, the sphere of influence around any planet is zero, and from the point of view of any planet, its own sphere of influence is infinite. Because of the huge size of the sun as compared to the planets and the distance between the planets and the sun, this assumption does not cause any great errors in the calculations and eliminates the need to know the radius of the sphere of influence around any particular planet.

2. BASIC MISSION ANALYSIS EQUATIONS

In this section, equations for specific angular momentum and specific energy for objects in orbit are derived in order to determine the path those objects will take under the gravitational influence of a massive central body. In deriving these equations, it will be assumed that Newton's laws of motion and Newton's law of gravitation hold and that other bodies which are either small (thus having little mass) or far away (thus exerting little force on the object of interest) are neglected. Under these assumptions, the analysis is reduced to a simple two-body problem which may be solved analytically. The derivation of an object's orbital-specific angular momentum begins by equating Newton's second law with his law of gravitation with the result being:

$$\vec{F} = -G\frac{m_1 m_2}{r^2}\frac{\vec{r}}{|r|} = -G\frac{m_1 m_2}{r^2}\hat{r} = m_2\frac{d^2\vec{r}}{dt^2} \Rightarrow \frac{d^2\vec{r}}{dt^2} = -G\frac{m_1}{r^3}\vec{r} \qquad (4.1)$$

where G = universal gravitational constant = 6.673×10^{-11} m^3/kg s^2; $\vec{F}$ = gravitational force between bodies "1" and "2"; m_i = mass of body "i"; and r = distance between the two bodies.

If $\mu = Gm_1$, where "μ" is the *standard gravitational constant*, then from Eq. (4.1) it is possible to obtain:

$$\frac{d^2\vec{r}}{dt^2} = -\frac{\mu}{r^3}\vec{r} \Rightarrow \frac{d^2\vec{r}}{dt^2} + \frac{\mu}{r^3}\vec{r} = 0 \qquad (4.2)$$

Using Eqs. (4.1) and (4.2), it is now possible to define an expression such that:

$$\vec{r} \times \vec{F} = \vec{r} \times m_2\frac{d^2\vec{r}}{dt^2} = \vec{r} \times \left(\frac{-\mu m_2}{r^3}\right)\vec{r} = \left(\frac{-\mu m_2}{r^3}\right)\vec{r} \times \vec{r} \qquad (4.3)$$

Because $\vec{r} \times \vec{r} = 0$, Eq. (4.3) may be rewritten as:

$$0 = \vec{r} \times m_2\frac{d^2\vec{r}}{dt^2} = m_2\frac{d}{dt}\left(\vec{r} \times \frac{d\vec{r}}{dt}\right) = m_2\frac{d}{dt}\left(\vec{r} \times \vec{V}\right) \Rightarrow \frac{\text{constant}}{m_2} = \vec{h} = \vec{r} \times \vec{V} \qquad (4.4)$$

where $\vec{h}$ = orbital-specific angular momentum (angular momentum per unit mass) = constant and $\vec{V}$ = relative velocity between the bodies.

Since h is a constant, Eq. (4.4) implies that objects in any given orbit have constant-specific angular momentum.

The analysis to derive an expression for an object's orbital-specific energy starts by using Eqs. (4.1) and (4.2) in conjunction with the definition for mechanical work to yield the following relationship:

$$dE = \vec{F} \cdot d\vec{r} = m_2\frac{d^2\vec{r}}{dt^2} \cdot \frac{d\vec{r}}{dt}dt = m_2\left(\frac{-\mu}{r^3}\right)\vec{r} \cdot d\vec{r} \qquad (4.5)$$

where E = orbital energy.

Using the chain rule and noting that:

$$m_2\frac{d^2\vec{r}}{dt^2} \cdot \frac{d\vec{r}}{dt}dt = m_2\frac{d}{dt}\left(\frac{d\vec{r}}{dt}\right) \cdot \frac{d\vec{r}}{dt}dt = m_2 d\left(\frac{d\vec{r}}{dt}\right) \cdot \frac{d\vec{r}}{dt} = m_2\frac{1}{2}d\left(\frac{d\vec{r}}{dt}\right)^2 = m_2 d\left(\frac{V^2}{2}\right) \qquad (4.6)$$

2. BASIC MISSION ANALYSIS EQUATIONS

and noting that:

$$m_2\left(\frac{-\mu}{r^3}\right)\vec{r}\cdot d\vec{r} = m_2\left(\frac{-\mu}{r^3}\right) r\, dr = -m_2\left(\frac{\mu}{r^2}\right) dr = m_2 d\left(\frac{\mu}{r}\right) \tag{4.7}$$

as well as using the results from Eqs. (4.6) and (4.7) in Eq. (4.5) then yields:

$$m_2 d\left(\frac{V^2}{2}\right) - m_2 d\left(\frac{\mu}{r}\right) = 0 \tag{4.8}$$

Integrating Eq. (4.8) yields an equation of the form:

$$\frac{V^2}{2} - \frac{\mu}{r} = \frac{\text{constant}}{m_2} = E_o \Rightarrow V = \sqrt{2E_o + \frac{2\mu}{r}} \tag{4.9}$$

where E_o = orbital-specific energy (energy per unit mass) = constant.

Since E_o is a constant, Eq. (4.9) implies that objects in any given orbit have constant-specific energy. The specific energy can be seen to be the total energy per unit mass of an object in orbit with the first term in Eq. (4.9) representing the object's kinetic energy and the second term in Eq. (4.9) representing the object's potential energy. The next few steps in the derivation are devoted to determine the value of E_o in terms of other orbital parameters.

Writing the vector equations for position, velocity, and acceleration in plane polar coordinates it is found that:

$$\vec{r} = r\hat{r} \tag{4.10a}$$

$$\frac{d\vec{r}}{dt} = V = \frac{dr}{dt}\hat{r} + r\frac{d\phi}{dt}\hat{\phi} \tag{4.10b}$$

$$\frac{d^2\vec{r}}{dt^2} = \vec{a} = \left[\frac{d^2\vec{r}}{dt^2} - r\left(\frac{d\phi}{dt}\right)^2\right]\hat{r} + \left[r\frac{d^2\phi}{dt^2} + 2\frac{dr}{dt}\frac{d\phi}{dt}\right]\hat{\phi} \tag{4.10c}$$

Noting that the gravitational force acts only in the radial direction and that the tangential force is zero, the vector component of the radial acceleration term in Eqs. (4.10) may be written as:

$$-\frac{\mu}{r^2} = \frac{d^2\vec{r}}{dt^2} - r\left(\frac{d\phi}{dt}\right)^2 \tag{4.11}$$

and the vector component of the tangential acceleration term in Eq. (4.10) may be written as:

$$0 = r\frac{d^2\phi}{dt^2} + 2\frac{dr}{dt}\frac{d\phi}{dt} = \frac{1}{r}\frac{d}{dt}\left(r^2\frac{d\phi}{dt}\right) \Rightarrow r^2\frac{d\phi}{dt} = h \Rightarrow \frac{d\phi}{dt} = \frac{h}{r^2} \tag{4.12}$$

As expected, Eq. (4.12) again shows that the specific angular momentum of objects in orbit is equal to a constant. Now, using the chain rule on Eq. (4.11) and incorporating the results of Eq. (4.12), it is possible to eliminate the time component from the results such that:

$$-\frac{\mu}{r^2} = \frac{d^2\vec{r}}{dt^2} - r(\sin(\alpha_d)V_{si})^2 = \frac{d}{dt}\left(\frac{d\vec{r}}{dt}\right) - r\left(\frac{d\phi}{dt}\right)^2 = \left(\frac{d\phi}{dt}\right)^2 \frac{d}{d\phi}\left(\frac{d\vec{r}}{d\phi}\right) - r\left(\frac{d\phi}{dt}\right)^2$$

$$= \left(\frac{h}{r^2}\right)^2 \frac{d}{d\phi}\left(\frac{d\vec{r}}{d\phi}\right) - r\left(\frac{h}{r^2}\right)^2 = \frac{h}{r^2}\frac{d}{d\phi}\left(\frac{h}{r^2}\right)\frac{d\vec{r}}{d\phi} - \frac{h^2}{r^3} \tag{4.13}$$

To solve Eq. (4.13), it will prove useful to make a change of variables such that $\psi = 1/r$. Incorporating this variable change into Eq. (4.13) then yields:

$$-\mu\psi^2 = h\psi^2 \frac{d}{d\phi}\left[h\psi^2\left(\frac{1}{\psi^2}\frac{d\psi}{d\phi}\right)\right] - h^2\psi^3 \Rightarrow \frac{\mu}{h^2} = \frac{d^2\psi}{d\phi^2} + \psi \quad (4.14)$$

Differential equation (4.14) can now easily be solved analytically to obtain:

$$\psi(\phi) = A\sin(\phi) + B\cos(\phi) + \frac{\mu}{h^2} \Rightarrow \frac{d\psi}{d\phi} = A\cos(\phi) - B\sin(\phi) \quad (4.15)$$

If an initial condition is assumed such that:

$$\frac{d\psi}{d\phi} = 0 \text{ at } \phi = 0 \Rightarrow \left.\frac{d\psi}{d\phi}\right|_{\phi=0} = A\cos(0) - B\sin(0) \Rightarrow A = 0 \quad (4.16)$$

Eq. (4.15) may now be rewritten using the definition for ψ stated earlier to yield:

$$\psi(\phi) = B\cos(\phi) + \frac{\mu}{h^2} = \frac{\mu}{h^2}[\epsilon\cos(\phi) + 1] \Rightarrow r(\phi) = \frac{h^2}{\mu[\epsilon\cos(\phi) + 1]} \quad (4.17)$$

where $\epsilon = \frac{Bh^2}{\mu}$ = constant and ϕ = *true anomaly*.

Eq. (4.17) expresses the equation of motion or trajectory of an object traveling under the gravitational influence of a massive central body and represents a conic section with an eccentricity "ϵ." This equation gives the distance from the central body "m_1" located at a focus of an orbital conic section as a function of the true anomaly angle "ϕ" and the specific angular momentum "h" of an object in orbit. In Fig. 4.1, the parameter "a" is called the major radius and the parameter "b" is called the minor radius. When the true anomaly is equal to 90 degree, the radial distance "r" to the central body (represented by the sun in Fig. 4.1) is equal to what is called the semilatus rectum which is equal to "h^2/μ." Fig. 4.1 illustrates how the various orbital parameters of an object under the gravitational influence of the central body change as a function of the orbital eccentricity. Note that when the eccentricity is less than one, the flight paths of objects in orbit are elliptical with circular orbits arising in the special case where the eccentricity is zero. In the cases where the eccentricity is greater than one, the trajectories traced by objects are hyperbolic rather than elliptical, and there is no actual orbit around the central body. Objects following hyperbolic trajectories do not remain gravitationally bound to the central bodies influencing them, but rather simply experience changes in the direction of their travel. In the special case where the orbital eccentricity is exactly equal to one the trajectory of the object is parabolic rather than hyperbolic.

The angle between the velocity vector of a departing spacecraft at the planetary sphere of influence boundary "V_{si}" and the planetary orbital velocity vector of the planet from which the spacecraft is departing "V_{pd}" is defined as the departure angle "α_d" of the spacecraft. The departure angle may be related to the true anomaly of the spacecraft around the sun by noting that:

$$V_r = \sin(\alpha_d)V_{si} = \frac{dr}{dt} \quad (4.18)$$

and that:

$$V_\phi = \cos(\alpha_d)V_{si} = r\frac{d\phi_d}{dt} \quad (4.19)$$

2. BASIC MISSION ANALYSIS EQUATIONS 35

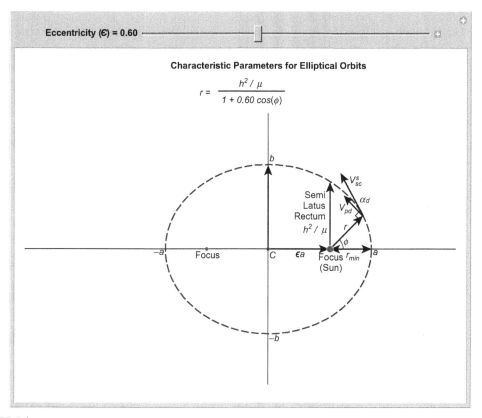

FIGURE 4.1

Orbital parameters.

Dividing Eq. (4.18) into Eq. (4.19) then yields:

$$\frac{V_r}{V_\phi} = \frac{\sin(\alpha_d)V_{si}}{\cos(\alpha_d)V_{si}} = \tan(\alpha_d) = \frac{1}{r}\frac{\frac{dr}{dt}}{\frac{d\phi_d}{dt}} = \frac{1}{r}\frac{dr}{d\phi_d} \quad (4.20)$$

Taking the derivative of the orbital equation of motion as expressed by Eq. (4.17) yields a relationship of the form:

$$\frac{dr}{d\phi_d} = \frac{h^2}{\mu} \frac{\epsilon \sin(\phi_d)}{[1+\epsilon \cos(\phi_d)]^2} \quad (4.21)$$

Substituting Eqs. (4.17) and (4.21) into Eq. (4.20) then yields an expression relating the angle of departure of a spacecraft on an interplanetary trajectory to the true anomaly of its orbit around the sun:

$$\tan(\alpha_d) = \frac{1}{r}\frac{dr}{d\phi_d} = \frac{\frac{h^2}{\mu}\frac{\epsilon \sin(\phi_d)}{[1+\epsilon \cos(\phi_d)]^2}}{\frac{h^2}{\mu}\frac{1}{[1+\epsilon \cos(\phi_d)]}} = \frac{\epsilon \sin(\phi_d)}{1+\epsilon \cos(\phi_d)} \quad (4.22)$$

Since the specific angular momentum and specific energy of an object in orbit are constant, their values may be determined from any convenient point in that orbit. It turns out that the derivation of the expression for the orbital-specific energy is simplified somewhat if the minimum value of the orbital radius (or periapsis) is chosen as the location to begin the calculations (note that the maximum value of the orbital radius is called the apoapsis). At periapsis, the spacecraft velocity vector and the radius vector are perpendicular to one another; and, it is at this point that the true anomaly is defined to be zero. From Eq. (4.4), therefore, the specific angular momentum can be represented by:

$$\vec{h} = \vec{r} \times \vec{V} \Rightarrow h = r_{min} V \sin\left(\frac{\pi}{2}\right) \Rightarrow h = r_{min} V \Rightarrow V = \frac{h}{r_{min}} \quad (4.23)$$

Substituting Eq. (4.23) into Eq. (4.9), the expression for the orbital-specific energy becomes:

$$E_o = \frac{h^2}{2r_{min}^2} - \frac{\mu}{r_{min}} \quad (4.24)$$

Incorporating the expression for the orbital radius from Eq. (4.17) into Eq. (4.24) under the assumption that the true anomaly is zero (eg, at $r = r_{min}$) one finds that:

$$E_o = \left\{\frac{h^2}{2}\left[\frac{\mu(\epsilon+1)}{h^2}\right]\right\}^2 - \frac{\mu[\mu(\epsilon+1)]}{h^2} = \frac{\mu}{2}\left[\frac{\mu(\epsilon^2-1)}{h^2}\right] \quad (4.25)$$

Examining Eq. (4.25), it should be noted that the total orbital energy varies as described in Table 4.1:

From Fig. 4.1, it can also be observed that $r_{min} = a - \epsilon a = a(1 - \epsilon)$. Using this relationship and Eq. (4.17) at a true anomaly of zero, it is possible to derive the following relationships:

$$r_{min} = a(1-\epsilon) = \frac{h^2}{\mu[1+\epsilon\cos(0)]} = \frac{h^2}{\mu(1+\epsilon)} \Rightarrow a(1-\epsilon^2) = \frac{h^2}{\mu} \Rightarrow a = \frac{h^2}{\mu(1-\epsilon^2)} \quad (4.26)$$

The relationships expressed in Eq. (4.26) may be used to rewrite the trajectory expression from Eq. (4.17) to yield:

$$r(\phi) = \frac{h^2}{\mu[\epsilon\cos(\phi)+1]} = \frac{a(1-\epsilon^2)}{\epsilon\cos(\phi)+1} \quad (4.27)$$

Note that for eccentricities greater than one (eg, hyperbolic trajectories), Eq. (4.26) yields values for "a" which are negative and for an eccentricity equal to one (eg, parabolic trajectory), Eq. (4.26)

Table 4.1 Orbital Energy Characteristics

Total Orbital Energy (E_o)	Eccentricity (ϵ)	Trajectory	Kinetic to Potential Energy
Negative	0	Circular	$|KE| < |PE|$
Negative	<1	Elliptical	$|KE| < |PE|$
Zero	=1	Parabolic	$|KE| = |PE|$
Positive	>1	Hyperbolic	$|KE| > |PE|$

yields a value "a" which is infinitely large. Using Eq. (4.26), it is possible now to rewrite the expression for the total orbital energy described by Eq. (4.25) such that:

$$E_o = -\frac{\mu}{2a} \tag{4.28}$$

Finally using Eq. (4.28), it is possible to cast the specific orbital energy described by Eq. (4.9) in a form such that only known orbital parameters are present:

$$\frac{V^2}{2} - \frac{\mu}{r} = -\frac{\mu}{2a} \tag{4.29}$$

As stated earlier, Eq. (4.29) relates the (varying) kinetic and potential energies of an object under the gravitational influence of a massive central body to its (constant) total energy. This workhorse equation, which is used in later sections, can be used to relate the velocity of the object to its radial position with respect to the central body. Thus rearranging Eq. (4.29), it is found that:

$$V = \sqrt{\frac{2\mu}{r} - \frac{\mu}{a}} \tag{4.30}$$

3. PATCHED CONIC EQUATIONS

The first step in calculating a transfer orbit between two planets is determining the properties of the heliocentric orbit that the spacecraft is to follow. These properties may be determined from the specific orbital energy formulation given in Eq. (4.30) and from the specific angular momentum formulation given in Eq. (4.4). These equations relate the velocity and departure angle of a spacecraft to its radial position with respect to the sun and a central planetary body. The degree to which a spacecraft can achieve a desired trajectory depends upon the specific impulse of the spacecraft's propulsion system and the amount of fuel relative to the weight of the spacecraft which can be carried onboard. In Fig. 4.2, the various orbital parameters which are necessary to specify the propulsive maneuvers required to effect a desired interplanetary trajectory are defined.

Depending upon the problem being analyzed the characteristics of the propulsion system may be used to determine the orbital parameters of which it is capable of achieving or, alternatively, by knowing the desired orbital parameters, it is possible to derive a set of requirements for the spacecraft propulsion system. In the particular analyses which follow, it is assumed that the desired orbital parameters are known ahead of time and that these parameters are used to determine the spacecraft propulsion system requirements. From a knowledge of the desired mission requirements, the necessary heliocentric planetary transfer orbit parameters may be determined from Eq. (4.30) which yields the velocity required by the spacecraft with respect to the sun once it leaves the departure planet's sphere of influence (eg, at the patch point). This velocity is given by:

$$V_{pd}^{hv} = \sqrt{\frac{2\mu_s}{R_{pd}} - \frac{\mu_s}{a_{sc}^s}} \tag{4.31}$$

where V_{pd}^{hv} = heliocentric velocity of the spacecraft after leaving the departure planet's sphere of influence; μ_s = solar standard gravitational parameter; R_{pd} = distance from the spacecraft (or departure planet) to the sun; and a_{sc}^s = major radius of the spacecraft trajectory with respect to the sun.

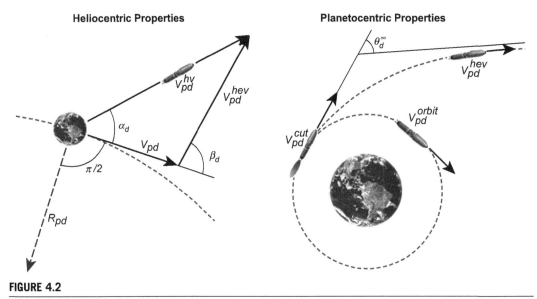

FIGURE 4.2

Orbital parameters for planetary departures.

The time required to complete the interplanetary orbital transfer along with the heliocentric velocity of the spacecraft after leaving the departure planet determines the required departure angle of the spacecraft with respect to the departure planet's orbital path around the sun. The spacecraft's heliocentric velocity and departure angle can then be used to determine its heliocentric-specific angular momentum from Eq. (4.4) such that:

$$\vec{h} = \vec{r} \times \vec{V} \Rightarrow h_s = R_{pd} V_{pd}^{hv} \sin\left(\alpha_d + \frac{\pi}{2}\right) \Rightarrow h_s = \cos(\alpha_d) R_{pd} V_{pd}^{hv} \qquad (4.32)$$

where α_d = spacecraft departure angle with respect to the planet and its orbital velocity vector around the sun and h_s = spacecraft heliocentric-specific angular momentum.

The time considerations mentioned in the preceding paragraphs are considered further in the next section. The spacecraft's heliocentric orbital velocity described in Eq. (4.31) is actually composed of two other vectors which include the heliocentric velocity vector of the departure planet and the planetocentric velocity vector of the spacecraft. These vectors which are illustrated in Fig. 4.2 may be related to one another through the use of the law of cosines with the result being:

$$\left(V_{pd}^{hev}\right)^2 = -2V_{pd} \cos(\alpha_d) V_{pd}^{hv} + V_{pd}^2 + \left(V_{pd}^{hv}\right)^2 \qquad (4.33)$$

where V_{pd}^{hev} = planetocentric velocity vector of the spacecraft and V_{pd} = heliocentric velocity vector of the planet of departure.

If the orbital radius of the planet of departure "R_{pd}" around the sun is assumed to be constant, the heliocentric velocity vector of the planet of departure in Eq. (4.33) may be calculated from Eq. (4.30) to yield:

$$V_{pd} = \sqrt{\frac{2\mu_s}{R_{pd}} - \frac{\mu_s}{R_{pd}}} = \sqrt{\frac{\mu_s}{R_{pd}}} \qquad (4.34)$$

The planetocentric velocity vector of the spacecraft defined as V_{pd}^{hev} is that velocity which the spacecraft possess once it leaves the planet's gravitational sphere of influence. This spacecraft velocity is also called the hyperbolic excess velocity and is proportional to the kinetic energy of the spacecraft after it has traveled infinitely far from the departure planet. The hyperbolic excess velocity can be related to the orbital-specific energy by noting from Eq. (4.9) that:

$$V_{pd}^{hev} = \lim_{H_{pd} \to \infty} \sqrt{2E_o + \frac{2\mu_{pd}}{r_{pd} + H_{pd}}} \Rightarrow 2E_o = \left(V_{pd}^{hev}\right)^2 \qquad (4.35)$$

where μ_{pd} = standard gravitational constant of the planet of departure; r_{pd} = radius of the planet of departure; and H_{pd} = height of spacecraft in orbit above the planet of departure.

If the spacecraft leaves the departure planet from a circular parking orbit at given distance "$r_{pd} + H_{pd}$" from the center of the planet, then Eq. (4.9) can again be used, this time in conjunction with Eq. (4.35), to calculate the spacecraft velocity at engine cutoff relative to the departure planet necessary to achieve the hyperbolic excess velocity required for insertion into the desired interplanetary transfer orbit.

$$V_{pd}^{cut} = \sqrt{2E_o + \frac{2\mu_{pd}}{r_{pd} + H_{pd}}} = \sqrt{\frac{2\mu_{pd}}{r_{pd} + H_{pd}} + \left(V_{pd}^{hev}\right)^2} \qquad (4.36)$$

where V_{pd}^{cut} = spacecraft velocity required to achieve the desired interplanetary transfer orbit.

Since the spacecraft is assumed to begin its interplanetary transfer from earth orbit, the nuclear engine only has to accelerate the spacecraft from its nominal earth orbit velocity to its required hyperbolic excess velocity. Using Eq. (4.30) again to determine the spacecraft's orbital velocity, it is found that:

$$V_{pd}^{orbit} = \sqrt{\frac{2\mu_{pd}}{r_{pd} + H_{pd}} - \frac{\mu_{pd}}{r_{pd} + H_{pd}}} = \sqrt{\frac{\mu_{pd}}{r_{pd} + H_{pd}}} \qquad (4.37)$$

The velocity increment required to be delivered by the spacecraft propulsion system may thus be determined by subtracting Eq. (4.37) from Eq. (4.36) yielding:

$$\Delta V_{pd} = V_{pd}^{cut} - V_{pd}^{orbit} = \sqrt{\frac{2\mu_{pd}}{r_{pd} + H_{pd}} + \left(V_{pd}^{hev}\right)^2} - \sqrt{\frac{\mu_{pd}}{r_{pd} + H_{pd}}} \qquad (4.38)$$

The semimajor axis "a_{pd}" of the hyperbolic planetary escape orbit can be evaluated by using Eq. (4.30) and the value for the engine cutoff velocity as determined from Eq. (4.36) such that:

$$V_{pd}^{cut} = \sqrt{\frac{2\mu_{pd}}{a_{pd}} - \frac{\mu_{pd}}{a_{pd}}} = \sqrt{\frac{\mu_{pd}}{a_{pd}}} \Rightarrow a_{pd} = \frac{\mu_{pd}}{\left(V_{pd}^{cut}\right)^2} \tag{4.39}$$

The eccentricity of the escape orbit can now be calculated from Eq. (4.26) to yield:

$$a_{pd} = \frac{h_s^2}{\mu_{pd}\left(1 - \epsilon_{pd}^2\right)} \Rightarrow \epsilon_{pd} = \sqrt{1 - \frac{h_s^2}{\mu_{pd}a_{pd}}} \tag{4.40}$$

The next few equations concern the patch conditions which relate the planetocentric orbital parameters to the heliocentric orbital parameters at the planet of departure. From Fig. 4.2 it may now be noted that:

$$V_{pd}^{hv}\sin(\alpha_d) = V_{pd}^{hv}\sin(\beta_d) \quad \text{and} \quad V_{pd}^{hv}\cos(\alpha_d) = V_{pd}^{hv}\cos(\beta_d) + V_{pd} \tag{4.41}$$

Using the geometric relationships defined by Eq. (4.41) to solve for "β_d" which is the angle between the planetocentric velocity vector of the departing spacecraft and the departure planet's heliocentric orbital velocity vector yields:

$$\tan(\beta_d) = \frac{V_{pd}^{hv}\sin(\alpha_d)}{V_{pd}^{hv}\cos(\beta_d) + V_{pd}} \tag{4.42}$$

Assuming that the hyperbolic planetary injection engine burn occurs parallel to the parking orbit vector (eg, resulting in a burn at periapsis where the planetary true anomaly is equal to zero), then the limiting planetocentric true anomaly θ_d^∞ at the edge of the planetary sphere of influence (or equivalently at $r \approx \infty$) may be determined from the hyperbolic trajectory profile given by Eq. (4.15) with the result being:

$$r = \frac{-h_s^2/\mu_{pd}}{1 + \epsilon_{pd}\cos(\theta_d^\infty)} \approx \infty \Rightarrow 0 = 1 + \epsilon_{pd}\cos(\theta_d^\infty) \Rightarrow \cos(\theta_d^\infty) = \frac{-1}{\epsilon_{pd}} \tag{4.43}$$

The angular position of the spacecraft around the departure planet where the engine burn must be initiated to allow it to achieve a desired interplanetary trajectory can now be determined as the angle between the planetocentric departure velocity vector of the spacecraft and the planet's heliocentric orbital velocity vector around the sun as expressed in Eq. (4.42) plus the limiting planetocentric true anomaly "θ_d^∞" determined from Eq. (4.43) such that:

$$\theta_d = \theta_d^\infty + \beta_d \tag{4.44}$$

where θ_d = angular position of the spacecraft around the departure planet where the departure burn must be initiated.

Upon arriving at the destination planet, the patch calculations for insertion in a desired arrival planet parking orbit are essentially the reverse of those described in the preceding paragraphs. Like the equations for planetary departures, these equations relate the velocity and arrival angle of a spacecraft to its radial position with respect to the sun and a central planetary body.

The heliocentric velocity of the spacecraft upon arriving at the destination planet may be determined from the conservation of total energy. Since the total specific energy of the spacecraft at the departure and destination planets is equal, Eq. (4.9) may be employed such that:

$$E_o = \frac{\left(V_{pd}^{hv}\right)^2}{2} - \frac{\mu_s}{R_{pd}} = \frac{\left(V_{pa}^{hv}\right)^2}{2} - \frac{\mu_s}{R_{pa}} \Rightarrow V_{pa}^{hv} = \sqrt{\left(V_{pd}^{hv}\right)^2 - 2\mu_s\left(\frac{1}{R_{pd}} - \frac{1}{R_{pa}}\right)} \quad (4.45)$$

where R_{pa} = distance from the spacecraft (or destination planet) to the sun and V_{pa}^{hv} = heliocentric velocity of the spacecraft upon arriving at the destination (eg, arrival) planet.

The true anomaly at the destination planet can be determined through the use of Eq. (4.27) by rearranging it to yield the following:

$$r_{pa} = \frac{a(1-\epsilon^2)}{\epsilon \cos(\phi_{pa}) + 1} \Rightarrow \phi_{pa} = \cos^{-1}\left[\frac{a(1-\epsilon^2) - r_{pa}}{\epsilon r_{pa}}\right] \quad (4.46)$$

The angle between the heliocentric velocity vector of the arriving spacecraft and the heliocentric orbital velocity vector of the destination planet can be determined from the conservation of angular momentum. Thus equating the angular momentum determined from Eq. (4.32) at the patch point of the departure planet with the angular momentum of the spacecraft at the patch point of the destination planet one finds that:

$$h_s = R_{pd} V_{pd}^{hv} \cos(\alpha_d) = R_{pa} V_{pa}^{hv} \cos(\alpha_a) \Rightarrow \alpha_a = \cos^{-1}\left[\frac{R_{pd} V_{pd}^{hv} \cos(\alpha_d)}{R_{pa} V_{pa}^{hv}}\right] \quad (4.47)$$

Fig. 4.3 illustrates how the heliocentric velocity vector of the destination planet and the planetocentric velocity vector of the spacecraft are related to one another at the patch point. Using the law of

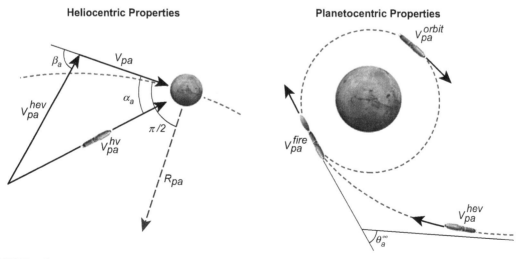

FIGURE 4.3

Orbital parameters for planetary arrivals.

cosines, an expression again may be determined for the velocity vector of the spacecraft with respect to the heliocentric velocity vector of the destination planet such that:

$$\left(V_{pa}^{hev}\right)^2 = \left(V_{pa}^{hv}\right)^2 + (V_{pa})^2 - 2V_{pa}^{hv}V_{pa}\cos(\alpha_a) \tag{4.48}$$

where V_{pa}^{hev} = spacecraft departure angle with respect to the planet and its orbital velocity vector around the sun; V_{pa} = heliocentric velocity vector of the destination planet; and α_a = angle between the spacecraft's and destination planet's heliocentric velocity vectors.

Fig. 4.3 also illustrates how the patch conditions allow for the determination of spacecraft angles of approach with respect to the planetary heliocentric velocity vector. These spacecraft approach angles (heliocentric and planetocentric) may be related to one another by noting that:

$$V_{pa}^{hv}\sin(\alpha_a) = V_{pa}^{hev}\sin(\beta_a) \quad \text{and} \quad V_{pa}^{hv}\cos(\alpha_a) = V_{pa}^{hev}\cos(\beta_a) + V_{pa} \tag{4.49}$$

Using the geometric relationships defined by Eq. (4.49) to solve for "β_a" which is the angle between the spacecraft's planetocentric arrival velocity vector and the planet's heliocentric orbital velocity vector around the sun, it is possible to derive an equation of the form:

$$\tan(\beta_a) = \frac{V_{pa}^{hv}\sin(\alpha_a)}{V_{pa}^{hv}\cos(\alpha_a) - V_{pa}} \tag{4.50}$$

Assuming that it is desired for the spacecraft to capture into a circular parking orbit of radius "r_{pa}" around the destination planet and following the same logic as presented earlier for deriving the equations related to planetary departures, the kinetic energy of the spacecraft at the planetary sphere of influence (or equivalently, infinitely far from the planet) can be related to the orbital-specific energy by again noting from Eq. (4.9) that:

$$V_{pa}^{hev} = \lim_{H_{pa}\to\infty}\sqrt{2E_o + \frac{2\mu_{pa}}{r_{pa}+H_{pa}}} \Rightarrow 2E_o = \left(V_{pa}^{hev}\right)^2 \tag{4.51}$$

As the spacecraft enters the sphere of influence of the destination planet possessing a hyperbolic excess velocity of "V_{pa}^{hev}," Eq. (4.9) can again be used in conjunction with Eq. (4.51), to calculate the velocity necessary to achieve planetary capture at the desired parking orbit radius.

$$V_{pa}^{fire} = \sqrt{2E_o + \frac{2\mu_{pa}}{r_{pa}+H_{pa}}} = \sqrt{\left(V_{pa}^{hev}\right)^2 + \frac{2\mu_{pa}}{r_{pa}+H_{pa}}} \tag{4.52}$$

Where V_{pa}^{fire} = velocity of the spacecraft when propulsive braking is initiated at the capture orbit height and H_{pa} = height of the capture orbit above the surface of the destination planet.

As the spacecraft approaches the destination planet, the semimajor axis a_{pa} of the inbound hyperbolic trajectory can be evaluated by using Eq. (4.30) and the value for the planetary capture velocity as determined from Eq. (4.52) to yield the following:

$$V_{pa}^{fire} = \sqrt{\frac{2\mu_{pa}}{a_{pa}} - \frac{\mu_{pa}}{a_{pa}}} \Rightarrow a_{pa} = \frac{\mu_{pa}}{\left(V_{pa}^{fire}\right)^2} \tag{4.53}$$

The eccentricity of the spacecraft inbound trajectory can now also be calculated using Eq. (4.26) such that:

$$a_{pa} = \frac{h_s^2}{\mu_{pa}\left(1 - \epsilon_{pa}^2\right)} \Rightarrow \epsilon_{pa} = \sqrt{1 - \frac{h_s^2}{\mu_{pa} a_{pa}}} \tag{4.54}$$

To achieve the most efficient planetary capture, the spacecraft must fire its engine perpendicular to the radius vector of the destination planet at its closest approach to the planet (eg, at periapsis). The radius at which this burn occurs is the radius of the spacecraft's parking orbit. The angle of approach between the spacecraft at the edge of the planetary sphere of influence (or equivalently at $r \approx \infty$) and periapsis (which is also the limiting planetocentric true anomaly, eg, θ_a^∞), where the planetary capture burn occurs may be determined from the hyperbolic trajectory profile given by Eq. (4.15) with the result being:

$$r = \frac{-h_{sc}^2/\mu_{pd}}{1 + \epsilon_{pa}\cos(\theta_a^\infty)} \approx \infty \Rightarrow 0 = 1 + \epsilon_{pa}\cos(\theta_a^\infty) \Rightarrow \cos(\theta_a^\infty) = \frac{-1}{\epsilon_{pa}} \tag{4.55}$$

The angular position of the spacecraft around the destination planet necessary for it to be captured into a desired parking orbit can now be determined as the angle between the spacecraft's planetocentric arrival velocity vector and the planet's heliocentric orbital velocity vector around the sun as expressed in Eq. (4.50) plus the limiting planetocentric true anomaly "θ_a^∞" determined from Eq. (4.54), thus:

$$\theta_a = \theta_a^\infty + \beta_a \tag{4.56}$$

where θ_a = angular position of the spacecraft around the destination planet where the capture burn must be initiated.

4. FLIGHT TIME EQUATIONS

In Section 3, expressions were derived which related a spacecraft's position and velocity to its true anomaly; that is, its angular displacement with respect to the periapsis of its trajectory. During the course of these derivations, the time variable was temporarily eliminated from the analysis. In this section, time is reintroduced so that transit times for interplanetary missions may be determined. The process of reintroducing the time variable begins by noting that Eq. (4.12) may be rewritten such that:

$$\frac{d\phi}{dt} = \frac{h}{r^2} \Rightarrow dt = \frac{r^2}{h} d\phi \tag{4.57}$$

If the trajectory expression from Eq. (4.17) is now incorporated into Eq. (4.57), it is possible to obtain:

$$dt = \frac{1}{h}\left\{\frac{h^2}{\mu[1 + \epsilon\cos(\phi)]}\right\}^2 d\phi = \frac{h^3}{\mu^2} \frac{d\phi}{[1 + \epsilon\cos(\phi)]^2} \tag{4.58}$$

By integrating Eq. (4.58) from a true anomaly of zero (periapsis) to some arbitrary true anomaly along the spacecraft trajectory, it is possible to obtain an expression for the spacecraft transit time between the two true anomaly points such that:

$$t = \frac{h^3}{\mu^2}\int_0^\phi \frac{d\phi'}{[1+\epsilon\cos(\phi')]^2} = \frac{h^3}{\mu^2}\frac{1}{(1-\epsilon^2)^{3/2}}\left\{2\tan^{-1}\left[\sqrt{\frac{1-\epsilon}{1+\epsilon}}\tan\left(\frac{\phi}{2}\right)\right] - \frac{\epsilon\sqrt{1-\epsilon^2}\sin(\phi)}{1+\epsilon\cos(\phi)}\right\}$$

(4.59)

Eq. (4.59) may be simplified considerably by introducing a new variable called the eccentric anomaly "E." Geometrically, this variable is illustrated in Fig. 4.4. In this figure, the orbital path of the spacecraft is inscribed within a circle having a radius equal to the trajectory's semimajor axis "a" and which just touches the orbital path of the spacecraft at its periapsis and apoapsis points. Fig. 4.4 also illustrates the nature of the relationship that exists between the true anomaly "ϕ" and the eccentric anomaly "E."

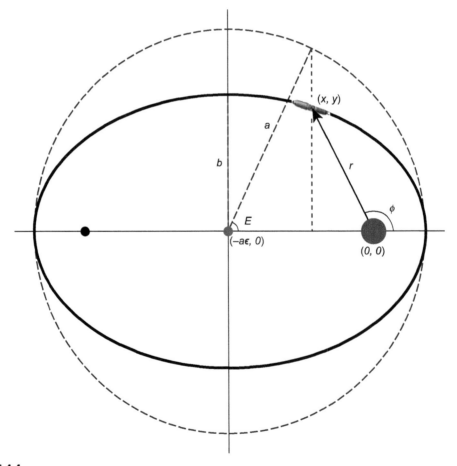

FIGURE 4.4

Geometric relationship between the true anomaly and the eccentric anomaly.

4. FLIGHT TIME EQUATIONS

At this point, it is useful to express the radial distance of the spacecraft from the focus of its trajectory in terms of the eccentric anomaly. First, the equation of the spacecraft trajectory is written in Cartesian coordinates with the origin at the primary focal point of the trajectory yielding:

$$\frac{(x+a\epsilon)^2}{a^2} + \frac{y^2}{b^2} = 1 \tag{4.60}$$

Referring to Fig. 4.4, also note that:

$$x = a\cos(E) - a\epsilon \tag{4.61}$$

Substituting Eq. (4.61) into Eq. (4.60) then yields:

$$1 = \frac{[a\cos(E) - a\epsilon + a\epsilon]^2}{a^2} + \frac{y^2}{b^2} = \cos^2(E) + \frac{y^2}{b^2} \tag{4.62}$$

From the mathematical description of an ellipse, it is found that the eccentricity may be represented by:

$$\epsilon = \sqrt{\frac{a^2 - b^2}{a^2}} \Rightarrow b = a\sqrt{1-\epsilon^2} \tag{4.63}$$

If Eq. (4.63) is incorporated into Eq. (4.62) and the terms rearranged it is found that:

$$1 = \cos^2(E) + \frac{y^2}{a^2(1-\epsilon^2)} \Rightarrow y^2 = a^2(1-\epsilon^2)[1-\cos^2(E)] = a^2(1-\epsilon^2)\sin^2(E) \tag{4.64}$$

Applying the Pythagorean theorem in conjunction with Eqs. (4.61) and (4.64) finally yields the radius of the spacecraft trajectory in terms of the eccentric anomaly such that:

$$r^2 = x^2 + y^2 = [a\cos(E) - a\epsilon]^2 + a^2(1-\epsilon^2)\sin^2(E)$$
$$= a^2[1 - \epsilon\cos(E)]^2 \Rightarrow r = a[1 - \epsilon\cos(E)] \tag{4.65}$$

By equating the trajectory radius expressed in terms of the true anomaly from Eq. (4.27) with the trajectory radius expressed in terms of the eccentric anomaly from Eq. (4.65), the mathematical relationship between the true anomaly and the eccentric anomaly may be shown to be of the form:

$$r = \frac{a(1-\epsilon^2)}{\epsilon\cos(\phi) + 1} = a[1 - \epsilon\cos(E)] \Rightarrow \cos(E) = \frac{\epsilon + \cos(\phi)}{1 + \epsilon\cos(\phi)} \tag{4.66}$$

Using trigonometric identities, it is also possible to establish from Eq. (4.61) that:

$$\sin(\phi) = \sqrt{1 - \cos^2(E)} = \frac{\sqrt{1-\epsilon^2}\sin(E)}{1 - \epsilon\cos(E)} \tag{4.67}$$

and:

$$\tan\left(\frac{\phi}{2}\right) = \frac{1 - \cos(\phi)}{\sin(\phi)} = \frac{1 - \epsilon\cos(E) - \cos(E) + \epsilon}{\sqrt{1-\epsilon^2}\sin(E)} = \frac{(\epsilon+1)[1-\cos(E)]}{\sqrt{1-\epsilon^2}\sin(E)}$$
$$= \sqrt{\frac{1+\epsilon}{1-\epsilon}}\frac{1-\cos(E)}{\sin(E)} = \sqrt{\frac{1+\epsilon}{1-\epsilon}}\tan\left(\frac{E}{2}\right) \tag{4.68}$$

If Eqs. (4.66)–(4.68) are now substituted into the interplanetary transit time Eq. (4.59), a somewhat simpler form of the transit time equation results in which the eccentric anomaly replaces the true anomaly such that:

$$t = \frac{h^3}{\mu_s^2} \frac{1}{(1-\epsilon^2)^{3/2}} \left\{ 2\tan^{-1}\left[\sqrt{\frac{1-\epsilon}{1+\epsilon}}\sqrt{\frac{1+\epsilon}{1-\epsilon}}\tan\left(\frac{E}{2}\right)\right] - \frac{\epsilon\sqrt{1-\epsilon^2}\frac{\sqrt{1-\epsilon^2}\sin(E)}{1-\epsilon\cos(E)}}{1+\epsilon\frac{\cos(E)-\epsilon}{1-\epsilon\cos(E)}} \right\}$$

$$= \frac{h^3}{\mu_s^2} \frac{E - \epsilon \sin(E)}{(1-\epsilon^2)^{3/2}} \tag{4.69}$$

Using Eq. (4.26), the transit time relationship of Eq. (4.69) can be simplified still further to yield:

$$t = \frac{h^3}{\mu_s^2} \frac{E - \epsilon \sin(E)}{(1-\epsilon^2)^{3/2}} = a^{3/2}(1-\epsilon^2)^{3/2} \frac{E - \epsilon \sin(E)}{\sqrt{\mu_s}(1-\epsilon^2)^{3/2}} = [E - \epsilon \sin(E)]\sqrt{\frac{a^3}{\mu_s}} \tag{4.70}$$

where t = transit time from an eccentric anomaly of 0 degree to an eccentric anomaly of E degree.

Eq. (4.70) is called Kepler's equation and is the expression generally used to calculate interplanetary transit times. Fig. 4.5 illustrates pictorially the orbital trajectory characteristics of various earth to Mars missions. The lowest-energy mission possible follows what is called a *Hohmann trajectory*. In a Hohmann trajectory, the apogee of the spacecraft transfer orbit is exactly equal to the destination planet's orbital radius around the sun and the perigee of the spacecraft transfer orbit is exactly equal to the departure planet's orbital radius around the sun. While a Hohmann interplanetary transfer is the lowest energy mission possible, it is also the slowest, requiring about 259 days to accomplish for an earth to Mars transfer.

The interactive version of Fig. 4.5 also illustrates how relative small velocity increases in excess of that required for a Hohmann transfer yields fairly large decreases in transit time. These transit time decreases are not primarily due to the added velocity increment available but rather are due to drops in the required transit distance. This decrease in transit distance gradually becomes smaller as the total mission velocity increases and asymptotically approaches a minimum equal to the semilatus rectum assuming all departure angles are tangent to the orbital velocity vector of the departure planet.

The sum of all the velocity changes necessary to accomplish a particular mission is called the *total mission velocity* and includes, not only the velocity increments required to escape from orbit around the departure planet and capture into orbit around the destination planet, but also all other required velocity changes such as adjusting the orbital plane of the transfer orbit, midcourse corrections, and so on. The propulsion system chosen for a particular interplanetary mission, therefore, must be capable of delivering this total mission velocity or the mission would be impossible to accomplish. This total mission velocity may be related to the vehicle mass fraction and engine-specific impulse through the application of the rocket equation as expressed by Eq. (2.8). The goal now is to determine for a particular interplanetary mission, some combination of mission velocity increments whose sum is equal to the maximum vehicle velocity from the rocket equation. These mission velocity increments are functions of the various orbital parameters (eg, true anomaly) and are chosen such that for a given total mission velocity, the interplanetary transit time is minimized. This calculation does not have a closed-form solution and must, therefore, be performed numerically. As an example, if one assumes

4. FLIGHT TIME EQUATIONS   47

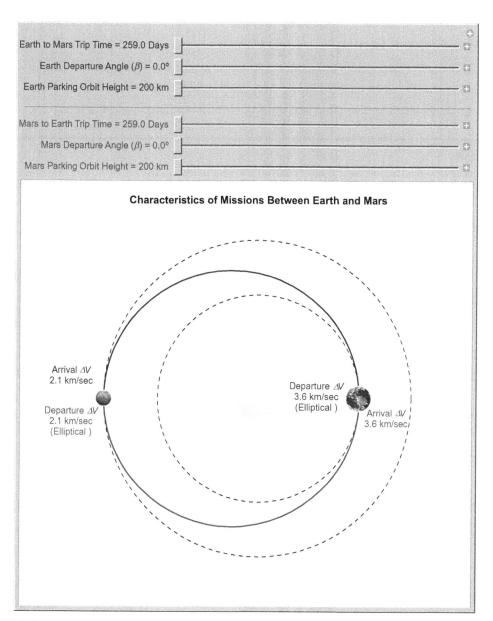

FIGURE 4.5

Earth/Mars mission characteristics.

that the mission in question is a one-way voyage which captures into orbit around a destination planet, then the problem would be set up in such a way that:

$$t_{min} = \text{Minimum}[t(a,\epsilon,\phi)] \text{ with the constraint that: } V_{tm}(I_{sp},f_m) = \Delta V_{pd}[a,\epsilon,\phi_d] + \Delta V_{pa}[a,\epsilon,\phi_a] \quad (4.71)$$

where: t_{min} = minimum total trip time and $t(a,\epsilon,\phi)$ = functional equivalent of trip time Eq. (4.70).

EXAMPLE

Determine the one-way trip time from the earth orbit to Mars orbit assuming that the transfer vehicle has a mass fraction of 0.25 and that it employs a nuclear thermal rocket engine having a specific impulse of 850 s. Assume that the transfer vehicle leaves earth from a 200 km high orbit in the same direction as the earth's velocity vector around the sun ($\beta_{earth} = 0° \Rightarrow \alpha_{earth} = 0° \Rightarrow \phi_{earth} = 0°$) and that it captures into a 100 km high Mars orbit. Also assume that:

Parameter	Value	Units
Distance from sun to earth	149,700,000	km
Radius of earth	6378	km
Earth standard gravitational constant	398,600	km³/s²
Distance from sun to Mars	228,000,000	km
Radius of Mars	3393	km
Mars standard gravitational constant	42,830	km³/s²
Sun standard gravitational constant	1.327×10^{11}	km³/s²

Note that this calculation does not minimize the earth to Mars transit time since for this problem, the spacecraft angle of departure has been fixed.

Solution

The first step in the calculation is to determine the target total mission velocity which is a function of the nuclear engine's specific impulse and the transfer vehicle's mass fraction. From the rocket equation as expressed by, the total velocity increment available from the vehicle for mission maneuvers may be calculated from Eq. (2.8) to be:

$$\Delta V_{vehicle} = -0.0098 I_{sp} \ln(f_m) = -0.0098 \frac{km}{s^2} \times 850 \, s \times \ln(0.25) = 11.548 \frac{km}{s} \quad (1)$$

The next step in the analysis is to determine the trajectory characteristics of the earth to Mars transfer maneuver. To perform this calculation, a knowledge of the solar orbital eccentricity is required. Since this parameter is not yet known, guesses are made for the solar orbital eccentricity until a value is found which yields a total mission velocity that matches the total velocity increment available from the vehicle. For the present, assume that the solar orbital eccentricity (ϵ) is 0.4. With this orbital eccentricity, the value for the major radius of the vehicle transfer orbit may be determined from Eq. (4.27) such that:

$$a = \frac{R_{earth}[1 + \epsilon \cos(\phi_{earth})]}{1 - \epsilon^2} = \frac{149,700,000 \, km[1 + 0.4 \times \cos(0)]}{1 - 0.4^2} = 249,500,000 \, km \quad (2)$$

4. FLIGHT TIME EQUATIONS

Knowing the major radius of the vehicle transfer orbit from Eq. (2), the heliocentric velocity of the spacecraft after leaving earth orbit may be calculated from Eq. (4.31) yielding:

$$V_{earth}^{hv} = \sqrt{\mu_s \left(\frac{2}{R_{earth}} - \frac{1}{a}\right)} = \sqrt{1.327 \times 10^{11} \frac{km^3}{s^2} \left(\frac{2}{149,700,000 \text{ km}} - \frac{1}{249,500,000 \text{ km}}\right)} = 35.23 \frac{km}{s} \quad (3)$$

In order to calculate the spacecraft hyperbolic excess velocity from earth, the velocity of earth around the sun must also be calculated. The velocity of earth around the sun may be calculated using Eq. (4.34) yielding:

$$V_{earth}^{hv} = \sqrt{\frac{\mu_s}{R_{earth}}} = \sqrt{\frac{1.327 \times 10^{11} \frac{km^3}{s^2}}{149700000 \text{ km}}} = 29.773 \frac{km}{s} \quad (4)$$

The hyperbolic excess velocity which is the velocity of the spacecraft after it has left the gravitational sphere of influence of the earth may now be calculated from Eq. (4.33) using the results from Eqs. (3) and (4) such that:

$$V_{earth}^{hev} = \sqrt{\left(V_{earth}^{hv}\right)^2 + V_{earth}^2 - 2V_{earth}^{hv}V_{earth}^{hv}\cos(\alpha_{earth})}$$

$$= \sqrt{35.23^2 \frac{km^2}{s^2} + 29.773^2 \frac{km^2}{s^2} - 2 \times 35.23 \frac{km}{s} \times 29.773 \frac{km}{s} \cos(0)} = 5.455 \frac{km^3}{s^2} \quad (5)$$

Knowing the hyperbolic excess velocity from Eq. (5), it is now possible to determine the spacecraft velocity at engine cutoff relative to earth required to insert the spacecraft into the desired Mars transfer orbit. From Eq. (4.36) the spacecraft velocity at engine cutoff may be determined to be:

$$V_{earth}^{cut} = \sqrt{\frac{2\mu_{earth}}{r_{earth} + H_{earth}} + \left(V_{earth}^{hev}\right)^2} = \sqrt{\frac{2,398,600 \frac{km^3}{s^2}}{6378 \text{ km} + 200 \text{ km}} + 5.455^2 \frac{km^2}{s^2}} = 12.286 \frac{km}{s} \quad (6)$$

To determine the velocity increment required by the spacecraft to insert the vehicle into the desired Mars transfer orbit from its earth parking orbit, the orbital velocity of the spacecraft must also be evaluated. Using Eq. (4.37) to calculate the spacecraft orbital velocity yields:

$$V_{earth}^{orbit} = \sqrt{\frac{\mu_{earth}}{r_{earth} + H_{earth}}} = \sqrt{\frac{398,600 \frac{km^3}{s^2}}{6378 \text{ km} + 200 \text{ km}}} = 7.784 \frac{km}{s} \quad (7)$$

Using the results from Eqs. (6) and (7), the velocity increment required to be delivered by the nuclear propulsion system to inject the spacecraft into its desired Mars transfer orbit is thus:

$$\Delta V_{earth} = V_{earth}^{cut} - V_{earth}^{orbit} = 12.286 \frac{km}{s} - 7.784 \frac{km}{s} = 4.502 \frac{km}{s} \quad (8)$$

Upon arriving at Mars, the calculations for inserting the spacecraft the desired parking orbit are the reverse of those used to describe the spacecraft leaving earth orbit. The first calculation required is the determination of the heliocentric velocity of the spacecraft upon arriving at Mars. This velocity may be determined from Eq. (4.45) and the results from Eq. (3) such that:

$$V_{Mars}^{hv} = \sqrt{\left(V_{earth}^{hv}\right)^2 + 2\mu_s\left(\frac{1}{R_{earth}} - \frac{1}{R_{Mars}}\right)}$$

$$= \sqrt{35.228^2 \frac{km^2}{s^2} + 2 \times 1.327 \times 10^{11} \frac{km^3}{s^2}\left(\frac{1}{149,700,000 \text{ km}} - \frac{1}{228,000,000 \text{ km}}\right)} = 25.143 \frac{km}{s} \quad (9)$$

In addition to the heliocentric arrival velocity of the spacecraft, it is also necessary to determine the angle at which the spacecraft arrives at Mars relative to the planet's orbital velocity vector. Using Eq. (4.47) to determine this angle yields:

$$\alpha_{Mars} = \cos^{-1}\left[\frac{R_{earth} V_{earth}^{hv} \cos(\alpha_{earth})}{R_{Mars} V_{Mars}^{hv}}\right] = \cos^{-1}\left[\frac{149,700,000 \text{ km} \times 35.228 \frac{km}{s} \times \cos(0)}{228,000,000 \text{ km} \times 25.143 \frac{km}{s}}\right] = 23.08° \quad (10)$$

In order to relate the spacecraft arrival velocity relative to Mars from its heliocentric arrival velocity it is necessary to first determine the orbital velocity of Mars around the sun. As before, use is made of Eq. (4.34) such that:

$$V_{Mars} = \sqrt{\frac{\mu_s}{R_{Mars}}} = \sqrt{\frac{1.327 \times 10^{11} \frac{km^3}{s^2}}{228,000,000 \text{ km}}} = 24.125 \frac{km}{s} \quad (11)$$

The spacecraft arrival velocity relative to Mars (which is equivalent to the hyperbolic escape velocity) can now be calculated using Eq. (4.48) yielding:

$$V_{Mars}^{hev} = \sqrt{\left(V_{Mars}^{hv}\right)^2 + V_{Mars}^2 - 2V_{Mars}^{hv} V_{Mars} \cos(\alpha_{Mars})}$$

$$= \sqrt{25.143^2 \frac{km^2}{s^2} + 24.125^2 \frac{km^2}{s^2} - 2 \times 25.125 \frac{km}{s} \times 24.125 \frac{km}{s} \times \cos(23.08°)} = 9.908 \frac{km}{s} \quad (12)$$

As the Mars gravity begins to affect the spacecraft, a propulsive braking maneuver is initiated at the height of the capture orbit above the Mars surface. The spacecraft velocity at the time the braking maneuver begins may be determined from Eq. (4.52) using the results from Eq. (12) such that:

$$V_{Mars}^{fire} = \sqrt{\left(V_{Mars}^{hev}\right)^2 + \frac{2\mu_{Mars}}{r_{Mars} + H_{Mars}}} = \sqrt{9.908^2 \frac{km^2}{s^2} + \frac{2 \times 42,830 \frac{km^3}{s^2}}{3393 \text{ km} + 100 \text{ km}}} = 11.077 \frac{km}{s} \quad (13)$$

To determine the velocity increment required by the spacecraft to capture into its Mars parking orbit, the orbital velocity of the spacecraft at the desired Mars capture height must also be evaluated. Using Eq. (4.37) to calculate the spacecraft orbital velocity at Mars then yields:

$$V_{Mars}^{orbit} = \sqrt{\frac{\mu_{Mars}}{r_{Mars} + H_{Mars}}} = \sqrt{\frac{42,830 \frac{km^3}{s^2}}{3393 \text{ km} + 100 \text{ km}}} = 3.502 \frac{km}{s} \quad (14)$$

4. FLIGHT TIME EQUATIONS

The velocity increment required to be delivered by the nuclear propulsion system to capture into its desired Mars parking orbit may be determined from Eqs. (13) and (14) such that:

$$\Delta V_{Mars} = V_{Mars}^{fire} - V_{Mars}^{orbit} = 11.077 \frac{km}{s} - 3.502 \frac{km}{s} = 7.575 \frac{km}{s} \quad (15)$$

The total velocity increment which must be delivered by the nuclear propulsion system to carry out the mission is the sum of the velocity increments to leave earth and capture at Mars. Using Eqs. (8) and (15), the total required mission velocity increment is thus:

$$\Delta V_{mission} = \Delta V_{earth} + \Delta V_{Mars} = 4.502 \frac{km}{s} + 7.575 \frac{km}{s} = 12.077 \frac{km}{s} \quad (16)$$

Comparing the $\Delta V_{vehicle}$ (11.548 km/s) available from the nuclear engine with the mission $\Delta V_{mission}$ (12.601 km/s), it can be seen that the solar orbital eccentricity of 0.4 was too low. A higher value for the orbital eccentricity of the Mars transfer orbit (implying a higher energy and thus quicker planetary transfer) will, therefore, be required to match the additional ΔV the nuclear engine is capable of delivering. Guessing several other values for the Mars transfer orbit eccentricity yields the following results in Table 1.

Table 1 Mars Transfer Orbit Characteristics as a Function of Orbit Eccentricity

ε	a (km)	$\Delta V_{mission}$ (km/s)	$\Delta V_{vehicle} - \Delta V_{mission}$ (km/s)
0.400	2.495×10^8	12.077	−0.529
0.300	2.139×10^8	9.127	2.421
0.350	2.303×10^8	10.655	−0.893
0.381	2.418×10^8	11.547	0.001 (close)

Knowing the characteristics of the spacecraft's earth to Mars transfer orbit, it is now possible to determine the travel time between the planets. Using the transfer orbit parameters from Table 1 in conjunction with Eq. (4.46), the true anomaly of the spacecraft as it arrives at Mars may be determined to be:

$$\phi_{Mars} = \cos^{-1}\left[\frac{a(1-\varepsilon^2) - R_{Mars}}{\varepsilon R_{Mars}}\right] = \cos^{-1}\left[\frac{241,800,000 \text{ km}(1 - 0.381^2) - 228,000,000 \text{ km}}{0.381 \times 228,000,000 \text{ km}}\right]$$

$$= 104.19° \quad (17)$$

Using the spacecraft's true anomaly at Mars from Eq. (17), it is also now possible to calculate the spacecraft's eccentric anomaly at Mars through the use of Eq. (4.66) such that:

$$E_{Mars} = \cos^{-1}\left[\frac{\varepsilon + \cos(\phi_{Mars})}{1 + \varepsilon \cos(\phi_{Mars})}\right] = \cos^{-1}\left[\frac{0.381 + \cos(104.19)}{1 + 0.381 \cos(104.19)}\right] = 1.420 \text{ rad} = 81.385° \quad (18)$$

Finally, using Eq. (4.70) the time required to travel between Earth and Mars may be determined. Since the departure engine burn was specified to occur at the spacecraft's transfer orbit perihelion (eg, $E_{earth} = 0$) as a result of the tangential departure burn at earth, only the eccentric anomaly at Mars from Eq. (18) is required to determine the travel time between the planets. The trip time to Mars is thus:

$$t = \sqrt{\frac{a^3}{\mu_s}}[E_{Mars} - \varepsilon \sin(E_{Mars})] = \frac{1}{86400}\frac{day}{s}\sqrt{\frac{(2.418 \times 10^8)^3 \text{km}^3}{1.327 \times 10^{11}\frac{km^3}{s^2}}}[1.420 - 0.381 \sin(81.385°)]$$

$$= 124.64 \text{ days} \quad (19)$$

52 CHAPTER 4 INTERPLANETARY MISSION ANALYSIS

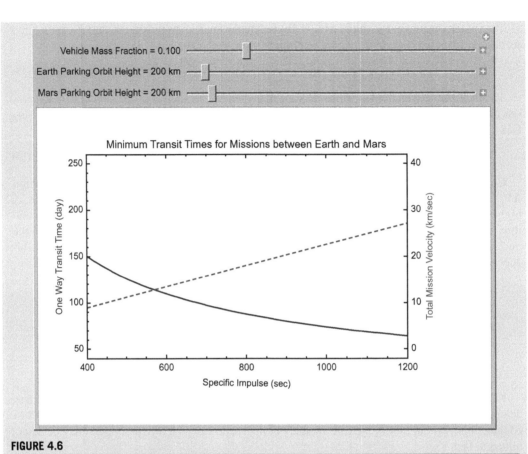

FIGURE 4.6

Minimum transit times between Earth and Mars.

In Fig. 4.6, the characteristics of a round-trip mission between earth and Mars are presented as a function of the performance characteristics of a spacecraft and its propulsion system. In these calculations, the round-trip time has been minimized such that the available total mission velocity budget is used as efficiently as possible. It should be pointed out that these calculations only account for the spacecraft departure and capture maneuvers with all other required velocity changes being ignored (eg, midcourse correction maneuvers, and contingencies). In addition, the orbits of earth and Mars are assumed to be perfectly circular at their average solar orbital radii, thus ignoring all the orbital eccentricities of the two planets. While these assumptions cannot be made for detailed mission analyses where high accuracy is critical for mission success, their neglect has been found to result in only small differences when compared to the more detailed mission studies.

While most interplanetary mission studies concentrate on voyages to Mars, nuclear or other advanced propulsion systems which possess a high-specific impulse can also be used on spacecraft to perform a wide variety of other planetary missions of scientific interest. Missions to the outer (or inner) planets which are difficult or impossible to perform with chemical propulsion systems suddenly become feasible when more efficient propulsion systems are employed. In Table 4.2, the trajectory characteristics of some of these other planetary missions are roughly estimated using the orbital relationships developed previously so that the propulsion system requirements necessary for their accomplishment may be determined. Again, the minimum trip times are calculated under the assumption that only the primary planetary departure and arrival propulsive maneuvers are required and that the planetary orbits are perfectly circular.

Table 4.2 Minimum Interplanetary Transit Times

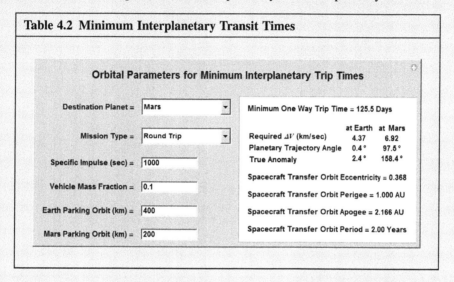

REFERENCE

[1] R.R. Bate, D.D. Mueller, J.E. White, Fundamentals of Astrodynamics, Dover Publications, Inc., New York, 1971, ISBN 0-486-60061-0.

CHAPTER 5

BASIC NUCLEAR STRUCTURE AND PROCESSES

The nucleus of the atom is normally thought to be simply an ensemble of bound nucleons composed primarily of protons and neutrons. While such a view is true, it represents only a partial picture of the nucleus. The nucleus actually contains a considerable amount of additional structure, the workings of which are still only partially understood as of the date of this writing. Fortunately, it is not necessary to fully understand the workings of the nucleus in order to exploit it for the purpose of releasing vast quantities of energy. By using the process of nuclear fission and perhaps in the future, nuclear fusion, energy from the nucleus can be extracted and used in such diverse areas as electric power production and rocket propulsion.

1. NUCLEAR STRUCTURE

Before getting started with our study of the nuclear reactor physics as it relates to nuclear rocket engines, it will be instructive to have a short qualitative discussion of the characteristics of the atomic nucleus itself.

The current picture of the structure of the atom and the nucleus in particular, was first formulated in 1911 when Ernest Rutherford, in a classic series of experiments, demonstrated that the atomic nucleus occupies only a small portion of the total volume of an atom. Since that time, later experiments have shown that classically the "diameter" of an atom is in the order of 10^{-8} cm, while the "diameter" of the nucleus is considerably less at roughly 10^{-12} cm. Even the nucleus can hardly be considered solid with the diameter of its constituent protons and neutrons having "diameters" of about 10^{-13} cm or about 10 times less than the diameter of the nucleus itself.

The nucleus of an atom under normal conditions is composed of subatomic particles called protons and neutrons (also called nucleons) which are of a class of elementary particles called hadrons. Hadrons are themselves composed of still more elementary particles called quarks. The quark model of elementary particles was first proposed in 1964 by Murray Gell-Mann and George Zweig to describe a rather large number of elementary particles discovered in particle accelerator experiments underway at the time. In the current quark model, quarks come in six flavors called up, down, charm, strange, top, and bottom. Quarks are theorized to possess fractional electric charges and also what is called color charge. Color charges can be either red, green, or blue and are related to the strong interaction which binds the quarks in the hadrons together. Any quark can possess any color charge. Color charges do not correspond in any way to actual colors, but rather constitute a useful tool with which to describe the strong interaction. All hadrons are color neutral, that is, all hadrons must consist of quark combinations whose color charges combine to yield white. Antiquarks can also exist and possess color charges of antired, antigreen, and antiblue. Table 5.1 presents some of the properties of the various quark flavors.

Table 5.1 Quark Flavor Properties

Quark	Symbol	Mass (MeV/c²)	Electric Charge
Up	u	1.7–3.1	+2/3
Down	d	4.1–5.7	−1/3
Charm	c	1180–1340	+2/3
Strange	s	80–130	−1/3
Top	t	172,000–173,800	+2/3
Bottom	b	4130–4370	−1/3

Only the up and down quarks are stable, and it is combinations of these quarks which constitute protons and neutrons. Protons and neutrons are part of a subclass of hadrons called baryons. Baryons consist of three quarks each having a different color charge. Protons having a net electric charge of +1, therefore, would consist of two up quarks and one down quark (uud). Neutrons, on the other hand, having no net electric charge would consist of two down quarks and one up quark (udd). The color charge between quarks in baryons is mediated by elementary particles called gluons which carry a color charge and an anticolor charge. Gluons act to exchange color charges between quarks and are the mechanism by which the strong interaction confines the quarks in baryons.

Besides protons and neutron, dozens of other baryons can exist under certain extreme conditions and are made up of combinations of the different quark flavors. A few examples of these baryons include the Λ (uds), the Ξ_c (dsc), and the Ω_{cb} (scb). All of these baryons are highly unstable and survive only briefly at extremely high energies, such as might be present in large particle accelerators or in exploding stars. Because these exotic baryons cannot be created in any of the more common nuclear interactions which will subsequently be discussed, they will receive no further mention here.

The other class of hadrons are called mesons. These elementary particles exist only fleetingly in the nucleus and consist of a quark and an antiquark. Similar to the baryons, mesons are bound together by the strong interaction, which is again mediated by gluons. Mesons act as the carrier of the strong nuclear force, which binds baryons together in the nucleus. The particular meson responsible for the strong interaction between protons and neutrons is the π meson or pion. Pions have about two-thirds of the mass of protons and neutrons and can be created only by violating the principle of conservation of mass and energy. This is permissible according to the Heisenberg uncertainty principle provided that the violation occurs for a sufficiently short amount of time, such that:

$$\Delta E \times \Delta t \leq \frac{h}{4\pi} \Rightarrow \Delta t \leq \frac{h}{4\pi \Delta E} \tag{5.1}$$

where ΔE = energy associated with the creation of the pion; Δt = time interval over which the pion can exist without violating the Heisenberg uncertainty principle; and h = Planck's constant.

Because the time during which the pions can exist is finite, the distance over which they can travel during their existence is also limited. Assuming that the pions travel at close to the speed of light and that the energy associated with their creation is governed by Einstein's famous mass/energy equivalence relationship ($E = mc^2$), Eq. (5.1) may be rewritten, such that:

$$\Delta t = \frac{d}{c} \lesssim \frac{h}{4\pi mc^2} \Rightarrow d \lesssim \frac{hc}{4\pi mc^2} \quad (5.2)$$

where d = distance traveled by the pion; m = pion mass; and c = speed of light.

Substituting in numerical values into Eq. (5.2) then yields:

$$d \lesssim \frac{1.24 \text{ eV } \mu\text{m}}{4\pi \times 135 \frac{\text{MeV}}{c^2} \times c^2 \times 10^6 \frac{\text{eV}}{\text{MeV}}} \approx 10^{-9} \ \mu\text{m} \quad (5.3)$$

Eq. (5.3), therefore, roughly represents the distance over which the strong nuclear force will be effective in binding nearby nucleons to one another and is of a similar order to the size of the protons and neutrons. A few of the possible quark interactions inside baryons via gluons and strong force coupling between baryons via pions is illustrated in Fig. 5.1.

The strength of the strong nuclear force between the pions and the baryons can be approximately described by what is called the Yukawa potential [1]. This potential arises when mediating particles having a nonzero mass such as the pion acts to create the strong nuclear force between baryons through pion particle exchange. The Yukawa potential has the form:

$$V = -g^2 \frac{e^{-kMr}}{r} \quad (5.4)$$

where V = Yukawa potential; g = coupling constant dependent upon the particular interaction under consideration; k = scaling constant dependent upon the particular interaction under consideration; M = mass of the force-mediating particle; and r = distance to the force-mediating particle.

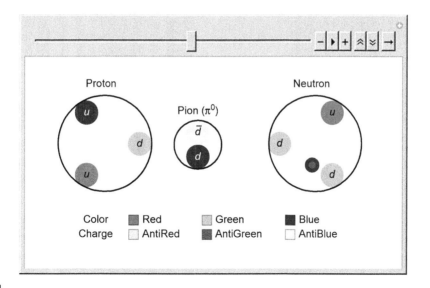

FIGURE 5.1

Quark–Gluon interactions in an atomic nucleus.

The coupling constant for the Yukawa potential in Eq. (5.4) is negative indicating that the force on the affected nucleons is attractive. If the mass of the mediating particle is equal to zero as is the case for the electrostatic potential where the mediating particle is a photon, the Yukawa potential reduces to the Coulomb potential, such that:

$$V_{coul} = g_{coul}^2 \frac{e^{-k(0)r}}{r} = g_{coul}^2 \frac{1}{r} \tag{5.5}$$

In Eq. (5.5), the coupling constant is positive indicating that the force between the nucleons is repulsive as would be the case in an atomic nucleus with multiple protons. Typically, the coupling constant for the Yukawa potential is such that the induced strong nuclear force is about 100 times stronger than the electrostatic force. Summing the Yukawa and Coulomb potentials yields a net potential which approximates what would be observed in the nucleus of an atom. This net potential is illustrated in Fig. 5.2 where "g" is proportional to the ratio of the coupling constant of the strong force to that of the electrostatic force. The scaling constant "k" is assumed to be equal to one. Note that since the Yukawa potential falls off much faster than the Coulomb potential, there exists a separation distance where the net nuclear potential experienced by the nucleons changes from attractive to repulsive. If it should happen that nucleus grows so large or becomes so distorted that the separation distance between

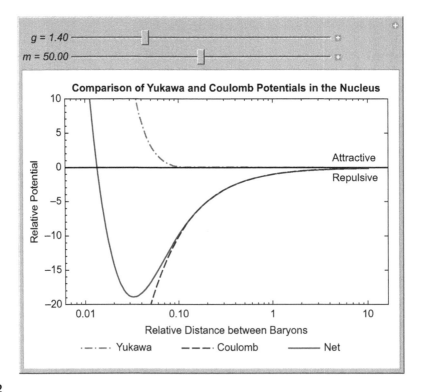

FIGURE 5.2

Yukawa and Coulomb potentials in an atomic nucleus.

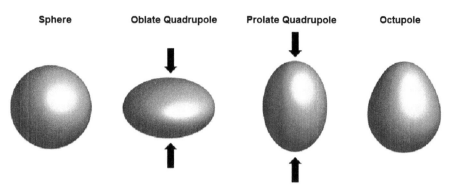

FIGURE 5.3

Excited quantum nuclear states [2].

nucleons is such that the net nuclear potential becomes repulsive, the electrostatic forces can come to dominate the nucleon—nucleon interactions and the nucleus will fly apart in a process called fission.

Generally speaking, the ensemble of nucleons in the nucleus of an atom can treated as a sphere and over time periods exceeding 10^{-16} s the nucleus indeed looks like a fuzzy spherical ball. If, however, the nucleus is observed on a time scale of about 10^{-18} s the nucleus can look slight ellipsoidal in shape. For large atoms, this deformation can lead to the nucleus spontaneously fissioning if the length of the ellipsoid is such that it causes the distance between nucleons to exceed the distance over which the nuclear potential becomes repulsive.

In artificially induced nuclear fission, free neutrons are captured in the nucleus of a large atom (such as ^{235}U) causing the nucleus to go into an excited quantum state similar to that which occurs in atoms when the electrons are put in excited quantum atomic states. Neutrons can easily penetrate into the nucleus of atoms since they are unaffected by the electrostatic force by virtue of the fact that they have no electric charge. Representations of some of these excited nuclear quantum states are illustrated in Fig. 5.3.

In these excited quantum nuclear states, the nucleus sometimes becomes deformed to such an extent that the critical separation distance between nucleons is exceeded resulting in a situation where the strong nuclear force can no longer hold the nucleus together. The electrostatic forces within the nucleus then quickly dominate causing the nucleus to fission within about 0.00005 s.

2. NUCLEAR FISSION

During the fission process, the nucleus nearly always splits into two pieces called fission fragments or fission particles and between one and three neutrons. These fission particles generally have unequal masses as can be seen in Figs. 5.4A and B, which show the fission yields as a function of atomic mass. The unequal mass split occurs as a result of certain stability factors having to do with the number of particles in the nucleus. The probability of occurrence of a particular fission product is generally designated by the symbol "γ."

During fission, it is found that the sum of the masses of these fission products is always less the mass of the original target nucleus plus the impacting neutron. This mass variance is termed the mass

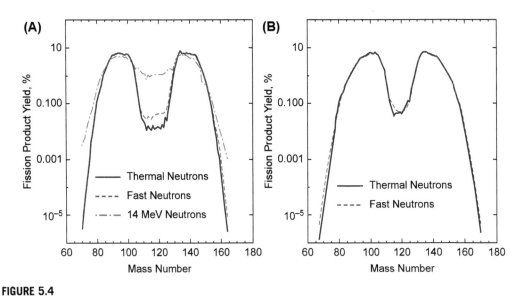

FIGURE 5.4

(A) ^{235}U fission product distribution. (B) ^{239}Pu fission product distribution.

defect and is the mass-energy equivalent of the difference between the total binding energies of the target nucleus and that of the fission products. The binding energy is defined as that energy required to break a nucleus into its constituent nucleons on a per nucleon basis. Fig. 5.5 illustrates how the binding energy for nucleons in the nucleus varies with the atomic mass of the nucleus. Note that for nuclear

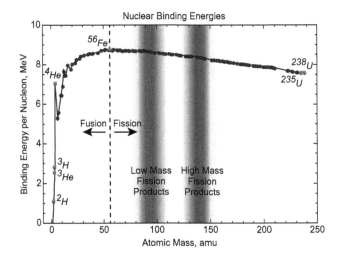

FIGURE 5.5

Binding energies as a function of atomic mass.

2. NUCLEAR FISSION

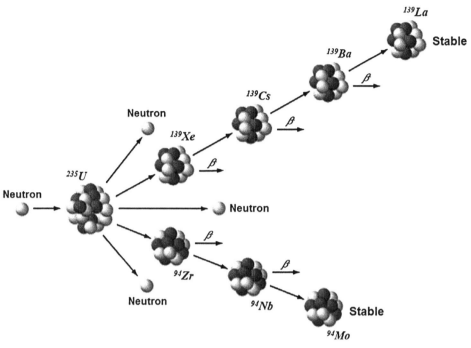

FIGURE 5.6

Typical fission reaction.

masses greater than that of iron, the binding energy favors fission reactions, while for nuclear masses less than that of iron, the binding energy favors fusion reactions.

The energy released during the fission process is truly phenomenal. Fig. 5.6 shows a fission reaction which is typical of the hundreds which are possible. In this reaction, a ^{235}U nucleus fissions, splitting into xenon and zirconium. These isotopes are highly unstable, however, having far too many neutrons for stability. As a result, the xenon and zirconium immediately have a series of β decays which yield, rather quickly, lanthanum and molybdenum. It is also interesting to note that for similar reasons having to do with the stability of the nucleus, almost all easily fissionable isotopes have odd numbers of nucleons.

EXAMPLE
Calculate the energy released by the fission reaction illustrated in Fig. 5.6.

Solution
The calculation of the energy release from this reaction will involve first determining the mass difference (or mass defect) in atomic mass units (amu) between the nuclides which existed before the reaction and the nuclides, which exist after the reaction.

CHAPTER 5 BASIC NUCLEAR STRUCTURE AND PROCESSES

Fission Reactants		Fission Products				
		^{139}La	138.90635			
^{235}U	235.04393	^{94}Mo	93.90509		236.05259	Fission reactants
n	1.00866	3 n	3.02599	$\Rightarrow$	−235.84017	Fission products
		5 B	0.00274		0.21242	Mass defect
	236.05259		235.84017			

Now using the famous Einstein matter to energy equivalence equation, $E = mc^2$, it is found that 1 amu of matter if converted completely to energy will yield 931 MeV. Therefore, the energy released from one fission reaction due to the mass defect is:

$$E = 931 \frac{\text{MeV}}{\text{amu}} \times 0.21242 \frac{\text{amu}}{\text{fission}} = 197.8 \frac{\text{MeV}}{\text{fission}} \approx 200 \frac{\text{MeV}}{\text{fission}} \quad (1)$$

The 200 MeV of energy released during the fission process is distributed approximately as follows:

Fission products ≈ 167 MeV β particles ≈ 8 MeV
Neutrons ≈ 5 MeV   Neutrinos ≈ 12 MeV
γ rays ≈ 8 MeV

Most of this energy can be captured in the form of heat as the various particles are slowed down through scattering interactions with their surroundings. The exception to this is the neutrino energy which is essentially lost. This unfortunate situation is due to the fact that since neutrinos have no charge and little or no mass, their ability to interact with surrounding matter is extremely slight. The energy distribution of the neutrons emitted during the fission process follows a fairly well-defined distribution called $\chi(E)$ distribution which may be represented quite well by the empirical formula given by

$$\chi(E) = 0.453 e^{-1.036E} \sinh\left(\sqrt{2.29E}\right) \quad (5.6)$$

A graph of the $\chi(E)$ distribution is shown in Fig. 5.7, where $\int_0^\infty \chi(E) \, dE = 1$.

To apply some context to the above calculations, it will be instructive to calculate the amount of energy which might be obtained from the complete fissioning of 1 g of ^{235}U:

$$E = 1 \text{ g} \times \frac{1}{235} \frac{\text{mol}}{\text{g}} \times 6.02 \times 10^{23} \frac{\text{atom}}{\text{mol}} \times (200 - 12) \frac{\text{MeV}}{\text{fission}} \times 1 \frac{\text{fission}}{\text{atom}} \times 1.6 \times 10^{-13} \frac{\text{W s}}{\text{MeV}}$$

$$= 7.7 \times 10^{10} \text{ W s}$$

From this equation, it is also possible to determine a rather useful conversion factor which is that about 1 J = 1 W s = 3×10^{10} fissions. This factor will be found useful quite often during the course of this study.

An interesting application may be made of the above relationship in application to a Mars mission using a nuclear thermal rocket (NTR). It was found from NERVA tests that a NTR engine producing 100,000 pounds of thrust will require a nuclear reactor producing about 2000 MW of power. For the entire Mars mission, it will be necessary to fire the rocket engine for about 90 min. Therefore, using the

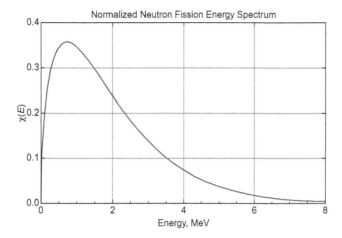

FIGURE 5.7

χ(E) fission energy distribution.

mass/energy fission relationship from above, the amount of ^{235}U which must be consumed to produce the energy required for the entire Mars mission is about:

$$\text{mass } ^{235}\text{U} = \frac{1}{7.7 \times 10^{10}} \frac{\text{g}}{\text{W s}} \times 2,000,000,000 \text{ W} \times 90 \text{ min} \times 60 \frac{\text{s}}{\text{min}} \approx 140 \text{ g}$$

From the equation mentioned previously, it can be seen that the entire Mars mission can be accomplished by fissioning only about 140 g of ^{235}U!

3. NUCLEAR CROSS SECTIONS

When a neutron encounters the nucleus of an atom, it does not always cause a fission. Depending on the nuclide, the neutron might also simply scatter off the nucleus or it might get absorbed by the nucleus, thereby creating a new isotope. The neutron could also cause the nucleus to go into an unstable excited state and emit two neutrons, or it might precipitate the occurrence of any of a number of other interactions. The probability of any particular nuclear interaction "x" depends on the type of nucleus, the neutron encounters, and the neutron energy. These probabilities are represented by a parameter called the microscopic neutron cross section (σ_x) which is measured in units of barns, where a barn is defined as 10^{-24} cm^2. A nuclear cross section, physically, is a measure of the effective area for interaction of a nucleus as seen by a neutron. By and large, except at very high energies, this area of interaction is independent of the actual size of the nucleus.

There are a number of neutron interaction cross sections which are of interest in a wide variety of fields of study; however, in this study, only a few types of cross sections are of importance. These cross sections are as follows:

σ_f = fission	σ_c = capture	σ_s = scattering
σ_{tr} = transport	σ_a = absorption: $\sigma_c + \sigma_f$	σ_t = total: $\sigma_a + \sigma_s + ...$

The variation of neutron cross sections with energy may typically be divided into three regions each exhibiting quite different behavior. These three regions may be described as follows.

3.1 1/V REGION

In this region, the cross section drops fairly smoothly and is proportional to the inverse of the neutron velocity. It is a low-energy characteristic of nuclide cross sections and depending upon the nuclide, the high end of the energy range can extend from fractions of an eV to several 1000 eV. Typically, thermal neutron cross sections for materials are quoted at a reference temperature of 20 °C which corresponds to a neutron energy of 0.025 eV or equivalently 2200 m/s. This reference cross section is useful in specifying the energy-dependent cross-sectional behavior in the low-energy range where the 1/V effect dominates. The 1/V cross-sectional behavior may be expressed in the following equation:

$$\sigma_{1/V}(E) = \sigma_{1/V}(2200)\frac{2200}{V} = \sigma_{1/V}(0.025)\sqrt{\frac{0.025}{E}} \tag{5.7}$$

3.2 RESONANCE REGION

In this region the cross section exhibits abrupt changes in magnitude as a result of various quantum mechanical effects. Generally speaking, these resonance peaks grow shorter and closer together as the neutron energy increases.

3.3 UNRESOLVED RESONANCE REGION OR FAST REGION

In this region, the resonances are so close together that they overlap one another creating a fairly flat curve exhibiting little change in magnitude as the neutron energy increases. The cross section in this region is roughly the same size as the "classical" cross-sectional size of the nucleus.

These various regions are illustrated clearly in the total microscopic cross section of ^{242}Pu as shown in Fig. 5.8.

Fission cross sections attain significant values in only a few isotopes of certain very heavy nuclides (called fissile nuclides) such as ^{235}U and ^{239}Pu. As mentioned before, stability considerations dictate that isotopes with large fission cross sections usually have odd numbers of nucleons in the nucleus, although this effect becomes less significant as the mass of the nucleus increases. This point is illustrated in Figs. 5.9 and 5.10, where for ^{235}U the fission cross section comprises the bulk of the total cross section, whereas for ^{238}U the fission portion of the total cross section comprises an insignificant portion of the total cross section except at very high neutron energies.

The most common nuclear fuel materials envisioned for use in NTR systems are ^{235}U and ^{239}Pu. These materials are fairly common and work well in most nuclear reactors. The use of ^{235}U is by far the most common fuel material used in reactor systems. Only a few test reactors systems contain plutonium fuel. Good as these materials are, however, they by no means have the highest fission cross sections. Fig. 5.11 shows that an isotope of americium, ^{242m}Am has a fission cross section an order of magnitude higher than either ^{235}U or ^{239}Pu. ^{242m}Am is, unfortunately, quite difficult to produce in quantity, however.

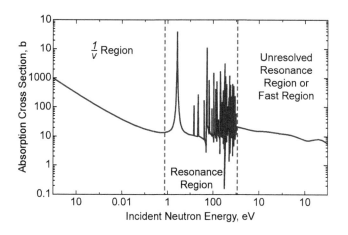

FIGURE 5.8

Typical cross-sectional variations with energy (^{242}Pu).

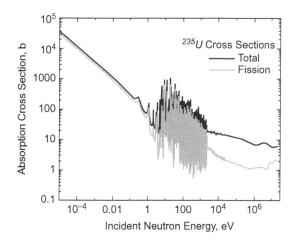

FIGURE 5.9

^{235}U cross sections.

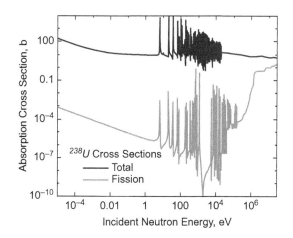

FIGURE 5.10

^{238}U cross sections.

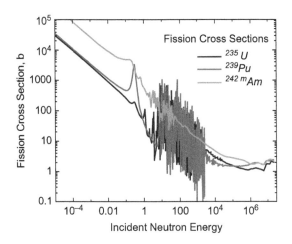

FIGURE 5.11

Fission cross sections of several nuclides.

Fig. 5.12 shows cross sections of several other materials which will almost certainly be used in NTR systems. Boron (actually ^{10}B) is a strong neutron absorber which would be useful in control elements to soak up excess neutrons produced in the chain reaction fission process. Graphite and beryllium are good reactor structural materials because they do not absorb neutrons easily, but rather scatter them such that the neutrons lose some of their energy. This loss of neutron energy due to scattering aids in the nuclear reaction process and can lead to a reduction in the amount of fissile material needed to construct the core. The reason why this is so will be discussed later in the course.

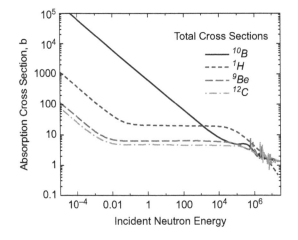

FIGURE 5.12

Total cross sections of NTR materials.

Hydrogen is also plotted in Fig. 5.12 because it is usually the propellant used in NTR systems, and because it has scattering characteristics which can greatly influence the behavior of the reactor.

The cross sections shown in Figs. 5.11 and 5.12 are obtained from the National Nuclear Data Center from the Evaluated Nuclear Data File (ENDF/B). The National Nuclear Data Center is a clearing house for all nuclear data and virtually any type of cross section for any nuclide can be found there.

In most calculations, the microscopic cross sections are not used by themselves but are usually multiplied by the atom density to yield what are called macroscopic cross sections as shown in the following:

$$\Sigma_x(E) \frac{1}{cm} = \sum_{j=1}^{\text{All Nuclides}} n_j \frac{\text{atom}}{b \, cm} \times \sigma_x^j(E) \frac{b}{\text{atom}} \tag{5.8}$$

The interpretation of macroscopic cross sections can best be understood by noting that the macroscopic cross section which is usually expressed in units of 1/cm can also be written as cm^2/cm^3. When viewed in this way, macroscopic cross section can be interpreted as representing the effective area for interaction "x" per cm^3. The reciprocal of the macroscopic cross section can also be understood as the average distance neutrons travel between interactions of type "x". In Eq. (5.8) the atom density has some rather odd, but quite useful units. These units may be determined as follows:

$$n \frac{\text{atom}}{b \, cm} = \rho \frac{g}{cm^3} \times \frac{1}{A} \frac{\text{mol}}{g} \times 6.02 \times 10^{23} \frac{\text{atom}}{\text{mol}} \times 10^{-24} \frac{cm^2}{b} = \frac{0.602}{A} \rho \tag{5.9}$$

where A = atomic weight.

4. NUCLEAR FLUX AND REACTION RATES

To find the rate at which the various neutron induced interactions proceed in a mixture of nuclides at a particular location in a reactor, it is necessary to determine a quantity called the neutron flux which is usually designated as "$\phi(r,E)$". To evaluate the neutron flux, one must first determine the number of neutrons of energy "E" crossing the surface of the differential volume element "dV" at location "r" as illustrated in Fig. 5.13.

The neutron flux may then be given by

$$\phi(r,E) \frac{\text{neut}}{cm^2 \, s} = \int_{\text{all } \Omega} n\left(r, E, \vec{\Omega}\right) u(E) \, d\vec{\Omega} \tag{5.10}$$

where $n\left(r, E, \vec{\Omega}\right)$ = neutron density at position "r" having an energy "E" traveling in direction "$\vec{\Omega}$"
and $u(E)$ = neutron velocity corresponding to energy "E."

Physically, the neutron flux in Eq. (5.10) can be interpreted as the total track length of neutrons per cm^3 within dE about "E." This interpretation can be made more clear if the units on neutron flux are rewritten as $\frac{\text{neut cm}}{cm^2 \, s}$. Reaction rates are proportional to the neutron flux level and the macroscopic reaction cross section, thus,

$$R_x(r,E) \frac{x}{cm^3 \, s} = \Sigma_x(r,E) \frac{1}{cm} \times \phi(r,E) \frac{\text{neut}}{cm^2 \, s} \times 1 \frac{x}{\text{neut}} \tag{5.11}$$

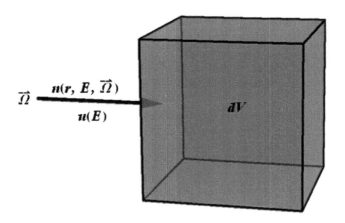

FIGURE 5.13

Flow of neutrons through a surface.

As an illustration, the power density at a particular location in a reactor is determined by using the fission cross section:

$$Q(r,E)\frac{W}{cm^3} = \Sigma_f(r,E)\frac{1}{cm} \times \phi(r,E)\frac{neut}{cm^2 \, s} \times 1\frac{fission}{neut} \times \frac{1}{3 \times 10^{10}}\frac{W \, s}{fission} \quad (5.12)$$

Also, the number of neutrons produced per unit volume can be determined by introducing the quantity "ν" which represents the number of neutrons on average emitted per fission. Typically, ν is about 2.5:

$$R_f \frac{neut}{cm^3 \, s} = \nu \frac{neut}{fission} \times \Sigma_f(r,E)\frac{1}{cm} \times \phi(r,E)\frac{neut}{cm^2 \, s} \times 1\frac{fission}{neut} \quad (5.13)$$

EXAMPLE

A neutron generator producing 1 eV monoenergetic neutrons is being used to irradiate a ^{239}Pu sample. This irradiation produces a fission rate in the sample equivalent to a power density typical of what it would experience in an operating nuclear rocket engine. Determine the neutron flux level required in the sample to achieve the desired power density.
Assume $\rho_{Pu239} = 19.84$ g/cm^3, $Q = 5$ kW/cm^3.

Solution

The first step in the solution is to determine the microscopic fission cross section at 1 eV for ^{239}Pu from the National Nuclear Data Center Evaluated Nuclear Data File (ENDF/B). From an examination of the plot, it may be determined that $\sigma_f^{Pu239}(1eV) = 34 \, b/atom$. The next step in the calculation involves putting ^{239}Pu in terms of atom density rather than mass density. From Eq. (5.9) the atom density may be determined to be

$$n_{Pu239} = \frac{0.602}{A_{Pu239}}\rho_{Pu239} = \frac{0.602}{239} 19.84 = 0.05 \frac{atom}{b \, cm} \quad (1)$$

The ^{239}Pu macroscopic fission cross section may now be determined from its microscopic fission cross section and its atom density determined from Eq. (1), such that:

$$\Sigma_f^{Pu239}(1\text{ eV}) = n_{Pu239}\frac{\text{atom}}{b\text{ cm}} \times \sigma_f^{Pu239}\frac{b}{\text{atom}} = 0.0534 = 1.7\frac{1}{\text{cm}} \tag{2}$$

From Eq. (5.12), the power density may be evaluated using the macroscopic fission cross section determined from Eq. (2) and the neutron flux, such that:

$$Q(1\text{ eV}) = \Sigma_f(1\text{ eV}) \times \phi(1\text{ eV}) \Rightarrow \phi(1\text{ eV}) = \frac{Q(1\text{ eV})\frac{W}{cm^3}}{\Sigma_f^{Pu239}(1\text{ eV})\frac{1}{cm}} = \frac{5000}{1.7} = 2940\frac{W}{cm^2} \tag{3}$$

The neutron flux calculated from Eq. (3) may be converted to more traditional units by using appropriate conversion factors, such that:

$$\phi(1\text{ eV}) = 2940\frac{W}{cm^2} \times 3 \times 10^{10}\frac{\text{fission}}{W\text{ s}} \times 1\frac{\text{neut}}{\text{fission}} = 8.82 \times 10^{13}\frac{\text{neut}}{cm^2\text{ s}} \tag{4}$$

5. DOPPLER BROADENING OF CROSS SECTIONS

The plots of nuclear cross sections presented earlier were shown solely as functions of the energy of the interacting neutrons. In reality, however, the energy at which the nuclear interactions occur are functions of the net velocity (or energy) between the approaching neutrons and the target nucleus. As a consequence, in order to eliminate the effects of thermal motion in the target nucleus, cross sections are normally presented at a temperature of 0 K. The effects of this thermal motion are most noticeable in low-lying resonance cross sections. Doppler broadening is a mechanism in which these resonance cross sections effectively broaden due to thermal vibrations in the target nucleus. These vibrations act so as to increase the energy interval within which a neutron of a given energy may be captured in a resonance. The effect of Doppler broadening is to increase the effective averaged absorption cross section of a nuclide at low neutron energies. It will be seen later that Doppler broadening is an extremely important mechanism in maintaining reactor stability.

To determine the degree to which these resonance cross sections broaden at elevated temperatures, it is necessary to integrate the resonance cross section over an appropriate distribution of particles as a function of the relative velocities between the neutron and the target nucleus. Normally, a Maxwellian particle distribution is used in the integration which, while strictly applicable only to matter in the gaseous state, has nevertheless, been found to adequately represent these temperature induced vibrational effects in solids without serious error. By integrating the cross section over a Maxwellian particle distribution characteristic of a particular temperature, the thermally averaged cross section as a function of energy and temperature may be determined, such that:

$$\overline{\Sigma}_c(E_n, T)\phi(E_n) = N_0\overline{\sigma}_{rc}(E_n, T)nV_n = \int_{-\infty}^{\infty} N(V_t)\sigma_{rc}(E_r)nV_r\, dV_r \tag{5.14}$$

where N_0 = atom density of the target nucleus; n = neutron density; E_n = neutron energy; V_n = neutron velocity; V_t = velocity of the target nucleus; V_r = relative velocity between the neutron and the target nucleus $= V_n - V_t$; $N(V_t) =$ Maxwellian atom density distribution as a function of V_t; $\sigma_{rc}(E_r) =$ microscopic resonance capture cross section at an energy "E_r" corresponding to "V_r"; and $T =$ temperature of the target nucleus.

The Maxwellian distribution described in the preceding paragraph for representing the number of particle at a given target nucleus velocity and temperature is defined by

$$N(V_t) = N_0 \sqrt{\frac{M}{2\pi kT}} e^{-\frac{MV_t^2}{2kT}} \tag{5.15}$$

where $k =$ Boltzmann constant and $M =$ mass of the target nucleus.

In Eq. (5.15), the velocity of the target nucleus (V_t) may be rewritten by noting that:

$$V_t = V_r - V_n \text{ and that } E_r = \frac{1}{2}\mu V_r^2 \Rightarrow V_r = \sqrt{\frac{2E_r}{\mu}} \text{ and } E_n = \frac{1}{2}\mu V_n^2 \Rightarrow V_n = \sqrt{\frac{2E_n}{\mu}} \tag{5.16}$$

where $\mu = \frac{mM}{m+M}$ reduced mass (for a center of mass coordinate system) $\approx m$ for $M \gg m$; $m =$ neutron mass; and $E_r =$ relative energy between the neutron and the target nucleus in a center of mass coordinate system.

By incorporating the definitions from Eq. (5.16) into the Maxwellian particle velocity distribution given in Eq. (5.15), it may be seen that:

$$N(E_r) = N_0 \sqrt{\frac{M}{2\pi kT}} e^{-\frac{M\left(\sqrt{\frac{2E_r}{\mu}} - \sqrt{\frac{2E_n}{\mu}}\right)^2}{2kT}} \approx N_0 \sqrt{\frac{M}{2\pi kT}} e^{-\frac{A(\sqrt{E_r} - \sqrt{E_n})^2}{kT}} \tag{5.17}$$

where $A = \frac{M}{m} =$ atomic weight of the target nucleus.

The next several steps are directed toward developing a generalized function to represent the Doppler broadening of a single isolated neutron capture resonance. First note that:

$$\sqrt{E_r} = \sqrt{E_n - E_n + E_r} = \sqrt{E_n \left(\frac{E_r}{E_n} - \frac{E_n}{E_n} + 1\right)} = \sqrt{E_n}\sqrt{\frac{E_r - E_n}{E_n} + 1} \tag{5.18}$$

Expanding the term in parentheses in Eq. (5.18) in a Taylor series then yields:

$$\sqrt{1 + \frac{E_r - E_n}{E_n}} = 1 + \frac{1}{2}\frac{E_r - E_n}{E_n} - \frac{1}{8}\left(\frac{E_r - E_n}{E_n}\right)^2 + \cdots \tag{5.19}$$

By truncating Eq. (5.19) to the first two terms (thus linearizing it) and then incorporating the result back into Eq. (5.18) yields an equation of the form:

$$\sqrt{E_r} \approx \sqrt{E_n}\left(1 + \frac{E_r - E_n}{2E_n}\right) \tag{5.20}$$

Substituting Eq. (5.20) into the expression for the exponent in Eq. (5.17) then yields:

$$-\frac{A}{kT}(\sqrt{E_r} - \sqrt{E_n})^2 \approx -\frac{A}{kT}\left[\sqrt{E_n}\left(1 + \frac{E_r - E_n}{2E_n}\right) - \sqrt{E_n}\right]^2 = -\frac{A}{4kE_nT}(E_r - E_n)^2 \tag{5.21}$$

The expression to the right of the equal sign in Eq. (5.21) may now be revised somewhat to obtain:

$$-\frac{A}{kT}(\sqrt{E_r} - \sqrt{E_n})^2 \approx -\frac{A}{4kTE_n}[(E_r - E_0) - (E_n - E_0)]^2$$

$$= -\frac{1}{4}\left(\frac{A\Gamma^2}{4kTE_n}\right)\left[\frac{2(E_r - E_0)}{\Gamma} - \frac{2(E_n - E_0)}{\Gamma}\right]^2 \quad (5.22)$$

where Γ = energy width of the resonance for all neutron induced reactions and E_0 = energy at the resonance peak.

The relationship given in Eq. (5.22) may now be incorporated into the Maxwellian particle velocity distribution given in Eq. (5.17) to obtain:

$$N(E_r) = N_0\sqrt{\frac{M}{2\pi kT}} e^{-\frac{1}{4}\left(\frac{A\Gamma^2}{4kTE_n}\right)\left[\frac{2(E_r-E_0)}{\Gamma} - \frac{2(E_n-E_0)}{\Gamma}\right]^2} \quad (5.23)$$

To represent the unbroadened shape of a single cross-sectional resonance in Eq. (5.14), it is necessary to present what is called the Breit–Wigner single-level cross-sectional formula [3]. This formula accurately describes the energy dependence of a single isolated neutron capture resonance. For more complicated situations, such as low-lying overlapping fission cross-sectional resonances, more complicated multilevel formulas are often required to accurately describe the resonance structure. For the particular example under consideration here, however, such multilevel formulas are not needed and are not discussed further. In any case, for a neutron capture cross section in the vicinity of a resonance the Breit–Wigner formula is

$$\sigma_{rc}(E_r) = \frac{\sigma_0 \Gamma_c}{\Gamma}\sqrt{\frac{E_0}{E_r}} \frac{1}{1 + \frac{4}{\Gamma^2}(E_r - E_0)^2} \quad (5.24)$$

where σ_0 = microscopic neutron capture cross section at E_0 and Γ_c = energy width of the resonance for neutron capture.

Substituting Eqs. (5.23) and (5.24) into Eq. (5.14) and rearranging terms then yields:

$$\bar{\sigma}_{rc}(E_n, T) = \frac{1}{N_0 n}\int_{-\infty}^{\infty} \frac{\sigma_0 \Gamma_c}{\Gamma}\sqrt{\frac{E_0}{E_r}} \frac{1}{1 + \frac{4}{\Gamma^2}(E_r - E_0)^2} N_0\sqrt{\frac{M}{2\pi kT}} e^{-\frac{1}{4}\left(\frac{A\Gamma^2}{4kTE_n}\right)\left[\frac{2(E_r-E_0)}{\Gamma} - \frac{2(E_n-E_0)}{\Gamma}\right]^2} n\frac{V_r}{V_n} d\left(\sqrt{\frac{2E_r}{m}}\right) \quad (5.25)$$

From the relationships of Eq. (5.16), it may be noted that:

$$\frac{V_r}{V_n} = \sqrt{\frac{E_r}{E_n}} \quad (5.26)$$

Substituting Eq. (5.26) into Eq. (5.25), rearranging terms and pulling those terms which do not depend on V_r outside the integral then gives for the temperature-dependent capture cross section:

$$\bar{\sigma}_{rc}(E_n, T) = \frac{\sigma_0 \Gamma_c}{\Gamma}\sqrt{\frac{E_0}{E_n}}\sqrt{\frac{M}{2\pi kT}}\sqrt{\frac{1}{2mE_r}} \int_{-\infty}^{\infty} \frac{e^{-\frac{1}{4}\left(\frac{A\Gamma^2}{4kTE_n}\right)\left[\frac{2(E_r-E_0)}{\Gamma} - \frac{2(E_n-E_0)}{\Gamma}\right]^2}}{1 + \frac{4}{\Gamma^2}(E_r - E_0)^2} dE_r \quad (5.27)$$

Eq. (5.27) can now be simplified somewhat by introducing the following variable changes:

$$\zeta = \sqrt{\frac{A\Gamma^2}{4kTE_n}}, \quad x = \frac{2(E_n - E_0)}{\Gamma}, \quad y = \frac{2(E_r - E_0)}{\Gamma} \Rightarrow dy = \frac{2}{\Gamma}dE_r \quad (5.28)$$

Note that the integral in Eq. (5.27) is approximately zero except for when E_r and E_n are near E_0. Using this fact and the definitions of Eq. (5.28), it is possible to rewrite Eq. (5.27), such that:

$$\bar{\sigma}_{rc}(E_n, T) = \frac{\sigma_0 \Gamma_c}{\Gamma}\sqrt{\frac{E_0}{E_n}}\sqrt{\frac{M}{2\pi kT}}\sqrt{\frac{1}{2mE_0}}\frac{\Gamma}{2}\int_{-\infty}^{\infty}\frac{e^{-\frac{1}{4}\zeta^2(y-x)^2}}{1+y^2}dy = \frac{\sigma_0\Gamma_c}{\Gamma}\sqrt{\frac{E_0}{E_n}}\frac{\zeta}{2\sqrt{\pi}}\int_{-\infty}^{\infty}\frac{e^{-\frac{1}{4}\zeta^2(y-x)^2}}{1+y^2}dy \quad (5.29)$$

From the results of Eq. (5.29), it will prove useful to define a new function called the Doppler integral, such that:

$$\psi(\zeta, x) = \frac{\zeta}{2\sqrt{\pi}}\int_{-\infty}^{\infty}\frac{e^{-\frac{1}{4}\zeta^2(y-x)^2}}{1+y^2}dy \quad (5.30)$$

The Doppler function cannot be explicitly evaluated, however, it has been widely tabulated and there are a number of very good numerical approximations [4] in the literature. Rewriting Eq. (5.29) in terms of the Doppler function thus yields the temperature-adjusted resonance neutron capture cross section for a single isolated resonance as a function of the temperature and the neutron energy:

$$\bar{\sigma}_{rc}(E_n, T) = \frac{\sigma_0 \Gamma_c}{\Gamma}\sqrt{\frac{E_0}{E_n}}\psi(\zeta, x) \quad (5.31)$$

To illustrate how Doppler broadening affects the shape of a neutron capture cross section as the temperature of the material is varied, the large capture resonance in ^{238}U at 6.67 eV is plotted in Fig. 5.14 which consists of the average resonance capture cross section as expressed by Eq. (5.31) added to the 1/V base capture cross section as expressed by Eq. (5.7):

6. INTERACTION OF NEUTRON BEAMS WITH MATTER

If a neutron beam is directed toward a piece of material having a total neutron interaction cross section of $\Sigma_t(E)$, it will be observed that the neutron beam attenuates as it passes through the material. In the analysis which follows, it will be assumed that neutrons are lost to the system after undergoing a single interaction.

As illustrated in Fig. 5.15, the number of nuclear interactions in a differential volume ($dV = A\,dx$) of material can be represented by the number of neutrons entering the differential volume through the left face minus the number of neutrons emerging from the differential volume through the right face, such that:

$$R(E)dV = \Sigma_t(E)\phi(E)dV = \underbrace{\Sigma_t(E)n(x,E)u(E)A\,dx}_{\text{neutrons lost due to interactions}}$$

$$= \underbrace{n(x,E)u(E)A}_{\text{neutrons entering dV from left face}} - \underbrace{n(x+dx,E)u(E)A}_{\text{neutrons entering dV through right face}} \quad (5.32)$$

where A = cross-sectional area intercepted by the neutron beam.

6. INTERACTION OF NEUTRON BEAMS WITH MATTER   73

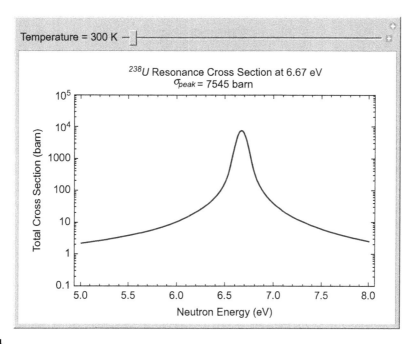

FIGURE 5.14

Cross Section Doppler broadening in the vicinity of a resonance.

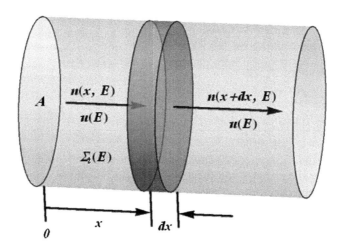

FIGURE 5.15

Interaction of a beam of neutrons in matter.

Table 5.2 Neutron Mean Free Paths in Various Materials

Substance	Σ_t (0.025 eV)	λ_{mfp} (0.025 eV) (cm)
H_2O	3.47	0.3
H_2	(Liquid)	1.60
H_2 (STP)	0.002	500
B	103	0.01
C	0.385	2.6
^{238}U	0.764	1.3
Stainless steel	~1.5	0.7

Making the assumption that:

$$n(x+dx,E) = n(x,E) + \frac{dn(x,E)}{dx}dx \qquad (5.33)$$

Eq. (5.32) may be rewritten, such that:

$$\Sigma_t(E)n(x,E)u(E)A\,dx = u(E)\left[n(x,E)-n(x,E)-\frac{dn(x,E)}{dx}dx\right]A \Rightarrow \Sigma_t(E)dx = -\frac{dn(x,E)}{n(x,E)} \qquad (5.34)$$

If (x,E) at $x=0$ is $n_0(E)$, then Eq. (5.34) may be integrated to yield:

$$\Sigma_t(E)\int_0^x dx = -\int_{n_0(E)}^{n(x,E)} \frac{dn'(x,E)}{n'(x,E)} \Rightarrow n(x,E) = n_0(E)e^{-\Sigma_t(E)x} \qquad (5.35)$$

The average distance traveled by neutron before it interacts with the material through which it is traveling is known as the neutron mean free path. The mean free path is determined by calculating the number of neutrons removed from the beam between "x" and "$x+dx$" and multiplying that number by the distance "x". This value represents the number of neutron centimeters removed from the beam in the region "dx". If all such contributions are added together, one gets the total number of centimeters traveled by all the neutrons removed from the beam. If this value representing the total number of centimeters traveled by all the neutrons in the beam is divided by the total number of neutrons initially in the beam, we are left with the average distance traveled by the neutrons in the beam, thus:

$$\lambda_{mfp}(E) = \frac{\int_0^\infty [x\Sigma_t(E)u(E)n(x,E)\,dx]dSdE}{n_0(E)u(E)dSdE} = \frac{\int_0^\infty x\Sigma_t(E)u(E)n_0(E)e^{-\Sigma_t(E)x}\,dx}{n_0(E)u(E)} = \frac{1}{\Sigma_t(E)} \qquad (5.36)$$

In Table 5.2, the mean free path for 2200 m/s (0.025 eV) neutrons in various materials is presented. Because neutrons are neutral particles, the mean free paths can be quite significant in many materials, ranging from fractions of a centimeter to several centimeters.

7. NUCLEAR FUSION

Besides nuclear fission, there is one other nuclear reaction that bears special mention with regard to the application of nuclear processes to space propulsion. That process is nuclear fusion. In nuclear fusion, the nuclei of two small light atoms merge with one another to form the nucleus of a somewhat heavier atom releasing vast quantities of energy in the process. Fig. 5.16 illustrates a fairly commonly considered fusion reaction in which the reactants are deuterium and tritium.

While it is relatively easy to initiate nuclear fission events by the absorption of neutrons into large, heavy atomic nuclei, nuclear fusion is much more difficult to initiate. The difficulty in initiating nuclear fusion lies in the fact that the positively charged nuclei of the atoms which are to undergo fusion must be brought close enough together that the strong nuclear force between their constituent nucleons is able to overcome the natural electrostatic repulsive force between the protons in the respective nuclei. The activation energy required by the nuclei to overcome this electrostatic force is considerable and requires that the nuclei have energy equivalent temperatures in the order of hundreds of millions of degrees. While the energy required by the fusing atoms to initiate the fusion reaction is considerable, the energy released by the fusion reaction is considerably greater.

Comparing the energy output from fusion reactions to fission reactions reveals that on a unit mass basis, fusion reactions are much more energetic. The reason fusion reactions are more energetic than fission reactions is illustrated in Fig. 5.5 which shows the large binding energies differences between reactants and products in the common fusion reactions as compared with typical fission reactions. Fusion reactions often considered for power production or as the basis for a fusion powered rocket engine generally reference the use of isotopes of hydrogen and helium, primarily deuterium, tritium and helium-3. A few examples of these commonly considered reactions include:

$$^{2}_{1}H + ^{3}_{1}H \rightarrow ^{4}_{2}He(3.5 \text{ MeV}) + ^{1}_{0}n(14.1 \text{ MeV}) \qquad (5.37)$$

$$^{2}_{1}H + ^{2}_{1}H \xrightarrow{50\%} \begin{array}{l} ^{3}_{1}H(1.01 \text{ MeV}) + ^{1}_{1}p(3.02 \text{ MeV}) \\ \\ ^{3}_{2}He(0.82 \text{ MeV}) + ^{0}_{1}n(2.45 \text{ MeV}) \end{array} \qquad (5.38)$$

$$^{2}_{1}H + ^{3}_{2}He \rightarrow ^{4}_{2}He(3.6 \text{ MeV}) + ^{1}_{1}p(14.7 \text{ MeV}) \qquad (5.39)$$

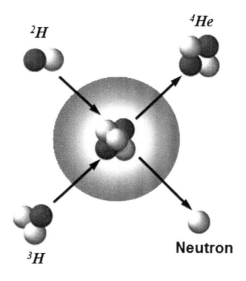

FIGURE 5.16

Fusion of deuterium and tritium.

If the fusion cross sections for the reactions shown in Eq. (5.37)–(5.39) are averaged over Maxwellian velocity distributions characteristic of different temperatures, it is possible to obtain the temperature-averaged fusion reaction rates illustrated in Fig. 5.17 [5]. Note that the reaction rates of the various fusion reactions are significant only at temperatures in the hundreds of millions of degrees due to the high activation energies required to initiate the reactions. At these temperatures, all the electrons are stripped from the atoms leaving the reactants in a fully ionized gaseous plasma state consisting of bare nuclei and free floating electrons.

In order for a fusion reaction to be self-sustaining, the fusion plasma must be confined for a sufficiently long period of time that the fusion energy released by the charged particles is sufficient to heat the reactants to the temperatures required to keep the fusion reaction rate constant. The power produced by neutrons is normally not considered in the power balance since they generally contribute very little to heating the plasma due to their lack of an electrical charge and consequent long mean free path. The fusion energy produced by a 50/50 mixture of the atoms which are to undergo fusion may be given by

$$E_f = n_1 n_2 \langle \sigma v \rangle E_c \tau = \frac{n^2}{4} \langle \sigma v \rangle E_c \tau \qquad (5.40)$$

where E_f = energy produced from fusion by the charged particles; n_1 and n_2 = fusion atom densities for reactant species one and 2; n = total fusion atom density (eg, $n_1 + n_2$); $\langle \sigma v \rangle$ = temperature-dependent fusion reaction rate per particle (from Fig. 5.17); τ = particle confinement time; and E_c = total charged particle energy release per fusion reaction.

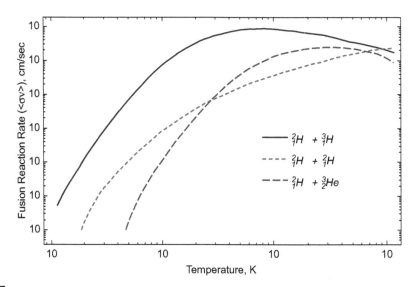

FIGURE 5.17

Fusion cross sections averaged over a Maxwellian temperature distribution.

From the kinetic theory of gases, the average kinetic energy of the particles in the gas as a function of temperature may be represented by

$$E_{KE} = 3nkT \tag{5.41}$$

where E_{KE} average kinetic energy of the particles in a gas; k = Boltzmann constant; and T = gas temperature.

To be self-sustaining, the energy produced by the fusioning plasma at a particular temperature must exceed the energy required by the plasma to reach that temperature. The point at which the fusion plasma is self-sustaining is called breakeven. Therefore, as a minimum, from Eqs. (5.40) and (5.41) breakeven requires that:

$$E_f > E_{KE} \Rightarrow \frac{n^2}{4} \langle \sigma v \rangle E_c \tau > 3nkT \tag{5.42}$$

Rearranging Eq. (5.42) to solve for the product of the plasma density times the confinement time yields what is known as the Lawson criterion [6]. This criteria states that at a particular plasma temperature, there exists a minimum value of the product of the plasma density times the confinement time necessary to achieve a self-sustaining fusion plasma:

$$n\tau > \frac{12kT}{\langle \sigma v \rangle E_c} \tag{5.43}$$

Although Eq. (5.43) ignores several important energy loss mechanisms in the plasma, most notably radiation, it nevertheless gives some insight into the conditions necessary to achieve a self-sustaining fusion plasma. A plot of the Lawson criterion from Eq. (5.43) as a function of temperature is shown in Fig. 5.18. Note in the plot the deuterium/tritium ($_1^2H + {_1^3}H$) reaction has the lowest minimum value for the Lawson criteria at a $n\tau$ value of about 1.6×10^{14} cm^3/s and a temperature of about 300,000,000 K.

Since the temperatures required for fusion are so incredibly high, it is impossible contain the fusing plasma in any type of solid container. As a consequence, other techniques must be employed to confine the plasma. Generally speaking, there are two methods by which it is possible to confine the plasma for the requisite amount of time. One way to confine the plasma is in specially configured magnetic fields called magnetic bottles which interact with the charged particles in the plasma in such a way as to prevent the plasma from escaping from within the magnetic fields. Plasmas confined in this manner usually have a low particle density and require confinement times in the order of seconds or even minutes. Another method by which the fusing plasma may be confined is through what is called inertial confinement. In this confinement method, small specially constructed pellets of the fusion reactants are super compressed and raised to fusion temperatures by converging laser or ion beams. Because of the inertia of the particles undergoing fusion, it is possible to confine the high density plasma for the extremely short amount of time required to achieve significant amounts of fusion energy. The general regions of operation for the two types of confinement methods are illustrated in Fig. 5.19 which plots the minimum values of $n\tau$ for the reaction types shown in Fig. 5.18.

Although numerous magnetic and inertial confinement configurations for fusion power and propulsion systems have been proposed over the years, none as of the date of this writing, have achieved breakeven [7]. Many of the proposed fusion propulsion designs are conceptual only with little experimental information available to support their ultimate feasibility. Practical fusion propulsion

CHAPTER 5 BASIC NUCLEAR STRUCTURE AND PROCESSES

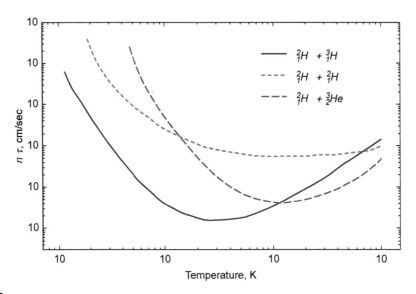

FIGURE 5.18

Lawson criterion for a self-sustaining fusion plasma.

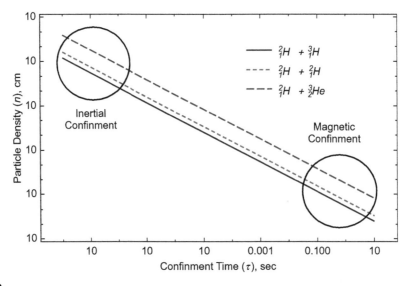

FIGURE 5.19

Operational regions of inertial and magnetic confinement fusion.

systems will require that the energy produced by the fusioning plasma exceed by some amount the energy required to sustain the fusion reactions. This excess energy over that required for breakeven is necessary to compensate for the various processes by which energy is lost from the propulsion system. These loss processes include parasitic radiation, electrical conversion inefficiencies, plasma lost in the exhaust, etc. The fraction of the power produced by the fusioning plasma that exceeds the power lost from the fusioning plasma is defined as the fusion energy gain factor or "Q" factor. Fusion gain factors in the range of five or greater will probably be required for practical fusion propulsion systems.

If a practical fusion propulsion systems could be constructed, the implications with regard to interplanetary travel would be enormous. Virtually the entire solar system would be opened for human space travel. Fusion-based propulsion systems would, for all practical purposes, eliminate mission window constraints and effectively allow unlimited manned exploration of the planets. The following example illustrates potentially just how great the specific impulse could be with fusion propulsion systems.

EXAMPLE

Determine the specific impulse and propellant mass flow rate for a fusion propulsion system using deuterium and helium-3 as propellants. Assume that the specific heat ratio of the propellant is 1.67 and the average molecular weight of the propellant is 3. Also assume that the engine produces 100,000 N of thrust.

Solution

The first step in the analysis of the fusion engine is to estimate the specific impulse of the rocket. From Fig. 5.18 which is annotated in Fig. 1, the temperature corresponding to the minimum $n\tau$ value of 4.1×10^{14} cm^3/s is found to be about 1.3×10^9 K.

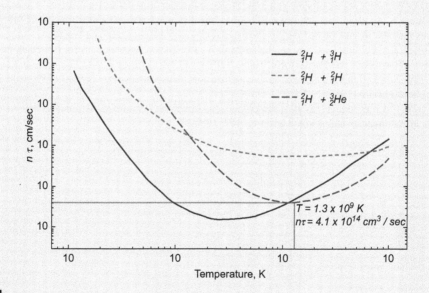

FIGURE 1
Optimal point of operation for the deuterium/helium-3 fusion engine.

Incorporating the temperature corresponding to the minimum $n\tau$ value for the deuterium - helium-3 reaction into the expression for the specific impulse as a function of temperature from Eq. (2.20) then yields:

$$I_{sp} = \frac{1}{g_c}\sqrt{\frac{2\gamma}{\gamma-1}\frac{R_u}{M}T_c} = \frac{1}{9.8\frac{m}{s^2}}\sqrt{\frac{2 \times 1.67}{1.67-1} \times \frac{8314.5\frac{g\,m^2}{K\,mol\,s^2}}{3\frac{g}{mol}} \times 1.3 \times 10^9\,K} = 432,000\,s \quad (1)$$

To determine the mass flow rate of propellant, it is necessary to use the definition of specific impulse, such that:

$$I_{sp} = \frac{F}{g_c\dot{m}} \Rightarrow \dot{m} = \frac{F}{I_{sp}} = \frac{100,000\,N}{9.8\frac{m}{s^2} \times 432,000\,s} = 0.0236\frac{kg}{s} = 23.6\frac{g}{s} \quad (2)$$

A specific impulse of 432,000 s is truly astounding and is considerably higher than any of the fission propulsion systems currently envisioned. Unfortunately, because working fusion reactors have proved notoriously difficult to construct in practice, it is unlikely that a working fusion propulsion system will be available in the near future.

REFERENCES

[1] H. Yukawa, On the interaction of elementary particles, Proceedings of the Physico-Mathematical Society of Japan 17 (1935) 48–57.
[2] C.J. Lister, J. Butterworth, Nuclear physics: exotic pear-shaped nuclei, Nature 497 (May 9, 2013) 190–191.
[3] G. Breit, E. Wigner, Capture of slow neutrons, Physical Review 49 (1936) 519.
[4] D.A.P. Palma, S. Aquilino, A.S. Martinez, F.C. Silva, The derivation of the Doppler broadening function using Frobenius method, Journal of Nuclear Science and Technology 43 (6) (2006) 617–622.
[5] W.R. Arnold, J.A. Phillips, G.A. Sawyer, E.J. Stovall Jr., J.L. Tuck, Cross Sections for the reactions D(d, p)T, D(d, n)He^3, T(d, n)He^4, and He^3(d, p)He^4 below 120 kev, Physical Review 93 (1954) 483–497.
[6] J.D. Lawson, Some criteria for a power producing thermonuclear reaction, Proceedings of Physical Society of London, Section B 70 (1957) 1–6.
[7] T. Kammash (Ed.), Fusion energy in space propulsion, American Institute of Aeronautics and Astronautics, Progress in Astronautics and Aeronautics, vol. 167, 1995.

CHAPTER 6

NEUTRON FLUX ENERGY DISTRIBUTION

During fission an atomic nucleus generally splits into two fission products and between two and three neutrons. These neutrons are extremely energetic having energies in the low MeV range. These neutrons lose energy through scattering collisions, and if they are not captured by a nucleus will eventually reach thermal equilibrium at energies of a fraction of an eV. It is found that light nuclei such as hydrogen and beryllium are the most effective at slowing-down neutrons. During this slowing-down process, the neutron population will take on a spectrum of energies which will depend upon other things, the amount of fissionable material, the probability of the neutrons being scattered as opposed to being absorbed, and so on.

1. CLASSICAL DERIVATION OF NEUTRON-SCATTERING INTERACTIONS

An examination of the cross-sectional plots illustrated earlier reveals that the cross sections are typically higher at lower neutron energies. It is, therefore, advantageous to find some means to reduce the neutron energy from the fast MeV range where the neutrons are born due to fission to the few eV thermal range, where the neutrons are more likely to interact with the nuclides of interest. The method nearly always used to reduce the neutron energy is through multiple scattering interactions off nearby nuclei.

The detailed prediction of the probable scattering angles between neutrons and other nuclei requires complex quantum mechanics analyses coupled with experiments. These analyses are required since one typically does not know the degree of inelasticity present in the collision process. Nevertheless, despite some restrictions, several useful relationships between scattering angles and collision energies can be derived simply from the conservation of energy and conservation of linear momentum considerations.

In the derivations which follow, it is assumed that a moving neutron scatters off a stationary nucleus in what is called the laboratory (L) coordinate system. The actual derivations prove easier to perform; however, if they are carried out in a coordinate system called the center of mass (C or COM) system where the total linear momentum of the particles is zero. In this coordinate system, the COM of the system of particles is stationary as illustrated in Fig. 6.1.

The trajectory of the object is parabolic rather than hyperbolic.

The derivations begin by writing out the equations for the conservation of momentum of the particles:

$$mV_{mC} - MV_{MC} = 0 \tag{6.1}$$

$$mV'_{mC} - MV'_{MC} = 0 \tag{6.2}$$

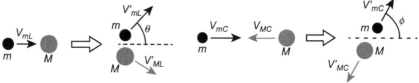

FIGURE 6.1

Coordinate system descriptions where m = neutron mass, M = mass of nucleus, and V = velocity.

and the equation for the conservation of energy of the particles:

$$\frac{1}{2}mV_{mC}^2 + \frac{1}{2}MV_{MC}^2 = \frac{1}{2}mV_{mC}'^2 + \frac{1}{2}MV_{MC}'^2 \tag{6.3}$$

Using Eqs. (6.1)–(6.3) to eliminate V_{MC} and V_{MC}':

$$\left[\frac{1}{2}m\left(\frac{M}{m}\right)^2 + \frac{1}{2}M\right]V_{mC}^2 = \left[\frac{1}{2}m\left(\frac{M}{m}\right)^2 + \frac{1}{2}M\right]V_{mC}'^2 \tag{6.4}$$

Therefore:

$$V_{mC}^2 = V_{mC}'^2 \quad \text{which also implies that} \quad V_{MC}^2 = V_{MC}'^2 \tag{6.5}$$

From Fig. 6.1 one can relate scattering in the COM system to scattering in the L system as:

$$V_{mC} = V_{mL} - V_{MC} \tag{6.6}$$

Using Eq. (6.1) in Eq. (6.6), one finds that:

$$V_{mC} = \frac{MV_{mL}}{M + m} \tag{6.7}$$

The vector relations between the neutron and the target nucleus after they collide with one another can be illustrated in Fig. 6.2. From these vector relations, all necessary scattering trigonometric relationships may be determined:

$$V_{mL}' \cos(\theta) = V_{mC}' \cos(\phi) + V_{MC} \tag{6.8}$$

$$V_{mL}' \sin(\theta) = V_{mC}' \sin(\phi) \tag{6.9}$$

The effort at this point is to derive a correlation which relates the scattering angle between the neutron and the nucleus and the neutron kinetic energy before and after the collision. Using the law of cosines:

$$V_{mL}'^2 = V_{mC}'^2 + V_{MC}^2 + 2V_{mC}'V_{MC}\cos(\phi) \tag{6.10}$$

Combining Eqs. (6.2), (6.5), (6.7), and (6.10), one can obtain:

$$V_{mL}'^2 = \left[\left(\frac{M}{m}\right)^2 + 1 + 2\frac{M}{m}\cos(\phi)\right]\left(\frac{mV_{mL}}{M + m}\right)^2 \tag{6.11}$$

1. CLASSICAL DERIVATION OF NEUTRON-SCATTERING INTERACTIONS

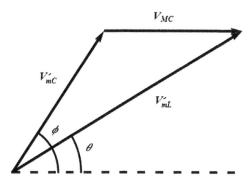

FIGURE 6.2

Vector relationships for scattering interactions.

If one defines $A = \frac{M}{m}$ as the mass ratio and notes that:

$$\frac{E'}{E} = \frac{\frac{1}{2}mV'^2_{mL}}{\frac{1}{2}mV^2_{mL}} = \frac{V'^2_{mL}}{V^2_{mL}} \tag{6.12}$$

using Eq. (6.11) in Eq. (6.12) then yields:

$$\frac{E'}{E} = \frac{A^2 + 2A\cos(\phi) + 1}{(1+A)^2} \tag{6.13}$$

Also from Eqs. (6.2), (6.5), (6.8), and (6.9), one can determine that:

$$\tan(\theta) = \frac{\sin(\phi)}{\cos(\phi) + \frac{1}{A}} \Rightarrow \cos(\theta) = \frac{A\cos(\phi) + 1}{\sqrt{A^2 + 2A\cos(\phi) + 1}} \tag{6.14}$$

The results obtained from Eq. (6.14) now be used to determine the cosine of the average scattering angle which is defined by:

$$\overline{\cos(\theta)} = \mu_0 = \frac{\int_0^{4\pi} \cos(\theta)\, d\theta}{\int_0^{4\pi} \theta\, d\theta} = \frac{2}{3A} \tag{6.15}$$

By making a change of variables, Eq. (6.13) may be rewritten as:

$$\frac{E'}{E} = \frac{1}{2}[(1+\alpha) + (1-\alpha)\cos(\phi)] \quad \text{where:} \quad \alpha \equiv \left(\frac{A-1}{A+1}\right)^2 \tag{6.16}$$

It proves useful to introduce a new variable "u" called *lethargy*, where $u \equiv \ln\left(\frac{E_0}{E}\right)$. E_0 is some arbitrary fixed energy at the top of the fission energy range which is usually taken to be around 10–15 MeV. Thus at the top of the energy range, neutrons start off with zero lethargy and as they lose energy through collisions with the surrounding media, the neutrons progressively gain lethargy.

Assuming equal scattering probabilities of $\cos(\phi)$ from -1 to $+1$ and where αE is the maximum possible energy loss during a collision, Eq. (6.16) can be used to determine the average logarithmic energy decrement per collision, "ξ." This value is also equal to the average gain in lethargy per collision:

$$\xi = \overline{\ln\left(\frac{E}{E'}\right)} = \frac{\int_{\alpha E}^{E} \ln\left(\frac{E}{E'}\right) dE'}{\int_{\alpha E}^{E} dE'} = \frac{-\int_{-1}^{1} \ln\{\frac{1}{2}[(1+\alpha) + (1-\alpha)\cos(\phi)]\} \, d[\cos(\phi)]}{\int_{-1}^{1} d[\cos(\phi)]}$$

$$= 1 + \frac{\alpha}{1-\alpha}\ln(\alpha) \qquad (6.17)$$

Table 6.1 Scattering Parameters for Various Nuclides

Element	A	α	ξ	N
H	1	0	1	18
Be	9	0.640	0.206	86
C	12	0.716	0.158	114
^{235}U	235	0.983	0.00084	2172

To calculate the average number of neutron-scattering events required to thermalize a fission-born neutron, one simply needs to divide ξ into the log of the ratio of the fast to thermal neutron energies. If $E_{\text{fast}} = 2$ MeV and $E_{\text{thermal}} = 0.025$ eV then:

$$N = \frac{\ln\left(\frac{E_{\text{fast}}}{E_{\text{thermal}}}\right)}{\xi} = \frac{\ln\left(\frac{2,000,000}{0.025}\right)}{\xi} = \frac{18.2}{\xi} \qquad (6.18)$$

In Table 6.1, the results of applying Eq. (6.18) to various nuclides are given, so as to determine the average number of scattering interactions required to thermalize fission-born neutrons.

2. ENERGY DISTRIBUTION OF NEUTRONS IN THE SLOWING-DOWN RANGE

Now that an expression has been developed describing how neutrons may be made to slow down through scattering interactions with other nuclei, it is instructive to use that description to develop a qualitative picture of the neutron energy distribution resulting those interactions. These multiple scattering interactions result in the neutrons being transported from the fast energy range where they are born in fission events to just above the thermal energy range where they begin to come into equilibrium with their surroundings. From Fig. 6.3, it can be seen that the number of neutrons scattered between E' and $E - dE'$ is on average the number of neutrons scattered into the interval between E and $E - dE$ assuming there is no neutron absorption.

If one lets the quantity "q" be the slowing-down density of neutrons at energy "E," then the number of neutrons scattering into dE from dE' will occur at a rate of:

$$q\frac{\ln(E) - \ln(E - dE)}{\xi} = \frac{q \, dE}{\xi E} \qquad (6.19)$$

One can also represent the neutron-scattering reaction rate within dE by:

$$R_s = \phi(E)\Sigma_s(E) \, dE \qquad (6.20)$$

2. ENERGY DISTRIBUTION OF NEUTRONS IN THE SLOWING-DOWN RANGE

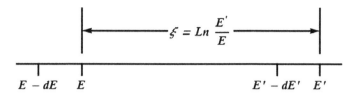

FIGURE 6.3

Slowing-down scattering interactions.

If the scattering processes are in steady state, Eq. (6.19) is equal to Eq. (6.20), therefore:

$$\Sigma_s(E)\phi(E)\,dE = \frac{q\,dE}{\xi E} \tag{6.21}$$

The energy-dependent neutron flux with no neutron absorption in the slowing-down region can now be determined by rearranging Eq. (6.21) to yield:

$$\phi(E) = \frac{q}{E\xi\Sigma_s(E)} \tag{6.22}$$

If neutron absorption takes place during the slowing-down process, the slowing-down density "q" is no longer independent of energy, but is reduced by the amount of absorption within dE. If there is only weak absorption, the amount of absorption may be represented by:

$$\frac{dq(E)}{dE}\,dE = \Sigma_c(E)\phi(E)\,dE \Rightarrow \frac{dq(E)}{dE} = \Sigma_c(E)\phi(E) \tag{6.23}$$

The rate at which neutrons leave dE if there is weak absorption can then be determined by modifying Eq. (6.20) such that:

$$R_{sa} = [\Sigma_s(E) + \Sigma_c(E)]\phi(E) \tag{6.24}$$

The rate at which neutrons enter dE is a function only of the rate at which they scatter out of dE' and is approximated by modifying Eq. (6.21) to give an expression of the form:

$$[\Sigma_s(E) + \Sigma_c(E)]\phi(E)\,dE \approx \frac{q}{\xi E}\,dE \tag{6.25}$$

Rearranging terms in Eq. (6.25) to solve for the neutron flux then yields:

$$\phi(E) \approx \frac{q(E)}{\xi E[\Sigma_c(E) + \Sigma_s(E)]} \tag{6.26}$$

Substituting Eq. (6.26) into Eq. (6.23) and integrating will yield an expression for the energy-dependent slowing-down density of the form:

$$\int_E^{E_0} \frac{dq(E')}{q(E')} = \int_E^{E_0} \frac{\Sigma_c(E')}{\xi E'[\Sigma_s(E') + \Sigma_c(E')]}\,dE' \Rightarrow q(E) = q_0 e^{\int_E^{E_0} \frac{\Sigma_c(E')}{\xi E'[\Sigma_s(E') + \Sigma_c(E')]}\,dE'} \tag{6.27}$$

Substituting Eq. (6.27) into Eq. (6.26) will then yield the neutron flux as a function of energy in the slowing-down region when there is weak neutron absorption:

$$\phi(E) \approx \frac{q_0 e^{\int_E^{E_0} \frac{\Sigma_c(E')}{\xi E'[\Sigma_s(E') + \Sigma_c(E')]} dE'}}{\xi E[\Sigma_s(E) + \Sigma_a(E)]} \quad (6.28)$$

If one now makes the assumption that $\frac{\Sigma_c(E)}{\Sigma_c(E) + \Sigma_s(E)}$ is a weak function of energy, then the integral in Eq. (6.28) may be solved to further approximate the neutron flux as a function of energy in the slowing-down region when there is weak neutron absorption yielding:

$$\phi(E) = \frac{q_0 \left(\frac{E_0}{E}\right)^{\frac{E\Sigma_c}{\xi(E\Sigma_c + E\Sigma_s)}}}{\xi E[\Sigma_c(E) + \Sigma_s(E)]} \quad (6.29)$$

3. ENERGY DISTRIBUTION OF NEUTRONS IN THE FISSION SOURCE RANGE

The fission source range is defined as that range of neutron energies within which neutrons appear as a result of fission events. Generally speaking, this energy range extends from about 10 keV to about 10 MeV and corresponds to the $\chi(E)$ distribution discussed earlier. If one neglects scattering in this energy range, the neutron balance equation can be written as:

$$\text{Removal Rate} = \text{Production Rate}$$

The above-mentioned neutron balance equation may also be written as:

$$\Sigma_t(E)\phi(E) = \chi(E) \int_0^\infty \nu(E') \Sigma_f(E') \phi(E') dE' \quad (6.30)$$

where $\nu(E')$ is the number of neutrons emitted on average per fission. This function is actually a very weak function of energy and is, therefore, often treated as a constant. In Eq. (6.30) the integral evaluates to a constant, therefore:

$$\Sigma_t(E)\phi(E) = C\chi(E) \quad (6.31)$$

Using Eq. (6.31), the neutron flux may now be written as:

$$\phi(E) = \frac{C\chi(E)}{\Sigma_t(E)} \quad (6.32)$$

Since $\Sigma_t(E)$ is fairly constant at high energies, the neutron flux; therefore, approximately follows the $\chi(E)$ distribution in the fission source energy range.

4. ENERGY DISTRIBUTION OF NEUTRONS IN THE THERMAL ENERGY RANGE

The thermal energy range is characterized as those energies wherein neutron-scattering interactions can occasionally result in the neutrons gaining energy as well as losing energy as a consequence of the

4. ENERGY DISTRIBUTION OF NEUTRONS IN THE THERMAL ENERGY RANGE

thermal motion of the nuclei off which the neutrons scatter. If it is assumed that there is no neutron absorption and that there is no source of neutrons due to fission, the thermalized neutrons will follow a Maxwellian energy distribution:

$$n(v)\, dv = Cv^2 e^{-\frac{1}{2}\frac{mv^2}{kT}}\, dv \tag{6.33}$$

where C = constant; v = neutron velocity; $n(v)\, dv$ = number of neutrons between v and $v + dv$ per unit volume; m = neutron mass; T = temperature of the scattering medium; and k = Boltzmann constant.

In the above-mentioned equation, $n(v)\, dv$ also corresponds to the number of neutrons between E and $E + dE$ per unit volume where $E = \frac{1}{2}mv^2$, therefore:

$$n(v)\, dv = n(E)\, dE = \frac{[n(E)v(E)]\, dE}{v(E)} = \frac{\phi(E)\, dE}{v(E)} \tag{6.34}$$

Also note that:

$$E = \frac{1}{2}mv^2 \Rightarrow dE = mv\, dv \Rightarrow dv = \frac{1}{mv}\, dE \tag{6.35}$$

If one now substitutes Eqs. (6.34) and (6.35) into Eq. (6.33) it is found that:

$$\frac{\phi(E)}{v(E)}\, dE = CEe^{\frac{-E}{kT}}\frac{1}{mv(E)}\, dE \Rightarrow \phi(E) = C'Ee^{\frac{-E}{kT}} \tag{6.36}$$

The most probable energy for a Maxwellian distribution occurs at $E = kT$. As was mentioned earlier, thermal neutron cross sections for materials are normally quoted at a reference temperature of 20°C which corresponds to a peak in the Maxwellian distribution at 0.025 eV or equivalently 2200 m/s. A Maxwellian distribution with a 20°C reference temperature is illustrated in Fig. 6.4.

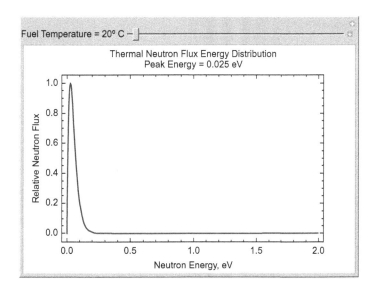

FIGURE 6.4
Maxwellian distribution.

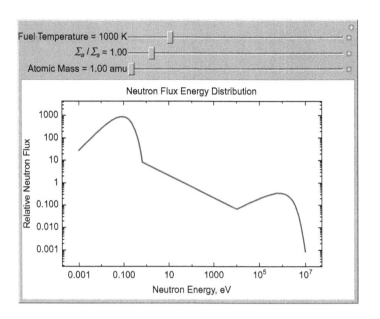

FIGURE 6.5

Neutron energy distribution.

5. SUMMARY OF THE NEUTRON ENERGY DISTRIBUTION SPECTRUM

To summarize the results from the previous sections, it is found that over all neutron energy ranges, the neutron flux level varies approximately as:

$$\text{Fission Source Range:} \quad \phi(E) = \frac{C\chi(E)}{\Sigma_t(E)} \quad 10\text{ keV} < E < 10\text{ MeV}$$

$$\text{Slowing Down Region:} \quad \phi(E) = \frac{q_0 \left(\frac{E_0}{E}\right)^{\frac{E\Sigma_c}{\xi(E\Sigma_c + E\Sigma_s)}}}{\xi E[\Sigma_c(E) + E\Sigma_s(E)]} \quad E_c < E < 10\text{ keV}$$

$$\text{Thermal Energy Range:} \quad \phi(E) = C_2 E e^{\frac{-E}{kT}} \quad 0\text{ eV} < E < E_c$$

In attempting to estimate the cut point energy "E_c" at which the thermal energy range transitions to the slowing-down energy range one quickly finds that defining a single energy cut point yields unreasonable results. This difficulty is due to the fact that the Maxwellian energy distribution which describes the thermal neutron energy range shifts significantly over the range of temperatures normally experienced during reactor operation. As a consequence, the cut point energy is usually made to vary with the temperature of the reactor core and is generally found to be between about 0.1 and 1.0 eV. In Fig. 6.5, a qualitative picture of the energy distribution of neutrons in the various energy ranges described previously is presented.

NEUTRON BALANCE EQUATION AND TRANSPORT THEORY

7

The time rate of change of the neutron population in a nuclear reactor depends upon the rate at which neutrons are produced in the reactor as compared to the rate at which neutrons are lost from the reactor. During steady-state operation, the production and loss rates in the reactor exactly cancel and the neutron population remains constant. The neutronics calculations necessary to determine the exact spatial dependence of the neutron population in the reactor generally require a calculational technique called transport theory. Transport theory is challenging from a calculational standpoint; however, an approximation to transport theory called diffusion theory is often used to perform the needed neutronics calculations. Diffusion theory has been found to be acceptable when high accuracy is not required or when the neutron population gradients in the reactor are not very large.

1. NEUTRON BALANCE EQUATION

In order to determine the neutron flux distribution within a nuclear reactor core, it is necessary to develop a relationship which describes the pointwise nuclear processes which occur within the core. This relationship is formally described by the neutron balance equation given below:

$$\frac{dn}{dt} = -\text{neutron leakage rate} + \text{neutron production rate} - \text{neutron loss rate} \stackrel{\text{steady state}}{=} 0 \qquad (7.1)$$

where n = neutron density.

1.1 LEAKAGE (L)

In order to calculate the neutron leakage, a new variable called *neutron current* must be defined. This quantity has the same units as the neutron flux but is a vector quantity. Physically, the neutron current can be interpreted as the total track length of neutrons per cm^3 within dE about E directed along $\vec{\Omega}$ about $d\Omega$ as shown in Fig. 7.1.

CHAPTER 7 NEUTRON BALANCE EQUATION AND TRANSPORT THEORY

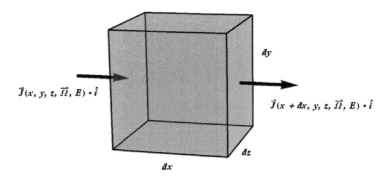

FIGURE 7.1

Neutron leakage through a differential element, where $\vec{J}(x,y,z,E,\vec{\Omega})$ = neutron current at position (x, y, z) having an energy "E" directed along the vector direction.

The net loss of neutrons due to leakage across the faces perpendicular to "x" (eg, $dy\,dz$) then is:

$$\left[\frac{\vec{J}(x,y,z,E,\vec{\Omega}) - \vec{J}(x+dx,y,z,E,\vec{\Omega})}{dx}\right] \cdot \hat{i}\, dx\, dy\, dz\, dE$$

$$= \left[\left(\hat{i}\frac{\partial}{\partial x}\right) \cdot \vec{J}(x,y,z,E,\vec{\Omega})\right] dx\, dy\, dz\, dE \tag{7.2}$$

The total net loss of neutrons due to leakage across all faces in $dV = dx\,dy\,dz$ may be determined by extending Eq. (7.2) to three dimensions:

$$L = \left[\left(\hat{i}\frac{\partial}{\partial x}\right) + \left(\hat{j}\frac{\partial}{\partial y}\right) + \left(\hat{k}\frac{\partial}{\partial y}\right)\right] \cdot \vec{J}(x,y,z,E,\vec{\Omega})\, dx\, dy\, dz\, dE = \vec{\nabla} \cdot \vec{J}(r,E,\vec{\Omega})\, dV\, dE \tag{7.3}$$

1.2 FISSION PRODUCTION RATE (P_f)

In order to determine the energy-dependent fission neutron production rate it is necessary to first calculate the total neutron production rate due to fissions by integrating the fission reaction rate over all nuclides "j" and all neutron energies. The total fission production rate is then scaled by $\chi(E)$ to yield the fission production rate between E and $E + dE$:

$$P_f = \chi(E) \int_0^\infty \nu\Sigma_f(r,E')\phi(r,E')\, dE'\, dV\, dE \tag{7.4}$$

1.3 SCATTERING PRODUCTION RATE (P_s)

The energy-dependent in-scattering neutron production rate is a result of neutron scattering into the energy range between E and $E + dE$ as a result of scattering interactions from all other energies. To evaluate this term, it is necessary to integrate the in-scattering interaction rate over all energies and all nuclides:

$$P_s = \int_0^\infty \Sigma_s\left(r, E' \to E\right) \phi\left(r, E'\right) dE' \, dV \, dE \tag{7.5}$$

1.4 ABSORPTION LOSS RATE (R_a)

The energy-dependent neutron loss rate is a result of the absorption of neutrons in the energy range between E and $E + dE$ by all nuclides. Any interaction which results in the loss of a neutron is accounted for with this term. Besides neutron capture, which is obviously accounted for in this term, neutron fission must also be accounted for since a neutron must be absorbed for the fission to take place. Neutron out-scattering could also be accounted for here; however, it is generally treated as a separate term:

$$R_a = [\Sigma_c(r, E) + \Sigma_f(r, E)]\phi(r, E) \, dV \, dE = \Sigma_a(r, E)\phi(r, E) \, dV \, dE \tag{7.6}$$

1.5 SCATTERING LOSS RATE (R_s)

The energy-dependent out-scattering neutron loss rate is a result of neutrons scattering out of the energy range between E and $E + dE$ into all other energies. To evaluate this term, it is necessary to integrate the out-scattering interaction rate over all energies and all nuclides:

$$R_s = \int_0^\infty \Sigma_s\left(r, E \to E'\right) \phi\left(r, E'\right) dE' \, dV \, dE = \Sigma_s(r, E)\phi(r, E) \, dV \, dE \tag{7.7}$$

1.6 STEADY-STATE NEUTRON BALANCE EQUATION

The steady-state neutron balance equation may now be determined by incorporating the neutron production and loss terms as described by Eqs. (7.3)–(7.7) into Eq. (7.1):

$$\begin{aligned} 0 &= -L + P_f + P_s - R_a - R_s \\ &= -\vec{\nabla} \cdot \vec{J}\left(r, E, \vec{\Omega}\right) + \chi(E) \int_0^\infty \nu^j \Sigma_f^j\left(r, E'\right) \phi\left(r, E'\right) dE' \\ &\quad + \int_0^\infty \Sigma_s^j\left(r, E' \to E\right) \phi\left(r, E'\right) dE' - \Sigma_a^j(r, E)\phi(r, E) - \Sigma_s^j(r, E)\phi(r, E) \end{aligned} \tag{7.8}$$

It should be noted that Eq. (7.8) has no approximations; however, it cannot be solved in its present form since it contains two unknown variables, namely. the neutron flux: $\phi(r,E)$ and the neutron current: $\vec{J}\left(r, E, \vec{\Omega}\right)$. What is needed to solve the neutron balance equation is another expression which relates the neutron flux to the neutron current. This neutron flux to neutron current relational expression is typically determined by applying a procedure called *transport theory* which is discussed in the next section.

2. TRANSPORT THEORY

In transport theory, angular dependencies are included when evaluating the expressions for the neutron flux and the neutron scattering cross section. In particular, a new parameter called the neutron directional flux density is introduced, which is defined as:

$$\psi\left(r, E, \vec{\Omega}\right) = u(E)n\psi\left(r, E, \vec{\Omega}\right) \tag{7.9}$$

Using Eq. (7.9) the scalar neutron flux and neutron current, respectively, can be expressed as:

$$\phi(r, E) = \int_{\text{all } \Omega} \psi\left(r, E, \vec{\Omega}\right) d\Omega = u(E) \int_{\text{all } \Omega} n\left(r, E, \vec{\Omega}\right) d\Omega \tag{7.10}$$

$$\vec{J}(r, E) = \int_{\text{all } \Omega} \vec{\Omega}\psi\left(r, E, \vec{\Omega}\right) d\Omega \tag{7.11}$$

A common transport theory approach (though not the only approach) to solving the expressions for the neutron flux and current is by the *spherical harmonics* method. Other transport theory methods include the *Fourier transform* approach, the *discrete ordinates* [1] technique, and the *Monte Carlo* [2,3] method. These methods, while quite powerful and useful in many situations, are not discussed here. The main advantage to the spherical harmonics method is that it is invariant to changes in axis orientation. This invariance to changes in axis orientation can be quite useful in that while the choice of orientation may help considerably in finding a solution to $\psi\left(r, E, \vec{\Omega}\right)$, its numerical values will remain unaltered regardless of the orientation chosen. The spherical harmonics method employs a procedure in which the neutron flux and current are expanded using Legendre polynomials. Legendre polynomials form the solution to the Legendre differential equation which occurs when solving Laplace's equation and related partial differential equations in spherical coordinates. The Legendre differential equation is expressed as follows:

$$(1 - x^2)\frac{d^2 P_l}{dx^2} - 2x\frac{dP_l}{dx} + l(l+1)P_l = 0 \tag{7.12}$$

The Legendre polynomials which form the solution set of Eq. (7.12) are defined by:

$$P_0(x) = 1, \quad P_1(x) = x, \quad (2l+1)P_l(x) = (l+1)P_{l+1}(x) + lP_{l-1}(x) \tag{7.13}$$

Functions may be expanded using Legendre polynomials in a manner similar to that employed in Fourier expansions wherein functions are expanded in sine and cosine series. In a Legendre expansion, functions may be represented as a sum of Legendre polynomials:

$$F(x) \equiv \sum_{l=0}^{\infty} A_l(2l+1)P_l(x) \tag{7.14}$$

The expansion coefficients "A_l" in Eq. (7.14) are found by making use of the orthogonality relationship for Legendre polynomials which is given by:

$$\frac{1}{2}(2l-1)\int_{-1}^{1} P_l(x)P_m(x)\,dx = \delta_{lm} \tag{7.15}$$

where δ_{lm} = Kronecker delta function = $\begin{cases} 0: l \neq m \\ 1: l = m \end{cases}$

2. TRANSPORT THEORY

Using the orthogonality relationship given by Eq. (7.15) the expansion coefficients are then determined by:

$$A_l \equiv \frac{1}{2} \int_{-1}^{1} F(x) P_l(x) \, dx \tag{7.16}$$

By applying Eq. (7.14) to a one-dimensional case where $\psi\left(r, E, \vec{\Omega}\right)$ depends only on "z" and "μ," the neutron directional flux density can be expanded such that:

$$\psi(r, E, \mu) = \sum_{l=0}^{\infty} (2l+1) \psi_l(r, E) P_l(\mu) \tag{7.17}$$

where $\psi_l(r, E) = \frac{1}{2} \int_{-1}^{1} \psi(r, E, \mu) P_l(\vec{\mu}) \, d\mu$ and $\mu = \vec{\Omega} \cdot \hat{k}$.

The neutron scattering cross section can be expanded such that:

$$\Sigma_s\left(r, E' \to E, \mu_0\right) = \sum_{l=0}^{\infty} (2l+1) \Sigma_{sl}\left(r, E' \to E\right) P_l(\mu_0) \tag{7.18}$$

where $\Sigma_{sl}\left(r, E' \to E\right) = \frac{1}{2} \int_{-1}^{1} \Sigma_s(r, E' \to E, \mu_0) P_l(\mu_0) \, d\mu$ and $\mu_0 = \vec{\Omega} \cdot \vec{\Omega}$.

To make the following analysis easier to follow, the neutron balance equation as presented by Eq. (7.8) is now rewritten in one-dimensional form with only one isotope to yield:

$$\frac{\partial \vec{J}(r, E)}{\partial z} + \Sigma_t(r, E) \phi(r, E) = \int_0^{\infty} \left[\chi(E) \nu \Sigma_f\left(r, E'\right) + \Sigma_s\left(r, E' \to E\right)\right] \phi\left(r, E'\right) dE' \tag{7.19}$$

If Eq. (7.19) is now again rewritten so as to incorporate the expressions for neutron flux and current as presented in Eqs. (7.10) and (7.11) one obtains:

$$\mu \frac{\partial}{\partial z} \psi(r, E, \mu) + \Sigma_t(r, E) \psi(r, E, \mu) = \frac{1}{2} \chi(E) \int_0^{\infty} \int_{-1}^{1} \nu \Sigma_f\left(r, E'\right) \psi\left(r, E', \mu'\right) d\mu' \, dE'$$
$$+ \frac{1}{4\pi} \int_0^{\infty} \int_0^{2\pi} \int_{-1}^{1} \Sigma_s\left(r, E' \to E, \mu_0\right) \psi\left(r, E', \mu\right) d\mu' \, d\varphi' \, dE' \tag{7.20}$$

The neutron balance relation as expressed in Eq. (7.20) is now expanded in spherical harmonics using the Legendre polynomial expressions presented in Eqs. (7.17) and (7.18) yielding:

$$\sum_{n=0}^{\infty} \left[(2n+1) \mu P_n(\mu) \frac{\partial}{\partial z} \psi_n(r, E) + \Sigma_t(r, E)(2n+1) P_n(\mu) \psi_n(r, E)\right]$$
$$= \frac{1}{2} \chi(E) \int_0^{\infty} \int_{-1}^{1} \nu \Sigma_f\left(r, E'\right) \sum_{n=0}^{\infty} (2n+1) P_n(\mu') \psi_n\left(r, E'\right) d\mu' \, dE'$$
$$+ \frac{1}{4\pi} \int_0^{\infty} \int_0^{2\pi} \int_{-1}^{1} \left[\sum_{n=0}^{\infty} \Sigma_{sn}\left(r, E' \to E\right)(2n+1) P_n(\mu_0)\right]$$
$$\times \left[\sum_{m=0}^{\infty} (2m+1) P_m(\mu') \psi_m\left(r, E'\right)\right] d\mu' \, d\varphi' \, dE' \tag{7.21}$$

To solve Eq. (7.22) it is necessary to express $P_n(\mu_0)$ in terms of μ, μ', φ, and φ'. This may be accomplished through the use of the law of angular addition of Legendre polynomials which can be stated as follows:

$$P_n(\mu_0) = P_n(\mu')P_n(\mu) + 2\sum_{m=1}^{n}\frac{(m-n)!}{(m+n)!}P_n^m(\mu')P_n^m(\mu)\cos\left[m(\varphi'-\varphi)\right] \quad (7.22)$$

where $P_n^m(\mu) \equiv (1-\mu^2)^{\frac{m}{2}}\frac{d^m P_n(\mu)}{d\mu^m} \equiv$ associated Legendre polynomial.

Substituting Eq. (7.22) into the neutron balance expression of Eq. (7.21) and applying the orthogonality condition of Legendre polynomials as expressed by Eq. (7.15) one obtains after a bit of algebra the following equation:

$$n\frac{\partial}{\partial z}\psi_{n-1}(r,E) + (n+1)\frac{\partial}{\partial z}\psi_{n+1}(r,E) + (2n+1)\Sigma_t(r,E)\psi_n(z,E)$$

$$= \delta_{0n}\chi(E)\int_0^\infty \nu\Sigma_f(r,E')\psi_0(r,E')\,dE' + (2n+1)\int_0^\infty \Sigma_{sn}(r,E'\to E)\psi_n(r,E')\,dE' \quad (7.23)$$

Eq. (7.23) represents an infinite set of coupled algebraic equations which are exactly equivalent to the integro-differential equation represented in Eq. (7.20). If Eq. (7.23) is truncated at $n = N$ and the subsequent $\frac{\partial}{\partial z}\psi_{n+1}(r,E)$ terms are neglected, it is possible to solve for the remaining ψ terms (ψ_0, ψ_1, ..., ψ_N) to yield what is the called "P_N" approximation to the spherical harmonics neutron transport equation. Since Legendre polynomial expansions normally converge quite quickly, P_3 is usually the highest order transport approximation routinely performed.

3. DIFFUSION THEORY APPROXIMATION

If the spherical harmonics equations are truncated at P_1, it is possible to obtain what is called the *diffusion theory* approximation to transport theory. This approximation permits the neutron balance equation to be solved in a much more tractable manner than would otherwise be possible. Using diffusion theory, analytical calculations of the neutron flux in space and energy can be performed allowing one to get an intuitive understanding as to how neutrons distribute themselves in a reactor. In the P_1 approximation only the zeroth- and first-order terms are retained. Rewriting Eq. (7.23) with $n = 0$ (zeroth-order term) yields:

$$\frac{\partial}{\partial z}\psi_1(r,E) + \Sigma_t(r,E)\psi_0(z,E) = \int_0^\infty \left[\chi(E)\nu\Sigma_f(r,E') + \Sigma_{s0}(r,E'\to E)\right]\psi_0(r,E')\,dE' \quad (7.24)$$

Rewriting Eq. (7.23) with $n = 1$ (first-order term) yields:

$$\frac{\partial}{\partial z}\psi_0(r,E) + 3\Sigma_t(r,E)\psi_1(r,E) = 3\int_0^\infty \Sigma_{s1}(r,E'\to E)\psi_1(r,E')\,dE' \quad (7.25)$$

From Eq. (7.17) one obtains for $n = 1$:

$$\psi(r,E,\mu) = \psi_0(r,E) + 3\mu\psi_1(r,E) \quad (7.26)$$

3. DIFFUSION THEORY APPROXIMATION

Using Eq. (7.17) in Eq. (7.10) an expression for the neutron flux may be determined as follows:

$$\phi(r,E) = \int_{\text{all }\Omega} \psi\left(r,E,\vec{\Omega}\right) d\Omega = \frac{1}{4\pi} \int_{-1}^{1} \int_{0}^{2\pi} [\psi_0(r,E) + 3\mu\psi_1(r,E)] \, d\varphi \, d\mu = \psi_0(r,E) \quad (7.27)$$

also noting that:

$$\mu = \vec{\Omega} \cdot \hat{k} \Rightarrow \mu\hat{k} = \vec{\Omega} \cdot \hat{k} \cdot \hat{k} = \vec{\Omega}$$

and using Eq. (7.17) in Eq. (7.11) an expression for the neutron current may be determined such that:

$$\vec{J}(r,E) = \int_{\text{all }\Omega} \vec{\Omega}\psi\left(r,E,\vec{\Omega}\right) d\Omega = \frac{1}{4\pi} \int_{-1}^{1} \int_{0}^{2\pi} \mu[\psi_0(r,E) + 3\mu\psi_1(r,E)]\hat{k} \, d\varphi \, d\mu = \psi_1(r,E)\hat{k} \quad (7.28)$$

Incorporating the definitions for the neutron flux and current as expressed by Eqs. (7.27) and (7.28), respectively, in Eq. (7.26) for the neutron directional flux density then results in:

$$\psi(r,E,\mu) = \phi(r,E) + 3\mu J(r,E) \quad (7.29)$$

Expanding the integral expression in Eq. (7.25) in a Taylor series using the definitions from Eq. (7.18) and retaining only the second term yields:

$$\int_{0}^{\infty} \Sigma_{s1}\left(r,E' \to E\right) \psi_1\left(r,E'\right) dE' \approx \mu_0 \Sigma_s(r,E)\psi_1(r,E) \quad (7.30)$$

Rewriting Eq. (7.25) using the results of Eqs. (7.28) and (7.30) then gives the expression:

$$\frac{\partial}{\partial z}\psi_0(r,E) + 3\Sigma_t(r,E)\vec{J}(r,E)\hat{k} = 3\mu_0\Sigma_s(r,E)\vec{J}(r,E)\hat{k} \quad (7.31)$$

Solving Eq. (7.31) for the neutron current using the results of Eq. (7.27) then yields:

$$\vec{J}(r,E)\hat{k} = \frac{-1}{3[\Sigma_t(r,E) - \mu_0\Sigma_s(r,E)]} \frac{\partial}{\partial z}\phi(r,E) = \frac{-1}{3\Sigma_{tr}(r,E)} \frac{\partial}{\partial z}\phi(r,E) = -D(r,E)\frac{\partial}{\partial z}\phi(r,E) \quad (7.32)$$

Eq. (7.32) is the diffusion theory approximation to transport theory and is an expression of *Fick's law* wherein $D(r,E)$ is the *diffusion coefficient*. This approximation is used almost exclusively in the chapters which follow as we analytically examine the neutron flux distributions in various geometric configurations. Extending Eq. (7.32) now to three dimensions yields:

$$\vec{J}(r,E) = -D(r,E)\left[\frac{\partial}{\partial x}\phi(r,E)\hat{i} + \frac{\partial}{\partial y}\phi(r,E)\hat{j} + \frac{\partial}{\partial z}\phi(r,E)\hat{k}\right] = -D(r,E)\vec{\nabla}\phi(r,E) \quad (7.33)$$

Eq. (7.33) is the three-dimensional diffusion theory approximation to transport theory and forms the theoretical basis for the use of multigroup diffusion theory in the calculation nuclear reactor behavior.

REFERENCES

[1] K.D. Parsons, "ANISN/PC Manual", EGG-2500, Idaho National Engineering Laboratory, April 2003.
[2] F. Brown, Fundamentals of Monte Carlo Particle Transport, LA-UR-05–4983, 2005.
[3] X-5 Monte Carlo Team, MCNP – A General N-Particle Transport Code, Version 5, Volume I: Overview and Theory, LA-UR-03–1987, 2003 (Updated 2005).

CHAPTER 8

MULTIGROUP NEUTRON DIFFUSION EQUATIONS

The integro-differential equation describing the spatially dependent neutron balance within a nuclear reactor, while formally correct, is extremely challenging to solve in most situations. Fortunately, this equation may be considerably simplified by using the diffusion theory approximation discussed earlier and by averaging the energy-dependent neutron cross sections over discrete energy regimes (or energy groups). The set of equations which result from these simplifications are called the multigroup neutron diffusion equations. These equations consist of coupled ordinary differential equations with constant coefficients which may be solved relatively easily using standard mathematical techniques. The solution to these equations yields spatially dependent neutron flux distributions for each discrete energy group plus an eigenvalue which describes the critical state of the reactor. By using these neutron flux distributions with the appropriate fission cross sections, detailed, fairly accurate power density distributions may be determined at all points throughout a nuclear reactor.

1. MULTIGROUP DIFFUSION THEORY

As was stated in Chapter 7 the neutron balance equation presented in Eq. (7.8) is formally correct, but unsolvable due to the fact that it contains both the neutron flux and the neutron current as unknowns. With the results of Eq. (7.33) an expression is now available by which the neutron flux and neutron current may be related to one another thereby enabling the neutron balance equation to be solved explicitly. Since the neutron current appears only in the leakage term of the neutron balance equation one may use Eq. (7.33) to eliminate the neutron current term from the leakage expression as given by Eq. (7.2):

$$L = \vec{\nabla} \cdot \vec{J}(r,E) = -D(r,E)\vec{\nabla} \cdot \vec{\nabla}\,\phi(r,E) = -D(r,E)\nabla^2 \phi(r,E) \tag{8.1}$$

Incorporating Eq. (8.1) into the energy-dependent neutron balance Eq. (7.8) for a reactor which is assumed to be in a steady-state condition then yields:

$$0 = D(r,E)\nabla^2\phi(r,E) + \chi(E)\int_0^\infty \nu\Sigma_f(r,E')\phi(r,E')\,dE' + \int_0^\infty \Sigma_s(r,E'\to E)\phi(r,E')\,dE'$$
$$- \Sigma_a(r,E)\phi(r,E) - \Sigma_s(r,E)\phi(r,E) \tag{8.2}$$

When appropriate boundary conditions are applied to the neutron balance equation as expressed by Eq. (8.2), the neutron flux as a function of space and energy may theoretically be

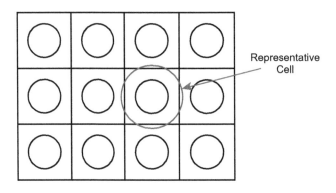

FIGURE 8.1

Square lattice reactor geometry.

determined. In practice, however, it is found that determining a solution to the neutron balance in its integro-differential form is quite difficult for all but the simplest geometric cases. This difficulty results from the fact that generally speaking, the neutron cross sections are very complicated functions of energy, and the reactor geometries themselves are often quite complex. As a consequence, the normal procedure for determining the neutron flux distribution in reactor involves a two-step process in which Eq. (8.2) is typically first solved using a higher-order P_n approximation in its full integro-differential form for a simple representative cell such as that illustrated in Fig. 8.1. The energy-dependent pointwise neutron fluxes which result from this calculation are then used to determine spatially smeared, cell-average cross sections for a few contiguous discrete energy ranges.

Second, once the flux averaged smeared cross sections have been determined, they are reincorporated into Eq. (8.2) to yield what are called the multigroup neutron diffusion equations. These neutron diffusion equations form a set of coupled ordinary differential equations with constant coefficients. The neutron diffusion equations are usually cast in matrix form and solved using appropriate boundary conditions to yield spatially dependent multienergy group neutron fluxes for the reactor configuration under consideration. The details of the flux distribution within the individual cells are lost in this procedure since only smeared average neutron fluxes and cross sections are used; however, for many situations this is acceptable. If the neutron flux details within individual cells are required, the smeared results can be scaled using the neutron fluxes from the cell calculation performed in the first step of the procedure when the smeared cross sections were first determined.

The number of energy groups used in any particular problem varies, but generally speaking, fewer groups (eg, 2–4) are used when analyzing *thermal reactors* (ie, reactors where the intent is to slow down most of the neutrons produced in fission to thermal energy levels) and more energy groups (eg, 8–10 or more) are used when analyzing *fast reactors* (ie, reactors where the intent is to minimize the number of neutrons reaching thermal energy levels). By convention, the highest energy group is the first energy group, and the energy group number designations increase as the energies decrease as shown in Table 8.1.

Table 8.1 Neutron Energy Group Structure

Group	Energy Interval
1	$E_1 - \infty$
2	$E_2 - E_1$
3	$E_3 - E_2$
$\vdots$	$\vdots$
G	$0 - E_{G-1}$

The flux averaged cross sections are determined by integrating the product of the cross section under consideration and the neutron flux over a specific energy interval (eg, the reaction rate over that energy interval), and dividing the result by the integral of the neutron flux over that interval (eg, the total neutron flux over that interval). The flux averaged cross sections for isotope "j" may thus be represented as

$$\phi^g(r) = \int_{E_g}^{E_{g-1}} \phi(r, E)\, dE$$

$$\sigma_c^{g,j}(r) = \frac{\int_{E_g}^{E_{g-1}} \sigma_c^j(r, E)\phi(r, E)\, dE}{\int_{E_g}^{E_{g-1}} \phi(r, E)\, dE} \quad \text{yielding} \quad \Sigma_c^g(r) = \sum_{j=1}^{J_{\max}} n^j(r)\sigma_c^{g,j}(r)$$

$$\sigma_f^{g,j}(r) = \frac{\int_{E_g}^{E_{g-1}} \chi(E) \left[\int_0^\infty \sigma_f^j(r, E')\phi(r, E')\, dE' \right] dE}{\int_{E_g}^{E_{g-1}} \phi(r, E)\, dE} \quad \text{yielding} \quad \Sigma_f^g(r) = \sum_{j=1}^{J_{\max}} n^j(r)\sigma_f^{g,j}(r)$$

$$\sigma_s^{g' \to g,j}(r) = \frac{\int_{E_g}^{E_{g-1}} \left[\int_{E_{g'}}^{E_{g'-1}} \sigma_s^j(r, E' \to E)\phi(r, E')\, dE' \right] dE}{\int_{E_g}^{E_{g-1}} \left[\int_{E_{g'}}^{E_{g'-1}} \phi(r, E')\, dE' \right] dE} \quad \text{yielding} \quad \Sigma_s^{g' \to g}(r) = \sum_{j=1}^{J_{\max}} n^j(r)\sigma_s^{g' \to g,j}(r)$$

$$\sigma_{tr}^{g,j}(r) = \frac{\int_{E_g}^{E_{g-1}} \sigma_{tr}^j(r, E)\phi(r, E)\, dE}{\int_{E_g}^{E_{g-1}} \phi(r, E)\, dE} \quad \text{yielding} \quad D^g(r) = \frac{1}{3}\sum_{j=1}^{J_{\max}} \frac{1}{n^j(r)\sigma_{tr}^{g,j}(r)}$$

(8.3)

If the averaged energy group cross sections in Eq. (8.3) are now incorporated into the steady-state neutron balance equation as expressed in Eq. (8.2), it is possible to obtain the neutron diffusion equations in their multigroup form:

$$0 = D^g(r)\nabla^2 \phi^g(r) + \chi^g \sum_{g'=1}^{G} \nu\Sigma_f^{g'}(r)\phi^{g'}(r) + \sum_{g'=1}^{G} \Sigma_s^{g' \to g}(r)\phi^{g'}(r) - \Sigma_a^g(r)\phi^g(r) - \sum_{g'=1}^{G} \Sigma_s^{g \to g'}(r)\phi^g(r) \quad (8.4)$$

where $\chi^g = \int_{E_g}^{E_{g-1}} \chi(E)\, dE$ and $\Sigma_c^g(r) = \Sigma_f^g(r) + \Sigma_c^g(r)$.

As an example in the use of Eq. (8.4), if "G" is set equal to 3, and the "r" dependency of the cross sections is dropped for the time being for clarity, three coupled differential equations are obtained to which if appropriate boundary conditions are applied can be solved for the position-dependent three energy group neutron fluxes:

$$\left(D^1\nabla^2 + \chi^1\nu\Sigma_f^1 - \Sigma_a^1 - \Sigma_s^{1 \to 2} - \Sigma_s^{1 \to 3}\right)\phi^1 + \left(\chi^1\nu\Sigma_f^2 + \Sigma_s^{2 \to 1}\right)\phi^2 + \left(\chi^1\nu\Sigma_f^3 + \Sigma_s^{3 \to 1}\right)\phi^3 = 0$$

$$\left(\chi^2\nu\Sigma_f^1 + \Sigma_s^{1 \to 2}\right)\phi^1 + \left(D^2\nabla^2 + \chi^2\nu\Sigma_f^2 - \Sigma_a^2 - \Sigma_s^{2 \to 1} - \Sigma_s^{2 \to 3}\right)\phi^2 + \left(\chi^2\nu\Sigma_f^3 + \Sigma_s^{3 \to 2}\right)\phi^3 = 0$$

$$\left(\chi^3\nu\Sigma_f^1 + \Sigma_s^{1 \to 3}\right)\phi^1 + \left(\chi^3\nu\Sigma_f^2 + \Sigma_s^{2 \to 3}\right)\phi^2 + \left(D_3\nabla^2 + \chi^3\nu\Sigma_f^3 - \Sigma_a^3 - \Sigma_s^{3 \to 1} - \Sigma_s^{3 \to 2}\right)\phi^3 = 0 \quad (8.5)$$

In Eq. (8.5), the grayed out terms are typically zero in three group calculations; however, in calculations involving "many" energy groups, it is often the case that some of the terms in the thermal energy range will be nonzero. Note that the grayed out terms are all either scattering terms where the neutrons gain energy through collisions (up-scatter terms) or terms involving the $\chi(E)$ function where the neutrons born in fission appear in the lower-energy groups.

2. ONE GROUP, ONE REGION NEUTRON DIFFUSION EQUATION

With the multigroup neutron diffusion equations in hand, it will prove useful now to analyze a very simple reactor configuration consisting of a one-dimensional slab with spatially independent cross sections in one energy group as shown in Fig. 8.2. This configuration illustrates the general shape of the neutron flux distribution in nuclear reactor systems as well as introduces a number of concepts which are used in subsequent analyses.

Applying Eq. (8.4) using only one energy group and dropping unneeded superscripts then yields:

$$D\frac{d^2\phi}{dz^2} + \nu\Sigma_f\phi - \Sigma_a\phi = 0 \quad (8.6)$$

Eq. (8.6) is now rewritten combining the constant cross-sectional terms, such that:

$$\frac{d^2\phi}{dz^2} + \frac{\nu\Sigma_f - \Sigma_a}{D}\phi = \frac{d^2\phi}{dz^2} + B_m^2\phi = 0 \quad (8.7)$$

where $B_m = \sqrt{\frac{\nu\Sigma_f - \Sigma_a}{D}}$.

2. ONE GROUP, ONE REGION NEUTRON DIFFUSION EQUATION

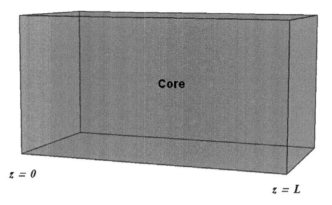

FIGURE 8.2

One-dimensional bare reactor.

In Eq. (8.7), B_m is known as the *materials buckling* for the reactor. This designation was chosen because B_m provides a measure of the "buckling" of the neutron flux throughout the reactor core in a manner similar to that observed in beams which buckle when subjected to end loading. The neutron diffusion differential equation represented by Eq. (8.7) is easily solved by standard methods to yield:

$$\phi(z) = A \cos(B_m z) + C \sin(B_m z) \qquad (8.8)$$

To solve for the arbitrary constants "A" and "C," appropriate boundary conditions must be applied to Eq. (8.8). As a first guess, one might expect that the neutron flux should go to zero at the reactor boundary. While the assumption of a zero boundary flux condition is a quite logical assumption to make and is often quite acceptable especially for large reactors, it turns out that in actual fact, it is not quite true. The problem stems from the fact that a finite number of neutrons escape the reactor by "leaking" out the sides and ends thus creating a finite neutron flux at the boundary. What is true at the boundary is that neutrons which leak out of the core do not return, that is to say that the neutron current is zero in the inward direction to the core. With this thought in mind, an expression is now sought which better represents the boundary condition at the edge of the reactor. Recall from the P_1 approximation of Eq. (7.29) that the neutron directional flux density was given by

$$\psi(r, E, \mu) = \phi(r, E) + 3\mu J(r, E) \qquad (8.9)$$

For all directions inward to the core at the core boundary, $\psi(0, E, \mu) = 0$ (assuming $r = 0$ is the core boundary):

$$\int_0^{-1} \mu \psi(0, E, \mu) \, d\mu = \int_0^{-1} \mu \phi(0, E) + 3\mu^2 J(0, E) \, d\mu = -\frac{1}{4}\phi(0, E) + \frac{1}{2}J(0, E) = 0$$

$$\Rightarrow \phi(0, E) = 2J(0, E) \qquad (8.10)$$

Recalling the one-dimensional diffusion theory approximation given by Eq. (7.32):

$$J(0,E) = -D(0,E)\frac{\partial \phi(0,E)}{\partial z} \qquad (8.11)$$

If Eq. (8.10) is incorporated into Eq. (8.11), one arrives at the following relationship at the reactor boundary:

$$\phi(0,E) = -2D(0,E)\frac{\partial \phi(0,E)}{\partial z} \qquad (8.12)$$

With the results from Eq. (8.12) a relationship between the slope of the neutron flux and the neutron flux level at the reactor boundary has now been determined. This relationship makes it possible to derive a straight line extrapolation to a fictitious point outside the reactor where the extrapolated flux goes to zero. By using the location of this fictitious point outside the reactor as a boundary condition where the neutron flux goes to zero, it is possible to obtain a more accurate representation of the core-wide neutron flux, especially at points near the reactor boundary as illustrated in Fig. 8.3.

From Fig. 8.3, the straight-line extrapolation equation for the neutron flux beyond the reactor boundary is thus observed to be

$$\phi(z,E) = \frac{\partial \phi(0,E)}{\partial z} z + b \qquad (8.13)$$

At the extrapolation distance "d," the flux goes to zero, therefore:

$$0 = \frac{\partial \phi(0,E)}{\partial z} d + b \Rightarrow b = -d\frac{\partial \phi(0,E)}{\partial z} \qquad (8.14)$$

Substituting Eq. (8.13) into Eq. (8.14), it is found that:

$$\phi(z,E) = \frac{\partial \phi(0,E)}{\partial z}(z-d) \Rightarrow \phi(0,E) = -d\frac{\partial \phi(0,E)}{\partial z} \qquad (8.15)$$

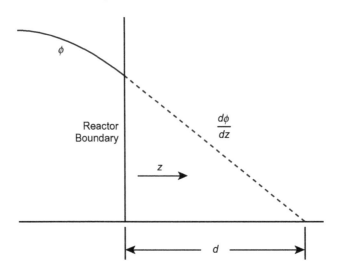

FIGURE 8.3

Neutron flux extrapolation at the reactor boundary.

Rearranging and equating Eqs. (8.12) and (8.15), it is found that:

$$\phi(0,E) = -d\frac{\partial \phi(0,E)}{\partial z} = -2D(0,E)\frac{\partial \phi(0,E)}{\partial z} \Rightarrow d = 2D(0,E) = \frac{2}{3\Sigma_{\text{tr}}(0,E)} \qquad (8.16)$$

Eq. (8.16) shows that the extrapolation length is equal to twice the value of the diffusion coefficient. Numerically, the diffusion coefficient is typically a few centimeters in length so, as was said earlier for large cores (eg, with dimensions in the order of meters), the extrapolation length is relatively unimportant. For small, highly enriched cores, however (eg, with dimensions in the order of a meter or less), the flux readjustments resulting from the incorporation of the extrapolation length in the reactor calculations can be quite significant.

For the one region reactor configuration described the preceding paragraphs, applying the extrapolation boundary conditions results in the neutron flux going to zero at "$-d$" and "$L+d$," therefore Eq. (8.8) is modified, such that:

$$\phi(z) = A\cos[B_{\text{m}}(z+d)] + C\sin[B_{\text{m}}(z+d)] \qquad (8.17)$$

Applying the boundary condition at $z = -d$ to Eq. (8.17) then yields:

$$0 = A\cos(0) + C\sin(0) \Rightarrow A = 0 \Rightarrow \phi(z) = C\sin[B_{\text{m}}(z+d)] \qquad (8.18)$$

Applying the boundary condition at $z = L+d$ to Eq. (8.18) results in the following expression:

$$0 = C\sin[B_{\text{m}}(L+2d)] \Rightarrow B_{\text{m}}(L+2d) = n\pi, \quad \text{where } n = 1,2,3,\ldots \qquad (8.19)$$

Rearranging Eq. (8.19) above reveals that the buckling must be equal to

$$B_{\text{m}} = \frac{n\pi}{L+2d} = \frac{n\pi}{L^*} \stackrel{?}{=} B_{gn} \qquad (8.20)$$

where $L^* = L + 2d$.

In Eq. (8.20) a change in the designation for the buckling has been made where B_{gn} is referred to as the *geometric buckling*. It is clear from this equation that one is faced with a dilemma. The dilemma arises from the fact that the problem is overspecified. There is no reason why the materials buckling (which depends solely upon characteristics of the materials which make up the reactor) should necessarily be equal to the geometric buckling (which depends solely upon the geometric configuration of the reactor). This dilemma is addressed by introducing a new variable λ_n into the equation for the materials buckling where "λ_n" are the eigenvalues for the problem. The question now arises as to what restrictions (if any) might need to be placed on values of "n." It turns out that indeed restrictions do have to be placed upon the acceptable values for "n." These restrictions are necessary to insure that the neutron flux not only matches the reactor boundary conditions, but also never takes on negative values within the reactor itself. As a consequence, "n" must be restricted to $n = 1$ which turns out also to yield the largest possible eigenvalue. *This largest possible eigenvalue (eg, "λ_1") is also called the k-effective or k_{eff} for the reactor*:

$$B_{gn}^2 = B_{\text{m}}^2 = \frac{\nu\Sigma_{\text{f}}/\lambda_n - \Sigma_{\text{a}}}{D} \Rightarrow \lambda_n = \frac{\nu\Sigma_{\text{f}}}{DB_{gn}^2 + \Sigma_{\text{a}}} \Rightarrow \lambda_1 = k_{\text{eff}} = \frac{\nu\Sigma_{\text{f}}}{DB_{\text{g}}^2 + \Sigma_{\text{a}}} = \frac{\text{Production Rate}}{\text{Loss Rate}} \qquad (8.21)$$

where $B_{\text{g}} = B_{g1}$.

An equation such as Eq. (8.21) which relates k_{eff} to the reactor geometry and material cross sections is called a *criticality equation*. It can be seen from this equation that *when $k_{\text{eff}} = 1$, the materials* buckling is equal to the geometric buckling and the neutron production rate equals the neutron loss rate. In this situation, the reactor is said to be operating in a critical steady-state condition. When k_{eff} is greater than 1, the reactor is said to be *supercritical* with the neutron production rate exceeding the neutron loss rate. A supercritical condition implies that the reactor power level is no longer in a steady-state condition, but is in fact increasing with time. Conversely, *when k_{eff} is less than 1, the reactor is said to be subcritical* with the neutron loss rate exceeding the neutron production rate. A subcritical condition also implies that the reactor power level is no longer in a steady-state condition, but in this case it is decreasing with time.

Note that k_{eff} is somewhat of an artificial quantity since if it takes on a value other than 1, the reactor is not in a steady-state condition. This condition is inconsistent with the neutron diffusion equation as expressed by Eq. (8.6) since that equation assumes that the reactor is in a steady-state condition. Nevertheless, k_{eff} is a useful quantity in that it does approximately represent the increase the neutron population from one generation to the next. If the reactor is assumed to be infinitely large (eg, $L^* = \infty \Rightarrow B_g = 0$) then one obtains an expression for k_{eff} which is independent of the geometry. This quantity is designated as the k-infinity or k_∞ for the reactor:

$$k_{eff}|_{L \to \infty} = k_\infty = \frac{\nu \Sigma_f}{\Sigma_a} = \frac{\text{Production Rate}}{\text{Absorption Rate}} \qquad (8.22)$$

Using the relationship for the geometric buckling as given by Eq. (8.20) in the expression for the neutron flux as given by Eq. (8.18), one finally obtains the following equation which has the form of a "chopped sine" distribution:

$$\phi(z) = C \sin[B_g(z+d)] = C \sin\left[\frac{\pi}{L^*}(z+d)\right] \qquad (8.23)$$

The question now remains how does one determine a value for the arbitrary constant? It turns out that "C" is actually a scaling factor which depends upon the power level of the reactor. The reactor power level is determined by integrating the neutron flux times fission cross section over the volume of the reactor; therefore, for a constant fission cross section, the reactor power level is

$$P = CA\Sigma_f \int_0^L \sin\left[\frac{\pi}{L^*}(z+d)\right] dz = \frac{CA\Sigma_f L^*}{\pi}\left\{\cos\left(\frac{\pi}{L^*}d\right) - \cos\left[\frac{\pi}{L^*}(d+L)\right]\right\} \qquad (8.24)$$

where A = reactor cross-sectional area and P = reactor power level.

Rearranging Eq. (8.24) to solve for "C" then yields:

$$C = \frac{\pi P}{A\Sigma_f L^*\left\{\cos\left(\frac{\pi}{L^*}d\right) - \cos\left[\frac{\pi}{L^*}(d+L)\right]\right\}} \qquad (8.25)$$

Also note from Eq. (8.21) that *if an appropriate expression for the geometric buckling is known, it is possible to calculate k_{eff} as an algebraic equation without solving the neutron diffusion differential equation* as represented by Eq. (8.6). Several expressions for the one region, one group geometric buckling for other reactor geometries are presented in Table 8.2. Be aware that the expressions presented in Table 8.2 are valid only for one region configurations where the cross sections are averaged over a single energy group. Configurations consisting of more than one geometric region or which use several neutron energy groups will have different expressions for the buckling.

2. ONE GROUP, ONE REGION NEUTRON DIFFUSION EQUATION

Table 8.2 One Region, One Group Geometric Buckling

Geometry	Buckling, B_g^2
Infinite slab	$\left(\frac{\pi}{L^*}\right)^2$
Rectangular box	$\left(\frac{\pi}{L_x^*}\right)^2 + \left(\frac{\pi}{L_y^*}\right)^2 + \left(\frac{\pi}{L_z^*}\right)^2$
Sphere	$\left(\frac{\pi}{R^*}\right)^2$
Cylinder	$\left(\frac{\pi}{H^*}\right)^2 + \left(\frac{2.405}{R^*}\right)^2$

As an example on the use of the geometric buckling, we now employ the quantity to derive an expression for the radius at which a bare sphere of fissionable material is exactly critical. Recall that when the reactor is exactly critical, $k_{\text{eff}} = 1$ and that $B_g = B_m$, therefore from Eq. (8.7):

$$0 = \frac{d^2\phi}{dz^2} + B_m^2\phi = -B_g^2\phi + B_m^2\phi \tag{8.26}$$

From Eq. (8.26) and using the definition for the geometric and materials buckling from Eq. (8.21):

$$\left(\frac{\pi}{R_c^*}\right)^2 = \frac{\nu\Sigma_f - \Sigma_a}{D} \Rightarrow R_c^* = R_c + 2D = \pi\sqrt{\frac{D}{\nu\Sigma_f - \Sigma_a}} \Rightarrow R_c = \pi\sqrt{\frac{D}{\nu\Sigma_f - \Sigma_a}} - 2D \tag{8.27}$$

With Eq. (8.27) giving us an expression for the critical radius "R_c," it is now possible to calculate a quantity which is often used in casual conversation, but generally little understood ... *critical mass*. By knowing the critical radius of a sphere, its volume may easily be calculated and from the density of the fissionable material the corresponding mass of the sphere (eg, the critical mass) may be calculated, thus:

$$M_c = \rho_f V_c = \rho_f \frac{4}{3}\pi R_c^3 = \rho_f \frac{4\pi}{3}\left(\pi\sqrt{\frac{D}{\nu\Sigma_f - \Sigma_a}} - 2D\right)^3 \tag{8.28}$$

where M_c = critical mass; V_c = critical volume; and ρ_f = density of the fissionable material.

Another example on the use of the geometric buckling illustrates how the buckling term can be used to account for neutron leakage in multiple dimensions. In this particular case, the critical height of a cylinder is calculated under the assumption that the critical radius is known. Again, recalling that when the reactor is exactly critical, $k_{\text{eff}} = 1$ and that $B_g = B_m$ and also using the definition for the geometric buckling for a cylinder from Table 8.2 yields:

$$\left(\frac{\pi}{H_c^*}\right)^2 + \left(\frac{2.405}{R_c^*}\right)^2 = \frac{\nu\Sigma_f - \Sigma_a}{D} \Rightarrow H_c^* = H_c + 2D = \frac{\pi}{\sqrt{\frac{\nu\Sigma_f - \Sigma_a}{D} - \left(\frac{2.405}{R_c^*}\right)^2}}$$

$$\Rightarrow H_c = \frac{\pi}{\sqrt{\frac{\nu\Sigma_f - \Sigma_a}{D} - \left(\frac{2.405}{R_c^*}\right)^2}} - 2D \tag{8.29}$$

3. ONE GROUP, TWO REGION NEUTRON DIFFUSION EQUATION

One of the problems with the bare core configuration analyzed in the preceding section is that neutrons which escape the core due to leakage are permanently lost to the reactor. To reduce the neutron leakage, it has been found that if a material having a high scattering cross section and low absorption cross section is placed around the outside of the core, it is possible to reflect many of the neutrons escaping the reactor core due to leakage back into the core thus raising the reactor k_{eff}. By reducing the neutron leakage, it is therefore possible to reduce the size of the core and thereby reduce the amount of fissionable material required to make the reactor critical.

The configuration that is analyzed to illustrate the effect of a neutron reflector is given in Fig. 8.4. In the analysis which follows, advantage is taken of the problem's geometric symmetry to restrict the study to two rather than three regions.

3.1 CORE

For the core region, it is found that for a three-dimensional box geometry in one energy group:

$$0 = D\nabla^2\phi + \frac{\nu\Sigma_f}{k_{eff}}\phi - \Sigma_a\phi = D\left(\frac{d^2\phi}{dx^2} + \frac{d^2\phi}{dy^2} + \frac{d^2\phi}{dz^2}\right) + \frac{\nu\Sigma_f}{k_{eff}}\phi - \Sigma_a\phi = D\left(\frac{d^2\phi}{dx^2} - B_{gy}^2\phi - B_{gz}^2\phi\right)$$

$$+ \frac{\nu\Sigma_f}{k_{eff}}\phi - \Sigma_a = \frac{d^2\phi}{dx^2} + \left(-B_{gy}^2 - B_{gz}^2 + \frac{\nu\Sigma_f/k_{eff} - \Sigma_a}{D}\right)\phi = \frac{d^2\phi}{dx^2} + \alpha^2\phi$$

(8.30)

where $\alpha^2 = -B_{gy}^2 - B_{gz}^2 + \frac{\nu\Sigma_f/k_{eff} - \Sigma_a}{D}$ = buckling in the core.

Solving Eq. (8.30) then yields:

$$\phi(z) = A\cos(\alpha x) + C\sin(\alpha x)$$

(8.31)

Noting that due to the symmetry of the problem, a boundary condition at the reactor centerline may be taken to be

$$\left.\frac{d\phi}{dx}\right|_{x=0} = 0$$

(8.32)

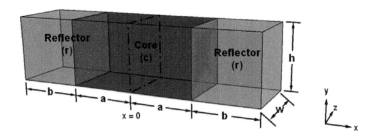

FIGURE 8.4

Two region, one-dimensional reactor.

3. ONE GROUP, TWO REGION NEUTRON DIFFUSION EQUATION

Upon applying boundary condition of Eq. (8.32) to the derivative of Eq. (8.31) then yields:

$$\frac{d\phi}{dx} = -A\sin(\alpha x) + C\cos(\alpha x) \Rightarrow 0 = -A\sin(0) + C\cos(0) \Rightarrow C = 0 \tag{8.33}$$

As a consequence, the neutron flux in the core from Eq. (8.33) becomes

$$\phi_c(x) = A\cos(\alpha x) \tag{8.34}$$

3.2 REFLECTOR

For the reflector region of the reactor, there are no fissionable materials, therefore:

$$0 = D\nabla^2\phi - \Sigma_a\phi = D\left(\frac{d^2\phi}{dx^2} + \frac{d^2\phi}{dy^2} + \frac{d^2\phi}{dz^2}\right) - \Sigma_a\phi = D\left(\frac{d^2\phi}{dx^2} - B_{gy}^2\phi - B_{gz}^2\phi\right) - \Sigma_a\phi$$

$$= \frac{d^2\phi}{dx^2} - \left(B_{gy}^2 + B_{gz}^2 + \frac{\Sigma_a}{D}\right)\phi = \frac{d^2\phi}{dx^2} - \beta^2\phi \tag{8.35}$$

where $\beta^2 = B_{gy}^2 + B_{gz}^2 + \frac{\Sigma_a}{D}$ = materials buckling in the reflector.

Solving the neutron diffusion differential equation then yields:

$$\phi(x) = E\cosh[\beta(a + b^* - x)] + F\sinh[\beta(a + b^* - x)] \tag{8.36}$$

where $b^* = b + 2D_r$ = extrapolated reflector thickness.

To solve for the arbitrary constants "E" and "F," it will again be assumed that the neutron flux goes to zero at the extrapolation distance just beyond the outer edge of the reflector (eg, $\phi = 0$ at $x = a + b^*$). Therefore, applying the zero flux boundary condition at the reflector extrapolation distance it is found from Eq. (8.36) that:

$$\phi(a + b^*) = 0 = E\cosh(0) + F\sinh(0) \Rightarrow E = 0 \tag{8.37}$$

Eq. (8.36), which represents the neutron flux in the reflector, now becomes:

$$\phi_r(x) = F\sinh[\beta(a + b^* - x)] \tag{8.38}$$

3.3 CORE + REFLECTOR

At the interface between the core and reflector the physics of the situation requires that the neutron flux be continuous, therefore using Eqs. (8.34) and (8.38) it is found that:

$$\phi_c(a) = \phi_r(a) \Rightarrow A\cos(\alpha a) = F\sinh(\beta b^*) \tag{8.39}$$

Also at the interface between the core and reflector, the physics of the situation requires that the neutron current too be continuous, therefore again using Eqs. (8.34) and (8.38) it is found that:

$$J_c(a) = J_r(a) \Rightarrow D_c\frac{d\phi_c}{dx}\bigg|_{x=a} = D_r\frac{d\phi_r}{dx}\bigg|_{x=a} \Rightarrow -AD_c\alpha\sin(\alpha x)|_{x=a}$$

$$= -FD_r\beta\cosh[\beta(a + b^* - x)]|_{x=a} \Rightarrow AD_c\alpha\sin(\alpha a) = FD_r\beta\cosh(\beta b^*) \tag{8.40}$$

If Eq. (8.39) is now divided into Eq. (8.40), it is possible to eliminate the arbitrary constants "A" and "F" to yield:

$$D_c \alpha \tan(\alpha a) = \frac{D_r \beta}{\tanh(\beta b^*)} \qquad (8.41)$$

Eq. (8.41) relates k_{eff} to all the other reactor parameters and so is the criticality equation for a one energy group, two region reactor configuration having the given boundary conditions. In Fig. 8.5, Eq. (8.41) is solved for the reactor dimensions "a" and "b*" using some typical material parameters (eg, cross sections) and a k_{eff} equal to one. Note that through the use of a reflector, the core size can be significantly reduced. The amount by which core thickness is reduced through the use of a reflector is termed the *reflector savings* for the reactor. Initially, even small increases in te thickness of the reflector can have a rather large effect on reducing the size of the core; however, as the thickness of the reflector continues to increase, its effect on the size of the core becomes progressively less significant. Eventually, it is found that even large increases in the thickness of the reflector have negligible effect on the size of the core and as a consequence negligible effect on k_{eff}. At this point, the reflector is said to be effectively infinite. In mathematical terms, the reason that additional increases in the size of the reflector have little effect on the size of the core—can be seen from an examination of Eq. (8.41). Note that even for fairly modest values of the argument, βb^* the function $\tanh(\beta b^*)$ closely approaches its asymptotic value of one. For example, note that $\tanh(4) =$

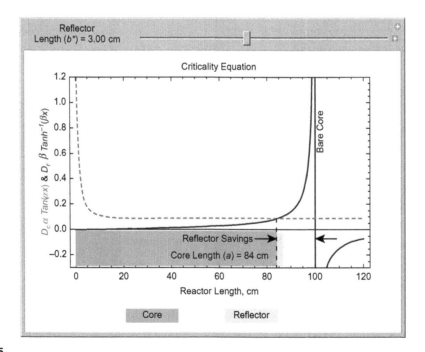

FIGURE 8.5

Reactor critical length as a function of reflector length.

0.99933 ≈ 1. Somewhat arbitrarily, therefore, *it will henceforth be assumed that reflectors having a thickness of:*

$$b^* \gtrsim \frac{4}{\beta} \tag{8.42}$$

will be treated as being effectively infinite.

Once the criticality relationship of Eq. (8.41) has been solved, it is possible to derive expression for the neutron flux in the core and reflector regions of the reactor. From Eq. (8.39) it is found that:

$$F = A \frac{\cos(\alpha a)}{\sinh(\beta b^*)} \tag{8.43}$$

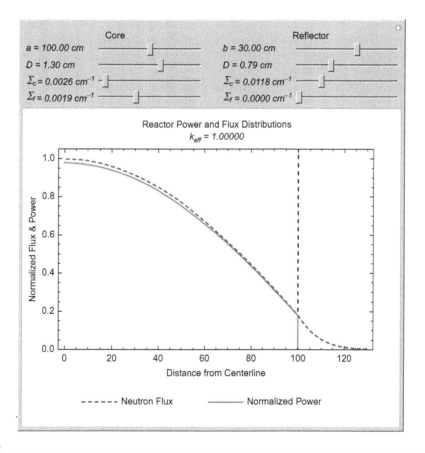

FIGURE 8.6

Two region, one group reactor power and flux distributions.

Substituting the results from Eq. (8.43) into Eq. (8.38) and recalling the results from Eq. (8.34) it is found that the neutron flux across the reactor is given by

$$\phi(x) = \begin{cases} A\cos(\alpha x): & 0 \le x \le a \\ A\dfrac{\cos(\alpha a)}{\sinh(\beta b^*)}\sinh[\beta(a+b^*-x)]: & a \le x \le a+b \end{cases} \quad (8.44)$$

In Fig. 8.6, the normalized pointwise reactor power density and neutron flux (eg, $A=1$) as determined from Eq. (8.44) are plotted for different geometric configurations and for various neutron cross-sectional values assuming $\nu = 2.5$.

4. TWO GROUP, TWO REGION NEUTRON DIFFUSION EQUATION

The one neutron energy group, two region reactor configuration described in Section 3 is very useful in giving a gross description of the spatial neutron flux distribution (and hence power distribution); however, by using additional neutron energy groups new spatial effects manifest themselves. It turns out that these spatial effects can have important thermal–hydraulic consequences as subsequently is seen. The configuration that is analyzed is identical to the configuration used in the one neutron energy group analysis shown in Fig. 8.4.

Using Eq. (8.4), with "G" set to 2 and assuming that the cross sections are constant within each region, one obtains the following two coupled differential equations:

$$\left(D_j^1 \nabla^2 + \frac{1}{\lambda}\chi^1 \nu \Sigma_{fj}^1 - \Sigma_{aj}^1 - \Sigma_{sj}^{1\to 2}\right)\phi^1(x) + \left(\frac{1}{\lambda}\chi^1 \nu \Sigma_{fj}^2 + \Sigma_{sj}^{2\to 1}\right)\phi^2(x) = 0$$

$$\left(\frac{1}{\lambda}\chi^2 \nu \Sigma_{fj}^1 + \Sigma_{sj}^{1\to 2}\right)\phi^1(x) + \left(D_j^2 \nabla^2 + \frac{1}{\lambda}\chi^2 \nu \Sigma_{fj}^2 - \Sigma_{aj}^2 - \Sigma_{sj}^{2\to 1}\right)\phi^2(x) = 0$$

(8.45)

where $j =$ "c" for the core region and $j =$ "r" for the reflector region.

If it is again assumed that the grayed out terms are zero and that bucklings are used to replace the second derivative terms in Eq. (8.45), and that $\chi^1 = 1$, the following expressions result:

$$\left\{-D_j^1\left[(B_j)^2 + (B_{gyz})^2\right] + \frac{1}{\lambda}\nu\Sigma_{fj}^1 - \Sigma_{aj}^1 - \Sigma_{sj}^{1\to 2}\right\}\phi^1(x) + \left(\frac{1}{\lambda}\nu\Sigma_{fj}^2\right)\phi^2(x) = 0$$

$$\Sigma_{sj}^{1\to 2}\phi^1(x) - \left\{D_j^2\left[(B_j)^2 + (B_{gyz})^2\right] + \Sigma_{aj}^2\right\}\phi^2(x) = 0$$

(8.46)

where $B_{gyz}^2 = B_{gy}^2 + B_{gz}^2$.

If Eq. (8.46) is now put in matrix form, it is found that:

$$\begin{pmatrix} -D_j^1\left[(B_j)^2 + (B_{gyz})^2\right] + \nu\Sigma_{fj}^1/\lambda - \Sigma_{aj}^1 - \Sigma_{sj}^{1\to 2} & \nu\Sigma_{fj}^2/\lambda \\ \Sigma_{sj}^{1\to 2} & -D_j^2\left[(B_j)^2 + (B_{gyz})^2\right] - \Sigma_{aj}^2 \end{pmatrix} \begin{pmatrix} \phi^1(x) \\ \phi^2(x) \end{pmatrix} = 0$$

(8.47)

4. TWO GROUP, TWO REGION NEUTRON DIFFUSION EQUATION

In order for the above matrix relationship to be true while at the same time requiring nonzero values for the neutron flux, it is required that the determinant of the matrix in Eq. (8.47) be equal to zero. The relationship for the determinant of the matrix given by the following equation where it is cast in terms of powers of the buckling "B":

$$0 = (B_j)^4 + (B_j)^2 \underbrace{\left[2(B_{gyz})^2 - \left(\frac{\nu \Sigma^1_{fj}/\lambda - \Sigma^1_{aj} - \Sigma^{1 \to 2}_{sj}}{D^1_j} \right) + \frac{\Sigma^2_{aj}}{D^2_j} \right]}_{P}$$

$$+ \underbrace{\left[(B_{gyz})^4 - (B_{gyz})^2 \left(\frac{\nu \Sigma^1_{fj}/\lambda - \Sigma^1_{aj} - \Sigma^{1 \to 2}_{sj}}{D^1_j} - \frac{\Sigma^2_{aj}}{D^2_j} \right) - \left(\frac{\Sigma^2_{aj} \left(\nu \Sigma^1_{fj}/\lambda - \Sigma^1_{aj} \right)}{D^1_j D^2_j} \right) + \frac{\Sigma^{1 \to 2}_{sj} \left(\nu \Sigma^2_{fj}/\lambda - \Sigma^2_{aj} \right)}{D^1_j D^2_j} \right]}_{Q}$$

$$= (B_j)^4 + P(B_j)^2 + Q \tag{8.48}$$

Solving Eq. (8.48) in terms of the buckling then yields:

$$(B_j)^4 + P(B_j)^2 + Q = 0 \Rightarrow (B_j)^2 = \frac{1}{2} \left(-P \pm \sqrt{P^2 - 4PQ} \right) \tag{8.49}$$

The relationship in Eq. (8.49) yields two buckling expressions which are represented as

$$\mu_j^2 = \frac{1}{2} \left(-P + \sqrt{P^2 - 4PQ} \right) \quad \cdots \quad \text{Can be positive or negative}$$

$$\rho_j^2 = \frac{1}{2} \left(-P - \sqrt{P^2 - 4PQ} \right) \quad \cdots \quad \text{Always negative (if no fissionable nuclides are present)}$$

$$\tag{8.50}$$

The two expressions for the buckling presented in Eqs. (8.50) lead to two differential equations describing the neutron flux in each region of the reactor:

$$\frac{d^2 [\phi^2(x)]}{dx^2} \pm \mu_j^2 \phi^2(x) = 0 \tag{8.51}$$

and

$$\frac{d^2 [\phi^2(x)]}{dx^2} - \rho_j^2 \phi^2(x) = 0 \tag{8.52}$$

As long as there is fissionable material present in the core region, values for λ (or more precisely k_{eff}) may be found which will cause the sign $\mu_j^2 = \mu_c^2$ to be positive in Eq. (8.51). In the reflector region, if it is assumed that there is no fissionable material present, the sign of $\mu_j^2 = \mu_r^2$ will always be negative in Eq. (8.51). To determine the ratio of the group 1 to the group 2 flux the second relationship in Eq. (8.46) is rearranged, such that:

$$\frac{\phi^1(x)}{\phi^2(x)} = \frac{D_j^2 \left[(B_j)^2 + (B_{gyz})^2 \right] + \Sigma^2_{aj}}{\Sigma^{1 \to 2}_{sj}} \tag{8.53}$$

Substituting the buckling parameters defined by Eq. (8.50) into Eq. (8.53) then yields for the flux ratios:

$$\alpha_j = \frac{D_j^2\left[(\mu_j)^2 + (B_{gyz})^2\right] + \Sigma_{aj}^2}{\Sigma_{sj}^{1\to 2}} \tag{8.54}$$

and

$$\beta_j = \frac{D_j^2\left[(\rho_j)^2 + (B_{gyz})^2\right] + \Sigma_{aj}^2}{\Sigma_{sj}^{1\to 2}} \tag{8.55}$$

The general equations for the group neutron fluxes may now be determined by solving the differential equations set forth in Eqs. (8.51) and (8.52) and using the neutron flux ratio relationships described by Eqs. (8.54) and (8.55):

$$\phi^2(x) = C_{1j}\sin(\mu_j x) + C_{2j}\cos(\mu_j x) + C_{3j}\sinh(\rho_j x) + C_{4j}\cosh(\rho_j x) \tag{8.56}$$

and

$$\phi^1(x) = C_{1j}\alpha_j \sin(\mu_j x) + C_{2j}\alpha_j \cos(\mu_j x) + C_{3j}\beta_j \sinh(\rho_j x) + C_{4j}\beta_j \cosh(\rho_j x) \tag{8.57}$$

Since the problem being examined is symmetric about the centerline of the reactor core where $x = 0$, it is possible to specify as boundary conditions that the first derivatives of the neutron flux will be zero at this location. Therefore, taking the derivatives of Eqs. (8.56) and (8.57) yields:

$$\frac{d\phi^2(x)}{dx} = C_{1c}\mu_c \cos(\mu_c x) - C_{2c}\mu_c \sin(\mu_c x) + C_{3c}\rho_c \cosh(\rho_c x) + C_{4c}\rho_c \sinh(\rho_c x) \tag{8.58}$$

and

$$\frac{d\phi^1(x)}{dx} = C_{1c}\mu_c\alpha_c \cos(\mu_c x) - C_{2c}\mu_c\alpha_c \sin(\mu_c x) + C_{3c}\rho_c\beta_c \cosh(\rho_c x) + C_{4c}\rho_c\beta_c \sinh(\rho_c x) \tag{8.59}$$

If the derivatives of Eqs. (8.58) and (8.59) are now set equal to zero at $x = 0$, it is found that:

$$0 = C_{1c}\mu_c \cos(0) - C_{2c}\mu_c \sin(0) + C_{3c}\rho_c \cosh(0) + C_{4c}\rho_c \sinh(0) = C_{1c}\mu_c + C_{3c}\rho_c \tag{8.60}$$

and

$$0 = C_{1c}\mu_c\alpha_c \cos(0) - C_{2c}\mu_c\alpha_c \sin(0) + C_{3c}\rho_c\beta_c \cosh(0) + C_{4c}\rho_c\beta_c \sinh(0) = C_{1c}\mu_c\alpha_c + C_{3c}\rho_c \tag{8.61}$$

In order for Eqs. (8.60) and (8.61) to be true, it follows that $C_{1c} = C_{3c} = 0$, and as a result, the equations for the neutron flux in the core region of the reactor (eg, $-a \le x \le a$) become from Eqs. (8.56) and (8.57):

$$\phi_c^2(x) = C_{2c}\cos(\mu_c x) + C_{4c}\cosh(\rho_c x) \tag{8.62}$$

and
$$\phi_c^1(x) = C_{2c}\alpha_c \cos(\mu_c x) + C_{4c}\beta_c \cosh(\rho_c x) \tag{8.63}$$

In the reflector region, it is possible to specify another boundary condition for the problem. In this case, the boundary condition is one in which the neutron flux is specified to go to zero at the extrapolation distance just beyond the outer edge of the reactor reflector. That is, $\phi^1 = 0$ at $x = a + b^{1*}$, where $b^{1*} = b + 2D_r^1$ and $\phi^2 = 0$ at $x = a + b^{2*}$, where $b^{2*} = b + 2D_r^2$. Using general flux expressions presented in Eqs. (8.56) and (8.57) and applying them to the reflector then yields:

$$\phi^2(x) = C_{1r}\sin[\mu_r(a+b^{2*}-x)] + C_{2r}\cos[\mu_r(a+b^{2*}-x)] + C_{3r}\sinh[\rho_r(a+b^{2*}-x)]$$
$$+ C_{4r}\cosh[\rho_r(a+b^{2*}-x)] \tag{8.64}$$

and

$$\phi^1(x) = C_{1r}\alpha_r \sin[\mu_r(a+b^{1*}-x)] + C_{2r}\alpha_r \cos[\mu_r(a+b^{1*}-x)] + C_{3r}\beta_r \sinh[\rho_r(a+b^{1*}-x)]$$
$$+ C_{4r}\beta_r \cosh[\rho_r(a+b^{1*}-x)]$$

$$\tag{8.65}$$

Applying the zero flux boundary conditions described in Eqs. (8.56) and (8.57) it is now found that:

$$0 = C_{1r}\sin(0) + C_{2r}\cos(0) + C_{3r}\sinh(0) + C_{4r}\cosh(0) = C_{2r} + C_{4r} \tag{8.66}$$

and

$$0 = C_{1r}\alpha_r \sin(0) + C_{2r}\alpha_r \cos(0) + C_{3r}\beta_r \sinh(0) + C_{4r}\beta_r \cosh(0) = C_{2r}\alpha_r + C_{4r}\beta_r \tag{8.67}$$

In order for Eqs. (8.66) and (8.67) to be true, it follows that $C_{2r} = C_{4r} = 0$, and as a result, the equations for the neutron flux in the reflector region of the reactor (eg, $a \leq x \leq a + b^{g*}$) become from Eqs. (8.64) and (8.65):

$$\phi_r^2(x) = C_{1r}\sin[\mu_r(a+b^{2*}-x)] + C_{3r}\sinh[\rho_r(a+b^{2*}-x)] \tag{8.68}$$

and

$$\phi_r^1(x) = C_{1r}\alpha_r \sin[\mu_r(a+b^{1*}-x)] + C_{3r}\beta_r \sinh[\rho_r(a+b^{1*}-x)] \tag{8.69}$$

To solve for the remaining "C" coefficients, it is necessary to impose a continuity condition on the neutron flux and current at the reactor core/reflector interface. As a consequence, equating the group 2 neutron flux from Eqs. (8.62) and (8.68) yields:

$$\phi_c^2(a) = \phi_r^2(a) \Rightarrow C_{2c}\cos(\mu_c a) + C_{4c}\cosh(\rho_c a) = C_{1r}\sin(\mu_r b^{2*}) + C_{3r}\sinh(\rho_r b^{2*}) \tag{8.70}$$

Equating the group 1 neutron flux from Eqs. (8.63) and (8.69) yields:

$$\phi_c^1(a) = \phi_r^1(a) \Rightarrow C_{2c}\alpha_c \cos(\mu_c a) + C_{4c}\beta_c \cosh(\rho_c a) = C_{1r}\alpha_r \sin(\mu_r b^{1*}) + C_{3r}\beta_r \sinh(\rho_r b^{1*})$$
$$\tag{8.71}$$

Using Eq. (8.58) to determine the group 2 neutron current in the core yields:

$$J_c^2(x) = D_c^2 \frac{d\phi^2(x)}{dx} = -C_{2c} D_c^2 \mu_c \sin(\mu_c x) + C_{4c} D_c^2 \rho_c \sinh(\rho_c x) \tag{8.72}$$

Using Eq. (8.59) to determine the group 1 current in the core yields:

$$J_c^1(x) = D_c^1 \frac{d\phi^1(x)}{dx} = -C_{2c} D_c^1 \mu_c \alpha_c \sin(\mu_c x) + C_{4c} D_c^1 \rho_c \beta_c \sinh(\rho_c x) \tag{8.73}$$

Taking the first derivative of Eq. (8.68) to determine the group 2 current in the reflector yields:

$$J_r^2(x) = D_r^2 \frac{d\phi^2(x)}{dx} = -C_{1r} D_r^2 \mu_r \cos\left[\mu_r(a + b^{2*} - x)\right] - C_{3r} D_r^2 \rho_r \cosh\left[\rho_r(a + b^{2*} - x)\right] \tag{8.74}$$

Taking the first derivative of Eq. (8.69) to determine the group 1 current in the reflector yields:

$$J_r^1(x) = D_r^1 \frac{d\phi^1(x)}{dx} = -C_{1r} D_r^1 \mu_r \alpha_r \cos\left[\mu_r(a + b^{1*} - x)\right] - C_{3r} D_r^1 \rho_r \beta_r \cosh\left[\rho_r(a + b^{1*} - x)\right] \tag{8.75}$$

Equating the group 2 neutron currents at the core reflector interface from Eqs. (8.72) and (8.74) yields:

$$J_c^2(a) = J_r^2(a) \Rightarrow -C_{2c} D_c^2 \mu_c \sin(\mu_c a) + C_{4c} D_c^2 \rho_c \sinh(\rho_c a)$$
$$= -C_{1r} D_r^2 \mu_r \cos(\mu_r b^{2*}) - C_{3r} D_r^2 \rho_r \cosh(\rho_r b^{2*}) \tag{8.76}$$

Equating the group 1 neutron currents at the core reflector interface from Eqs. (8.73) and (8.75) yields:

$$J_c^1(a) = J_r^1(a) \Rightarrow -C_{2c} D_c^1 \mu_c \alpha_c \sin(\mu_c a) + C_{4c} D_c^1 \rho_c \beta_c \sinh(\rho_c a)$$
$$= -C_{1r} D_r^1 \mu_r \alpha_r \cos(\mu_r b^{1*}) - C_{3r} D_r^1 \rho_r \beta_r \cosh(\rho_r b^{1*}) \tag{8.77}$$

Eqs. (8.70), (8.71), (8.76), and (8.77) constitute a set of four equations in four unknowns. Putting these equations in matrix form to solve for the "C" parameters yields:

$$\begin{pmatrix} \cos(\mu_c a) & \cosh(\rho_c a) & -\sin(\mu_r b^{2*}) & -\sinh(\rho_r b^{2*}) \\ \alpha_c \cos(\mu_c a) & \beta_c \cosh(\rho_c a) & -\alpha_r \sin(\mu_r b^{1*}) & -\beta_r \sinh(\rho_r b^{1*}) \\ -D_c^2 \mu_c \sin(\mu_c a) & D_c^2 \rho_c \sinh(\rho_c a) & D_r^2 \mu_r \cos(\mu_r b^{2*}) & D_r^2 \rho_r \cosh(\rho_r b^{2*}) \\ -D_c^1 \mu_c \alpha_c \sin(\mu_c a) & D_c^1 \rho_c \beta_c \sinh(\rho_c a) & D_r^1 \mu_r \alpha_r \cos(\mu_r b^{1*}) & D_r^1 \rho_r \beta_r \cosh(\rho_r b^{1*}) \end{pmatrix} \begin{pmatrix} C_{2c}^1 \\ C_{4c}^1 \\ C_{1r}^1 \\ C_{3r}^1 \end{pmatrix} = 0$$

(8.78)

Since Eq. (8.78) relates "λ" to all of the other reactor parameters, it constitutes the criticality equation for the two region, two neutron energy group reactor configuration. In order to have a nontrivial solution (eg, all "C" coefficients equal to zero), "λ" or some other reactor parameter is adjusted until the determinant of the matrix equals zero. As was mentioned in previous sections, only

4. TWO GROUP, TWO REGION NEUTRON DIFFUSION EQUATION

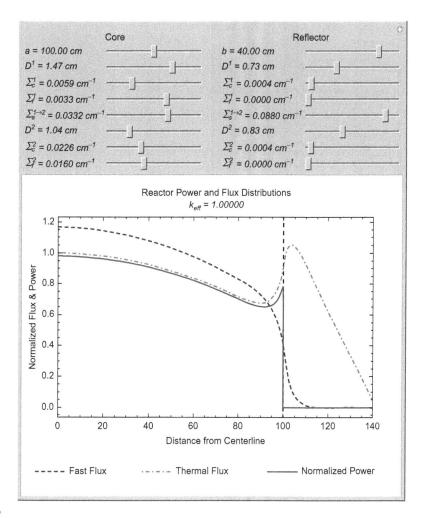

FIGURE 8.7

Two region, two group reactor power and flux distributions.

the largest value for "λ" results in physically meaningful (eg, no negative) values for the neutron flux distributions.

Once the criticality condition has been solved by determining a set of parameters which cause the determinant of Eq. (8.78) to be equal to zero, three of the equations comprising the matrix of Eq. (8.78) may be used to determine three of the "C" coefficients in terms of the fourth. The fact that only three equations are needed to determine the "C" coefficients is a direct result of the fact that the determinant of the matrix is equal to zero since the matrix equations are all no longer linearly independent.

Assuming that Eqs. (8.70), (8.71), and (8.76) are put in matrix form to solve for C_{4c}, C_{1r}, and C_{3r} in terms of C_{2c} then yields:

$$\begin{pmatrix} \cosh(\rho_c a) & -\sin(\mu_r b^{2*}) & -\sinh(\rho_r b^{2*}) \\ \beta_c \cosh(\rho_c a) & -\alpha_r \sin(\mu_r b^{1*}) & -\beta_r \sinh(\rho_r b^{1*}) \\ D_c^2 \rho_c \sinh(\rho_c a) & D_r^2 \mu_r \cos(\mu_r b^{2*}) & D_r^2 \rho_r \cosh(\rho_r b^{2*}) \end{pmatrix} \begin{pmatrix} C_{4c} \\ C_{1r}^1 \\ C_{3r}^1 \end{pmatrix} = C_{2c} \begin{pmatrix} -\cos(\mu_c^1 a) \\ -\alpha_c \cos(\mu_c^1 a) \\ D_c^2 \mu_c \sin(\mu_c^1 a) \end{pmatrix}$$

(8.79)

In Fig. 8.7, the normalized pointwise reactor power density and neutron flux (eg, $C_{2c}^1 = 1$) as determined from Eq. (8.78) using the "C" coefficients calculated from Eq. (8.79) are plotted for different geometric configurations and for various neutron cross-sectional values again assuming $\nu = 2.5$.

CHAPTER 9

THERMAL FLUID ASPECTS OF NUCLEAR ROCKETS

Because nuclear rockets generally operate near the thermal limits of their constituent materials, it is important that accurate evaluations be made of the temperature distributions within the engine's nuclear reactor. Since the heating rate due to fission at different locations in the reactor can vary significantly and because the propellant temperature increases tremendously as it travels through the reactor, the temperatures encountered by the fuel can easily vary over thousands of degrees between various portions of the reactor. In order to make accurate calculations of these temperature distributions, it is necessary, therefore, to not only have a good knowledge of the power distributions in the reactor, but it is also necessary to have a good knowledge of the propellant flow characteristics within the fuel elements and of the thermal characteristics of the fuel itself. For nuclear systems designed for power generation, where heat rejection is an issue, radiator designs which maximize the heat rejection for a given radiator mass are also important.

1. HEAT CONDUCTION IN NUCLEAR REACTOR FUEL ELEMENTS

It has been noted in earlier sections that the efficiency of most nuclear rocket engine systems is constrained by the fact that they are power limited rather than energy limited. Typically, the energy potentially available from the fissionable materials of the nuclear rocket engine reactor core is considerably in excess of that required to execute almost any conceivable interplanetary mission. The problem for nuclear rocket engine designers, therefore, becomes one of determining the best way to maximize the rate at which the available energy in the nuclear fuel can be extracted from the reactor core. From heat transfer considerations, the rate at which energy may be extracted from the core is limited primarily by the maximum fuel operating temperature and by the available surface area over which heat transfer can occur. The particle bed reactor (PBR) described earlier, for example, achieved high heat transfer rates through a unique fuel element design which used small fuel particles to greatly increase the surface area available for heat transfer. Nuclear engine for rocket vehicle application (NERVA) reactors achieved high heat transfer rates (though not as high as PBRs) through the use of a large number of small holes drilled axially through the prismatic fuel elements. New fuel compositions having higher melting temperatures can also be effective in increasing the engine efficiency, and research in this area is also ongoing. The topic of fuel materials and their properties are discussed at greater length in the chapter on nuclear materials.

Determining the fuel temperature distribution in a nuclear rocket core is usually quite complicated due to the fact that the power generated by the core is nonuniform, and the temperature of the coolant (which is usually the propellant) within the fuel elements varies considerably as it traverses through the reactor. In spite of these complications, however, it is possible to determine analytic expressions which, at least for simple geometries and constant material properties, can give a qualitative picture of the manner in which the fuel and propellant temperatures distributions vary throughout a nuclear rocket core.

The analysis to determine the core wide temperature distributions begins by writing a one-dimensional expression for the heat change within a differential volume "$\Delta V = \Delta x\, \Delta y\, \Delta z$" in the reactor over a short period of time "Δt" as illustrated in Fig. 9.1.

Performing the heat balance over the differential element illustrated in Fig. 9.1 then yields an equation of the form:

$$\underbrace{Q_t - Q_{\Delta t+t}}_{\text{change in heat content}} = \underbrace{q_{x+\Delta x}\Delta t \Delta y \Delta z}_{\text{heat out}} - \underbrace{q_x \Delta t \Delta y \Delta z}_{\text{heat in}} + P\Delta t \overbrace{\Delta x \Delta y \Delta z}^{\Delta V} \quad (9.1)$$

where Q = heat content; q = heat flux; and P = power density.

Rearranging Eq. (9.1) and taking the limit for small Δt then yields for the rate of heat transfer through the differential element a relationship of the form:

$$\lim_{\Delta t \to 0} \frac{Q_t - Q_{\Delta t+t}}{\Delta t} = \frac{dQ}{dt} = \frac{q_x - q_{x+\Delta x}}{\Delta x}\Delta V + P\Delta V \quad (9.2)$$

If the dimensions of the differential volume are sufficiently small, the heat flux out of the differential volume may be approximated by a linear function such that:

$$q_{x+\Delta x} = \frac{dq_x}{dx}\Delta x + q_x \quad (9.3)$$

If Eq. (9.3) is now incorporated into Eq. (9.2) it is found that:

$$\frac{dQ}{dt} = \frac{dq_x}{dx}\Delta V + P\Delta V \quad (9.4)$$

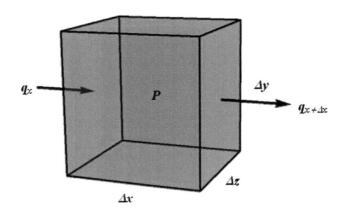

FIGURE 9.1

Heat flow through a differential element.

1. HEAT CONDUCTION IN NUCLEAR REACTOR FUEL ELEMENTS

From heat transfer theory, the *Fourier equation* is used to relate the heat transfer rate through a material to a temperature gradient. The one-dimensional form of this equation can be expressed as

$$q_x = k\frac{dT}{dx} \tag{9.5}$$

where k = thermal conductivity of the material through which the heat is being transferred and T = temperature.

Incorporating Eq. (9.5) into Eq. (9.4) then yields:

$$\frac{dQ}{dt} = \frac{d}{dx}\left(k\frac{dT}{dx}\right)\Delta V + P\Delta V = k\frac{d^2T}{dx^2}\Delta V + P\Delta V = \left(k\frac{d^2T}{dx^2} + P\right)\Delta V \tag{9.6}$$

From thermodynamics principles, the heat content of a quantity of material may be related to its temperature by an equation of the form:

$$Q = mTC_p \tag{9.7}$$

where m = mass of the material and C_p = specific heat of the material.

By substituting Eq. (9.7) into Eq. (9.6) and rearranging terms the one-dimensional form of the *general heat conduction equation* may be found such that:

$$\frac{dQ}{dt} = \frac{d}{dt}(mC_pT) = mC_p\frac{dT}{dt} = \left(k\frac{d^2T}{dx^2} + P\right)\Delta V \Rightarrow \frac{m}{\Delta V}\frac{C_p}{k}\frac{dT}{dt} = \frac{\rho C_p}{k}\frac{dT}{dt} = \frac{1}{\alpha}\frac{dT}{dt} = \frac{d^2T}{dx^2} + \frac{P}{k} \tag{9.8}$$

where ρ = density of the material and α = thermal diffusivity of the material.

Extending the general heat conduction equation (Eq. (9.8)) to three dimensions then yields:

$$\frac{1}{\alpha}\frac{dT}{dt} = \nabla^2 T + \frac{P}{k} \tag{9.9}$$

For steady-state cases, the time-dependent term in the general heat conduction equation described by Eq. (9.9) reduces to what is called *Poisson's equation*:

$$\nabla^2 T + \frac{P}{k} = 0 \tag{9.10}$$

Poisson's equation is now used in the following analysis to estimate the fuel temperature in a NERVA fuel element. The analysis consists of analyzing a single equivalent propellant flow channel within a fuel element as illustrated in Fig. 9.2. By approximating the hexagonal fuel portion of the flow

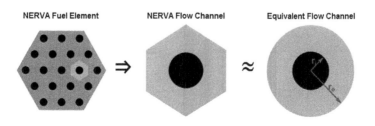

FIGURE 9.2

NERVA fuel element equivalent flow channel.

channel cell by a circular annulus region, all angular temperature dependencies are eliminated from the problem, thus allowing all terms containing an angular temperature component to be dropped. If it is also assumed that the length of the channel is long compared to the radial dimension (normally, a very good assumption), the axial temperature component of the problem can also be dropped. Thus by eliminating the angular and axial temperature variables, only the radial temperature component remains and the problem is simplified considerably. Under normal circumstances, only a modest loss of accuracy results from the application of these approximations.

The analysis begins by writing Eq. (9.10) in cylindrical coordinates and canceling out the angular and axial terms in the Laplacian operator such that:

$$\frac{d^2T}{dr^2} + \frac{1}{r}\frac{dT}{dr} + \underbrace{\frac{1}{r^2}\frac{dT}{d\theta}}_{0:\text{ No angular dependency}} + \underbrace{\frac{d^2T}{dz^2}}_{0:\text{ No axial dependency}} + \frac{P}{k} = 0 \Rightarrow \frac{1}{r}\frac{d}{dr}\left(r\frac{dT}{dr}\right) + \frac{P}{k} = 0 \quad (9.11)$$

Rearranging Eq. (9.11) and integrating then yields:

$$\frac{d}{dr}\left(r\frac{dT}{dr}\right) = -\frac{rP}{k} \Rightarrow r\frac{dT}{dr} = -\frac{P}{k}\int r\,dr = -\frac{r^2P}{2k} + C_1 \Rightarrow \frac{dT}{dr} = -\frac{rP}{2k} + \frac{C_1}{r} \quad (9.12)$$

Assuming that all the heat generated within a flow channel cell is removed by the propellant flow through the flow channel itself and that none of the heat is transferred across the cell boundary (eg, the flow cell boundary is adiabatic), then for the flow channel cell boundary condition using Fourier's equation, it is found that:

$$q_o = 0 = -kA\frac{dT}{dr}\bigg|_{r=r_o} \Rightarrow 0 = \frac{dT}{dr}\bigg|_{r=r_o} \quad (9.13)$$

Applying this zero temperature gradient boundary condition to Eq. (9.12) then allows the arbitrary constant "C_1" to be determined such that:

$$0 = -\frac{r_o P}{2k} + \frac{C_1}{r_o} \Rightarrow C_1 = \frac{r_o^2 P}{2k} \quad (9.14)$$

Incorporating the expression for the arbitrary constant from Eq. (9.14) into Eq. (9.12) and integrating a second time then yields:

$$\frac{dT}{dr} = -\frac{rP}{2k} + \frac{r_o^2 P}{2kr} \Rightarrow T - T_s = \frac{P}{2k}\int_{r_i}^{r}\left(-r' + \frac{r_o^2}{r'}\right)dr' = -\frac{P}{2k}\left[\frac{r^2 - r_i^2}{2} + r_o^2\text{Ln}\left(\frac{r_i}{r}\right)\right] \quad (9.15)$$

where T_s = fuel temperature at the surface of the propellant flow channel (eg, at $r = r_i$).

Since all the heat generated within the fuel portion of the flow channel cell must be transferred to the propellant flowing through the channel passageway, a simple heat balance leads to an expression of the form:

$$P\Delta V = P[\pi(r_o^2 - r_i^2)\Delta z] = h_c\Delta A(T_s - T_p) = h_c 2\pi r_i \Delta z(T_s - T_p) \quad (9.16)$$

where T_p = propellant temperature; h_c = heat transfer coefficient; ΔV = volume element; and ΔA = area element.

Rearranging Eq. (9.16) in terms of the fuel surface temperature then yields:

$$T_s = P\left[\frac{\pi(r_o^2 - r_i^2)\Delta z}{h_c 2\pi r_i \Delta z}\right] + T_p = \frac{P(r_o^2 - r_i^2)}{2h_c r_i} + T_p \qquad (9.17)$$

By substituting Eq. (9.17) into Eq. (9.15) and rearranging terms, an expression for temperature distribution within the fuel can be derived such that:

$$T = T_p + \frac{P(r_o^2 - r_i^2)}{2h_c r_i} - \frac{P}{2k}\left[\frac{r^2 - r_i^2}{2} + r_o^2 \text{Ln}\left(\frac{r_i}{r}\right)\right] \qquad (9.18)$$

The temperature distribution within the fuel due to changes in the various parameters, which comprise Eq. (9.18), can be observed in Fig. 9.3. Note especially how increasing the values for the heat transfer coefficient and the fuel thermal conductivity can serve to significantly reduce the peak fuel temperature assuming all the other parameters remain constant.

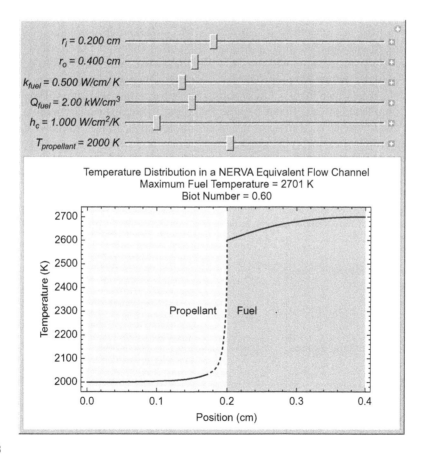

FIGURE 9.3

Temperature distribution in a NERVA equivalent flow channel.

The *Biot number* presented in Fig. 9.3 is a dimensionless group which gives an indication as to the flatness of the temperature distribution in the fuel and is often used in transient heat transfer calculations. The lower the Biot number, the flatter the temperature distribution. The Biot number is defined as

$$\text{Biot Number:} \quad Bi = \frac{h_c L}{k} = \frac{\text{external convective thermal resistance}}{\text{internal conductive thermal resistance}}$$

where L = characteristic length often set equal to $\frac{\text{fuel volume}}{\text{fuel surface area}}$.

2. CONVECTION PROCESSES IN NUCLEAR REACTOR FUEL ELEMENTS

With the exception of the heat transfer coefficient, all the quantities in the previous equations describing the temperature distribution within NERVA-type nuclear rocket fuel elements are easily measured or specified. Regarding the heat transfer coefficient, however, the situation is more complicated since the heat transfer coefficient itself is a function of many other variables. In addition, *the propellant flow in the reactor core is usually fairly turbulent* such that the propellant, while generally flowing in one direction from a macroscopic point of view, is nevertheless, quite random from a microscopic point of view with the flow stream having numerous small eddies and cross currents. Only rarely is the overall propellant flow laminar, wherein the fluid moves in well-defined layers or streamlines in which there is little intermixing between the fluid layers. From a heat transfer perspective, turbulent flow is desirable because the propellant mixing which occurs as a result of all the eddies and cross currents is an effective means of transferring heat from the flow channel walls to the bulk propellant.

Practically speaking, the random turbulent nature of the propellant flow generally precludes the use of analytically derived heat transfer coefficients. As a result, heat transfer coefficients along with the other parameters to which it is related are typically incorporated into several dimensionless groups and through experimentation, empirical relationships are derived which relate these dimensionless groups to one another. The dimensionless groups most often used to generate the convective heat transfer correlations include:

$$\text{Reynolds Number:} \quad Re = \frac{\rho V D}{\mu} = \frac{4\dot{m}}{\pi \mu D} = \frac{\text{inertial forces}}{\text{viscous forces}}$$

$$\text{Prandlt Number:} \quad Pr = \frac{c_p \mu}{k} = \frac{\text{viscous diffusion rate}}{\text{thermal diffusion rate}}$$

$$\text{Nusselt Number:} \quad Nu = \frac{h_c D}{k} = \frac{\text{convective heat transfer rate}}{\text{conductive heat transfer rate}}$$

$$\text{Stanton Number:} \quad St = \frac{h_c}{\rho V c_p} = \frac{\text{convective heat transfer rate into a fluid}}{\text{thermal capacity of the fluid}}$$

where μ = fluid viscosity; V = fluid velocity; and D = hydraulic diameter of the flow channel = $4 \frac{\text{Flow Channel Cross-Sectional Area}}{\text{Flow Channel Wetted Perimeter}}$.

The transition from laminar flow to turbulent flow takes place when the local flow velocity is sufficiently high so as to result in naturally occurring flow perturbations within the propellant stream

2. CONVECTION PROCESSES IN NUCLEAR REACTOR FUEL ELEMENTS

large enough to overcome the viscous damping forces that act to reduce these perturbations. The point at which this laminar to turbulent flow transition occurs can be approximated through the use of the Reynolds number which measures the ratio between the kinetic or inertial forces within the flow stream to the viscous forces. Experimentally, it has been shown that this laminar to turbulent flow transition occurs for a Reynolds number range roughly between 2300 and 10,000. If the propellant flow is such that there is laminar flow, the Nusselt number has been shown to be approximately constant and heat transfer to the propellant will be by conduction alone. In this case, it is found from theory and confirmed through experimentation that for laminar flow:

$$Nu = 3.66 \,(\text{constant wall temperature}) \quad \text{and} \quad Nu = 4.36 \,(\text{constant heat flux}) \quad (9.19)$$

Boundary conditions suggest that at the channel wall, the propellant flow velocity should be equal to zero since the channel wall is stationary. As one moves away from the wall, the flow velocity gradually increases until it finally reaches the free stream velocity. Consequently, there is a small region near the flow channel wall where the flow velocities are low and the local Reynolds number is sufficiently small so as to be below the threshold required for the initiation of turbulent flow resulting in a small region where the propellant flow is laminar. This small layer of laminar flow is called the *boundary layer* and its presence has a significant negative effect on the rate at which heat can be transferred to the bulk propellant. What is occurring is that as a result of the lack of turbulent fluid mixing between the flow streamlines in the boundary layer and the bulk fluid outside the boundary layer, fluid particles picking up heat near the channel wall have difficulty penetrating the boundary layer where they can deposit their energy to the bulk fluid. Fig. 9.4 illustrates the propellant velocity profile in and near the boundary layer at the channel wall. For a more detailed analysis of boundary layer behavior, the reader is referred to the classic work by Schlichting [1].

Noting that the Prandtl number relates the rate at which fluid particles penetrate the boundary layer through viscous processes to the rate at which heat is transported through the boundary layer through thermal processes, it should be possible using the Prandtl number in conjunction with the Reynolds number to develop functional relationships describing the rate at which heat can be transferred from the flow channel walls to the bulk fluid in turbulent flow.

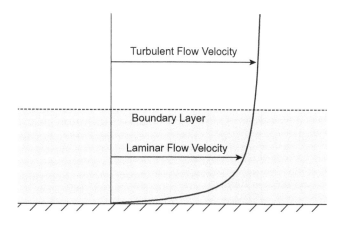

FIGURE 9.4

Velocity flow fields through a boundary layer.

Many such relationships have been developed over the years and only few are discussed here. The general form of these correlations were derived using boundary layer theory with some of the coefficients adjusted using curve fits to experimental data. These correlations have generally been quite successful in providing reasonable estimates of the heat transfer coefficient over a fairly wide range of conditions. The entire subject of convective heat transfer is quite large and has been widely described in numerous textbooks and articles. No attempt is made to justify the heat transfer equations presented. The correlations are simply described along with their ranges of applicability. If the reader desires more information, they are referred to any of the numerous textbooks on the subject such as those written by El Wakil [2] and Kreith [3].

Probably the most well known of these turbulent heat transfer correlations, which has been specifically designed for the heating of fluids flowing in smooth tubes, is the *Dittus–Boelter correlation* [4]. This correlation has been found to be good for conditions where the temperature difference between the channel wall and the bulk fluid is not very large and the Reynolds number is greater than about 10,000. Prandtl numbers for this correlation should be between about 0.7 and 120, and the physical properties should be evaluated at the bulk fluid temperature. The Dittus–Boelter correlation itself is given by

$$Nu = 0.023 Re^{0.8} Pr^{0.4} \Rightarrow h_c = 0.023 \frac{k}{D} Re^{0.8} Pr^{0.4} \quad (9.20)$$

For conditions in which there is a large temperature difference between the channel wall and the bulk fluid, the Dittus–Boelter correlation is modified to account for the viscosity differences between the fluid in contact with the channel wall and the bulk fluid which at times can be quite large. The resulting equation originally suggested by *Sieder and Tate* [5] can be represented by

$$Nu = 0.023 Re^{0.8} Pr^{0.3} \left(\frac{\mu_b}{\mu_w}\right)^{0.14} \Rightarrow h_c = 0.023 \frac{k}{D} Re^{0.8} Pr^{0.3} \left(\frac{\mu_b}{\mu_w}\right)^{0.14} \quad (9.21)$$

where μ_b = viscosity of the bulk propellant and μ_w = viscosity of the propellant at the channel wall.

This correlation has the same limits of applicability as the Dittus–Boelter correlation; however, it is somewhat more accurate for large temperature differences between the channel wall and the bulk fluid, although it generally requires an iterative solution since the propellant wall and bulk temperatures are not usually directly available beforehand.

Since the boundary layer is somewhat of a barrier to heat transfer, the thinner the boundary layer can be made, the easier it will be for heat to transfer into the bulk propellant flow. Often the walls of the propellant flow passages are intentionally roughened to enhance turbulence near the channel walls to decrease the boundary layer thickness, thus increasing the rate at which heat can be transferred into the bulk fluid. A later heat transfer correlation which accounted for channel roughening was developed by *Gnielinski* [6]. The Gnielinski correlation which is presented in the following is generally valid for Reynolds numbers greater than 3000 and less than 5×10^6 and Prandlt numbers between 0.5 and 2000:

$$Nu = \frac{\frac{f}{8} Pr(Re - 1000)}{1 + 12.7 \left(\frac{f}{8}\right)^{1/2} (Pr^{2/3} - 1)} \Rightarrow h_c = \frac{k}{D} \frac{\frac{f}{8} Pr(Re - 1000)}{1 + 12.7 \left(\frac{f}{8}\right)^{1/2} (Pr^{2/3} - 1)} \quad (9.22)$$

In the Gnielinski heat transfer correlation, an additional parameter called the *Darcy–Weisbach friction factor* "f" is used in its formulation. The Darcy–Weisbach friction factor is a dimensionless factor which is proportional to the pressure drop in the flow channel and is a function of the Reynolds number and the channel's *relative roughness* which is defined as the ratio of the average roughness height "ε" to the hydraulic diameter of the channel "D." To determine a numerical value for the Darcy–Weisbach friction factor, it is usually necessary to present two formulations, one for the laminar flow regime and the other for the turbulent flow regime. The formulation for the laminar portion of the friction factor can be determined exactly from boundary layer theory; however, the turbulent portion of the friction factor formulation must be determined from an experimentally derived empirical relationship such as that postulated by *Colebrook* [7]. For the entire range of flow conditions (eg, both laminar and turbulent), the Darcy–Weisbach friction factor is often presented in a form similar to:

$$\text{Darcy–Weisbach friction factor}(f) = \begin{cases} \text{Laminar Flow } (Re < 2300): \quad f = \dfrac{64}{Re} \\ \text{Turbulent Flow } (Re > 10,000): \quad \dfrac{1}{\sqrt{f}} = -2\,\text{Log}\left(\dfrac{\varepsilon/D}{3.7} + \dfrac{2.51}{Re\sqrt{f}}\right) \end{cases} \quad (9.23)$$

There actually have been a number of correlations generated over the years for the turbulent friction factor. One particular friction factor correlation generated by Wood [8] which expresses the friction factor as a closed form expression is useful in later analyses. Wood's correlation may be expressed by an equation of the form:

$$f = 0.094\left(\frac{\varepsilon}{D}\right)^{0.225} + 0.53\frac{\varepsilon}{D} + 88\left(\frac{\varepsilon}{D}\right)^{0.44}\frac{1}{Re^{1.62\left(\frac{\varepsilon}{D}\right)^{0.134}}} \quad (9.24)$$

Fig. 9.5 illustrates a Moody chart [9] which is generally used to graphically present the Darcy–Weisbach friction factor as a function of the Reynolds number and the relative roughness of the flow channel.

Determining which heat transfer correlation to use in any given situation requires an awareness of the limitations of the various correlations and a good knowledge of the flow regime in which the heat transfer processes are taking place. It should also be kept in mind that even if an appropriate heat transfer correlation is chosen for a particular analysis, the accuracies with which the correlation can be expected to predict the correct value for the heat transfer coefficient are still generally no better than about 10%. As a consequence, heat transfer calculations using these coefficients should be treated carefully and given appropriate margins when interpreting the results. A comparison of the three heat transfer correlations previously described is presented in Fig. 9.6. Note the large variations in the value for the Nusselt number for certain variable combinations. There are many other correlations in the literature for different flow situations other than flow through tubes. These situations include flow through packed beds, flow over exterior surfaces, flow over tube banks, flows using liquid metal coolants, and many other configurations. These situations, while interesting, are discussed here.

The heat transfer coefficients just described assume that there is no fluid mass transfer across the fixed wall bounding the flow stream. If there is fluid transfer across the wall (eg, the wall is porous), it is possible to significantly change the rate at which heat is transferred across the wall/fluid interface. This procedure of allowing the fluid to cross the wall interface is called *transpiration* and is often used

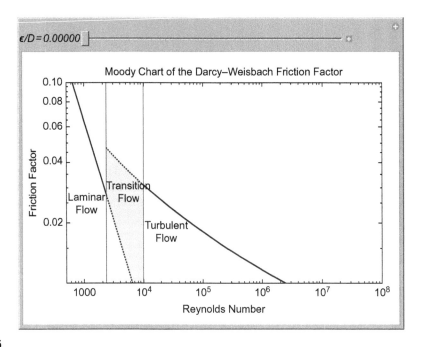

FIGURE 9.5

Moody chart.

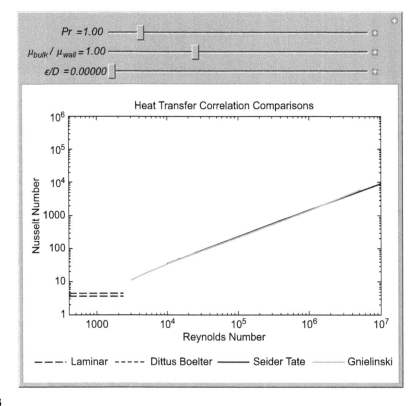

FIGURE 9.6

Heat transfer correlations.

2. CONVECTION PROCESSES IN NUCLEAR REACTOR FUEL ELEMENTS

in certain high heat flux locations such as the throat region of rocket nozzles to prevent overheating in that region. In essence, transpiration cooling works by changing the fluid boundary layer thickness in such a manner so as to promote or inhibit heat transfer at the wall. If fluid is expelled from the wall, the boundary layer thickens and heat transfer at the wall is diminished. If fluid is admitted into the walls, the boundary layer shrinks and heat transfer at the wall is enhanced. In the heat transfer correlation which follows [10], a ratio of the wall heat transfer coefficient with transpiration cooling to that without transpiration cooling is computed. This ratio is then multiplied by a heat transfer coefficient computed from one of the heat transfer correlations previously described to yield a heat transfer coefficient including transpiration cooling effects.

The transpiration correlation requires that a blowing parameter be defined such that:

$$B_h = \frac{\dot{m}''}{\rho V} \frac{1}{St} = \frac{\dot{m}''}{G_\infty} \frac{1}{St} \tag{9.25}$$

where B_h = heat transfer blowing parameter; G_∞ = propellant free stream mass flux; St = Stanton number associated with a transpiration cooled wall; and $\dot{m}''$ = transpiration propellant mass flux through wall.

The transpiration heat transfer correlation using the blowing parameter from Eq. (9.25) is then given by

$$\frac{St}{St_0} = \left[\frac{\text{Ln}(1+B_h)}{B_h}\right]^{\frac{5}{4}} (1+B_h)^{\frac{1}{4}} = \frac{St}{St_0} = \frac{h_c}{G_\infty c_p} \frac{G_\infty c_p}{h_{c0}} \Rightarrow h_c = h_{c0}\frac{St}{St_0} \tag{9.26}$$

where St_0 = Stanton number associated with a nontranspiration cooled wall and h_{c0} = heat transfer coefficient associated with a nontranspiration cooled wall.

Note that Eq. (9.26) is implicit in that the blowing parameter, B_h is not known a priori due to the fact that the correlation uses the Stanton number associated with the transpiration cooled wall. The correlation may easily be solved iteratively, however, as illustrated in the following example.

EXAMPLE

Using the Dittus–Boelter correlation, determine the heat transfer coefficient for a transpiration cooled tube having the following characteristics:

Parameter	Symbol	Value	Units
Free stream mass flux	$G_\infty = \rho_H V_\infty$	3	$\frac{gm}{cm^2 s}$
Tube diameter	D	2	cm
Viscosity	μ	0.0006	$\frac{gm}{cm\ s}$
Thermal conductivity	k	0.017	$\frac{W}{cm\ s}$
Prandlt number	Pr	0.7	
Wall transpiration mass flux	$\dot{m}''$	0.2	$\frac{gm}{cm^2 s}$

Solution
The first step in evaluating the transpiration heat transfer coefficient multiplier is to calculate the Reynolds number to determine whether the flow is laminar or turbulent:

$$Re = \frac{G_\infty D}{\mu} = \frac{3\frac{gm}{cm^2 s} \times 2\, cm}{0.0006 \frac{gm}{cm\, s}} = 10{,}000 \quad \therefore \text{turbulent} \tag{1}$$

Next, using the Reynolds number from Eq. (1), the Nusselt number is calculated using the Dittus–Boelter correlation:

$$Nu = 0.023 Re^{0.8} Pr^{0.4} = 0.02310000^{0.8} \times 0.7^{0.4} = 31.6 \tag{2}$$

From the Nusselt number, the free stream heat transfer coefficient is computed giving:

$$h_c = Nu\frac{k}{D} = 31.6 \frac{0.017 \frac{W}{cm\, K}}{2\, cm} = 0.27 \frac{W}{cm^2\, K} \tag{3}$$

The free stream Stanton number is now calculated from the Reynolds number and the Nusselt number, yielding:

$$St_0 = \frac{Nu}{Re Pr} = \frac{31.6}{10{,}000 \times 0.7} = 0.00451 \tag{4}$$

Computing the blowing parameter from Eq. (9.25) then gives:

$$B_h = \frac{\dot{m}''}{G_\infty}\frac{1}{St} = \frac{0.2\frac{gm}{cm^2 s}}{3\frac{gm}{cm^2 s}}\frac{1}{St} = \frac{1}{15 St} \tag{5}$$

Employing the blowing coefficient from Eq. (5) in the transpiration cooling correlation from Eq. (9.26), the transpiration Stanton number may now be iteratively determined, yielding:

$$\frac{St}{St_0} = \frac{St}{0.00451} = \left[\frac{Ln\left(1 + \frac{1}{15St}\right)}{\frac{1}{15St}}\right]^{\frac{5}{4}} (1 + St)^{\frac{1}{4}} \Rightarrow St = 0.00256 \tag{6}$$

Using the Stanton number assuming no transpiration cooling from Eq. (4), the Stanton number assuming transpiration cooling from Eq. (6), and the heat transfer coefficient assuming no transpiration cooling from Eq. (3), the heat transfer coefficient including transpiration effects may now be calculated such that:

$$h_c = h_{c0}\frac{St}{St_0} = 0.27 \frac{0.00256}{0.00451} = 0.153 \frac{W}{cm^2\, K} \tag{7}$$

3. NUCLEAR REACTOR TEMPERATURE AND PRESSURE DISTRIBUTIONS IN AXIAL FLOW GEOMETRY

It was noted earlier that determining the fuel temperature distribution in a nuclear rocket core is complicated by the fact that the power generated by the core is nonuniform and that the temperature of the propellant varies as it traverses through the reactor. Nevertheless, by using the core average power distributions derived earlier with suitable power peaking factors coupled with the expressions for the

3. NUCLEAR REACTOR TEMPERATURE AND PRESSURE DISTRIBUTIONS

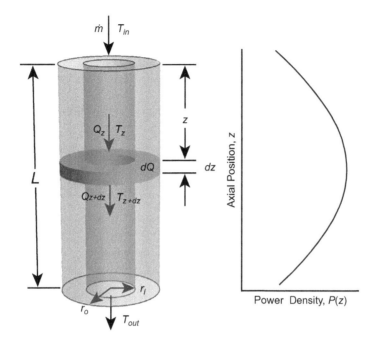

FIGURE 9.7

Axial representation of propellant channel flow cell.

temperature distribution in an equivalent flow channel cell, it is possible to obtain a qualitative picture of the overall temperature distribution throughout the nuclear rocket core. The analysis begins writing an equation for the heat balance over a differential slice of a flow channel cell such as would be found in a NERVA-type fuel element. A representation such as NERVA-type flow channel is illustrated in Fig. 9.7.

The heat balance can be expressed as

$$dQ = Q_{dz+z} - Q_z = \dot{m}C_p(T_{dz+z} - T_z) = \dot{m}C_p dT \qquad (9.27)$$

The amount of heat "dQ" generated by fission within the differential length "dz" is also equal to

$$dQ = P(z)\pi(r_o^2 - r_i^2)dz \qquad (9.28)$$

Combining Eqs. (9.27) and (9.28) then yields

$$dQ = \dot{m}C_p dT = P(z)\pi(r_o^2 - r_i^2)dz \qquad (9.29)$$

In order to solve Eq. (9.29) the functional relationship for the power density "$P(z)$" must be known. This relationship is determined by solving the nuclear criticality equation as expressed by Eq. (7.8) for the particular configuration under investigation. For the purpose of the present analysis, Eq. (7.8) is simplified assuming that the reactor core can be represented by a one energy group, two region diffusion theory model. Such a core model was previously derived, with the results being expressed by Eq. (8.44):

$$P(z) = A \, \text{Cos}(\alpha x) = A \, \text{Cos}\left[\alpha\left(\frac{L}{2} - z\right)\right] \tag{9.30}$$

where α = core buckling and $x = \left(\frac{L}{2} - z\right)$ = variable change to account for the coordinate system reference point shift between Eqs. (9.30) and (8.44).

Assuming that the average power density for a propellant flow cell is known, the arbitrary constant "A" may be determined integrating the power distribution function from Eq. (9.30) over the length of the core, dividing by the core length, and setting the result equal to the known power density such that:

$$P_{\text{ave}} = \frac{A}{L}\int_0^L \text{Cos}\left[\alpha\left(\frac{L}{2} - z\right)\right] dz = \frac{2A}{\alpha L}\text{Sin}\left(\frac{\alpha L}{2}\right) \Rightarrow A = \frac{\alpha P_{\text{ave}} L}{2\,\text{Sin}\left(\frac{\alpha L}{2}\right)} \tag{9.31}$$

Using the value for "A" from Eq. (9.31) in Eq. (9.30), the propellant flow cell power distribution becomes:

$$P(z) = \frac{\alpha P_{\text{ave}} L}{2\,\text{Sin}\left(\frac{\alpha L}{2}\right)} \text{Cos}\left[\alpha\left(\frac{L}{2} - z\right)\right] \tag{9.32}$$

Incorporating Eq. (9.32) into Eq. (9.29) and rearranging terms then yields:

$$dT = \frac{\alpha P_{\text{ave}} L}{2\,\text{Sin}\left(\frac{\alpha L}{2}\right)} \frac{\pi(r_o^2 - r_i^2)}{\dot{m}C_p} \text{Cos}\left[\alpha\left(\frac{L}{2} - z\right)\right] dz \tag{9.33}$$

By integrating Eq. (9.33) over the length of the reactor core under the assumption that the specific heat capacity of the propellant is a constant, it is possible to find an expression for the axial propellant temperature distribution of the form:

$$\int_{T_{\text{in}}}^{T_p} dT = \frac{\alpha P_{\text{ave}} L}{2\,\text{Sin}\left(\frac{\alpha L}{2}\right)} \frac{\pi(r_o^2 - r_i^2)}{\dot{m}C_p} \int_0^z \text{Cos}\left[\alpha\left(\frac{L}{2} - z'\right)\right] dz'$$

$$\Rightarrow T_p(z) = T_{\text{in}} + \frac{P_{\text{ave}} L}{2}\frac{\pi(r_o^2 - r_i^2)}{\dot{m}C_p}\left\{1 - \frac{\text{Sin}\left[\alpha\left(\frac{L}{2} - z\right)\right]}{\text{Sin}\left(\frac{\alpha L}{2}\right)}\right\} \tag{9.34}$$

where $T_p(z)$ = propellant temperature as a function of the axial position in the reactor core.

Now by substituting Eqs. (9.32) and (9.34) into Eq. (9.18) an expression may be found for the fuel temperature distribution at any axial position in the core such that:

$$T_f(z) = T_{\text{in}} + \frac{P_{\text{ave}} L}{2}\frac{\pi(r_o^2 - r_i^2)}{\dot{m}C_p}\left\{1 - \frac{\text{Sin}\left[\alpha\left(\frac{L}{2} - z\right)\right]}{\text{Sin}\left(\frac{\alpha L}{2}\right)}\right\} + \frac{\alpha P_{\text{ave}} L}{2}\frac{\text{Cos}\left[\alpha\left(\frac{L}{2} - z\right)\right]}{\text{Sin}\left(\frac{\alpha L}{2}\right)}$$

$$\times \left\{\frac{r_o^2 - r_i^2}{2h_c r_i} - \frac{1}{2k}\left[\frac{r^2 - r_i^2}{2} + r_o^2 \, \text{Ln}\left(\frac{r_i}{r}\right)\right]\right\} \tag{9.35}$$

If the radial position variable "r" in Eq. (9.35) is set to "r_i" the channel wall temperature distribution may be expressed in the form:

$$T_w(z) = T_{in} + \frac{P_{ave}L}{2}\frac{\pi(r_o^2 - r_i^2)}{\dot{m}C_p}\left\{1 - \frac{\text{Sin}\left[\alpha\left(\frac{L}{2} - z\right)\right]}{\text{Sin}\left(\frac{\alpha L}{2}\right)}\right\}$$
$$+ \frac{\alpha P_{ave}L(r_o^2 - r_i^2)}{4h_c r_i}\frac{\text{Cos}\left[\alpha\left(\frac{L}{2} - z\right)\right]}{\text{Sin}\left(\frac{\alpha L}{2}\right)} \quad (9.36)$$

If the derivative of the channel wall temperature distribution from Eq. (9.36) is now taken with respect to the axial position variable "z" and the result set equal to zero, an equation may be derived which locates the position where the channel wall temperature is a maximum such that:

$$\frac{dT_w}{dz} = \frac{\alpha P_{ave}L(r_o^2 - r_i^2)\{2\pi h_c r_i \text{ Cos}\left[\alpha\left(\frac{L}{2} - z\right)\right] + \alpha \dot{m}C_p \text{ Sin}\left[\alpha\left(\frac{L}{2} - z\right)\right]\}}{4h_c r_i \dot{m}C_p \text{ Sin}\left(\frac{\alpha L}{2}\right)} = 0 \quad (9.37)$$

Rearranging the terms in Eq. (9.37) then yields an equation for the axial position at which channel wall temperature is a maximum, where

$$z_w^{max} = \frac{L}{2} + \frac{1}{\alpha}\text{Tan}^{-1}\left(\frac{2\pi h_c r_i}{\alpha \dot{m}C_p}\right) \quad (9.38)$$

By inserting the value for "z_w^{max}" from Eq". (9.38) into channel wall temperature distribution from Eq. (9.36), the maximum channel wall temperature in the equivalent propellant channel flow cell may be determined.

In a similar manner, if the radial position variable "r" in Eq. (9.35) is set to "r_o" the peak fuel temperature distribution may be expressed by an equation of the form:

$$T_f^{max}(z) = T_{in} + \frac{P_{ave}L}{2}\frac{\pi(r_o^2 - r_i^2)}{\dot{m}C_p}\left\{1 - \frac{\text{Sin}\left[\alpha\left(\frac{L}{2} - z\right)\right]}{\text{Sin}\left(\frac{\alpha L}{2}\right)}\right\} + \frac{\alpha P_{ave}L}{2}\frac{\text{Cos}\left[\alpha\left(\frac{L}{2} - z\right)\right]}{\text{Sin}\left(\frac{\alpha L}{2}\right)}$$
$$\times \left\{\frac{r_o^2 - r_i^2}{2h_c r_i} - \frac{1}{2k}\left[\frac{r_o^2 - r_i^2}{2} + r_o^2 \text{ Ln}\left(\frac{r_i}{r_o}\right)\right]\right\} \quad (9.39)$$

Again, if the derivative of the peak fuel temperature distribution from Eq. (9.39) is now taken with respect to the axial position variable "z" and the result set equal to zero, an equation may be derived which locates the position where the fuel temperature is a maximum such that:

$$\frac{dT_f^{max}}{dz} = \frac{\pi \alpha P_{ave}L(r_o^2 - r_i^2)\text{Cos}\left[\alpha\left(\frac{L}{2} - z\right)\right]}{2\dot{m}C_p \text{ Sin}\left(\frac{\alpha L}{2}\right)}$$
$$+ \frac{\alpha^2 P_{ave}L\left[(2k - h_c r_i)(r_o^2 - r_i^2) - 2h_c r_i r_o^2 \text{ Ln}\left(\frac{r_i}{r_o}\right)\right]\text{Sin}\left[\alpha\left(\frac{L}{2} - z\right)\right]}{8h_c k r_i \text{ Sin}\left(\frac{\alpha L}{2}\right)} = 0 \quad (9.40)$$

Rearranging the terms in Eq. (9.40) then yields an equation for the axial position at which the fuel temperature is a maximum, where

$$z_f^{max} = \frac{L}{2} - \frac{1}{\alpha}\text{Tan}^{-1}\left\{\frac{4\pi k h_c r_i (r_o^2 - r_i^2)}{\alpha \dot{m} C_p \left[(h_c r_i - 2k)(r_o^2 - r_i^2) + 2h_c r_i r_o^2 \text{Ln}\left(\frac{r_i}{r_o}\right)\right]}\right\} \qquad (9.41)$$

By inserting the value for "z_f^{max}" from Eq. (9.41) into peak fuel temperature distribution from Eq. (9.39), the maximum fuel temperature in the equivalent propellant channel flow cell may be determined. In Fig. 9.8, the influence of the various parameters affecting the characteristics of the axial temperature distributions in an equivalent propellant channel flow cell may be examined.

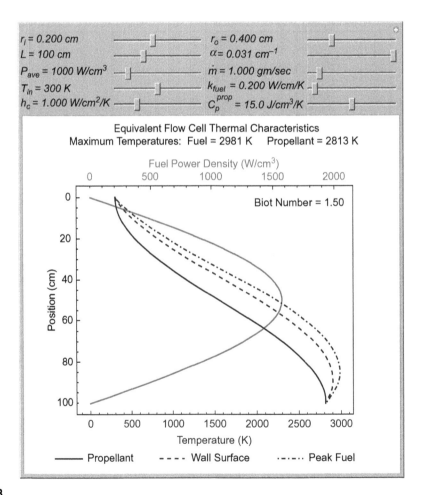

FIGURE 9.8

Reactor equivalent channel temperature profiles.

3. NUCLEAR REACTOR TEMPERATURE AND PRESSURE DISTRIBUTIONS

As the propellant gains heat during its traverse of the flow channel, it also experiences a pressure loss due to viscous friction effects resulting from the propellant's interactions with the channel wall. The greater the friction experienced by the propellant, the greater the pressure loss. This pressure loss represents additional work which must be supplied by the engine's pumping system. Consequently, one might suppose that it would be most advantageous to reduce the friction as much as possible so as to keep pressure losses to a minimum. However, it is precisely these frictional effects which cause the flow turbulence that enhances the heat transfer rate from the channel wall to the propellant. Thus, increased flow friction works to minimize peak fuel temperatures. As a result, the configuration of the propellant flow channels entails a design optimization which balances propellant flow pressure losses against increased fuel temperatures.

Pressure drop is often calculated through the use of the *Darcy formula* which is generally valid when compressibility effects can be neglected. In its differential form, the equation may be given by

$$dp = f \frac{dz}{2r_i} \frac{\rho V^2}{2} \qquad (9.42)$$

where f = Darcy–Weisbach friction factor; ρ = propellant density; and V = propellant velocity.

Recall that the *universal gas law* may be expressed as

$$p = \rho RT \Rightarrow \rho = \frac{p}{RT} \qquad (9.43)$$

where R = propellant gas constant.

Incorporating the universal gas law from Eq. (9.43) into the continuity expression from Eq. (2.26) then yields:

$$\dot{m} = \rho V A = \frac{pVA}{RT} \Rightarrow V = \frac{\dot{m}RT}{pA} \qquad (9.44)$$

Using the universal gas law from Eq. (9.43) and the expression for propellant velocity from Eq. (9.44) in the Darcy formula from Eq. (9.42) then yields for the differential pressure drop:

$$dp = f \frac{dz}{4r_i} \frac{p}{RT} \left(\frac{\dot{m}RT}{pA}\right)^2 = f \frac{dz}{4r_i} \frac{\dot{m}^2 RT}{pA^2} \qquad (9.45)$$

Rearranging Eq. (9.45) and replacing the channel area "A" with an equivalent relationship in terms of the channel radius then yields:

$$p\,dp = f \frac{dz}{4r_i} \frac{1}{RT} \left(\frac{\dot{m}RT}{A}\right)^2 = f \frac{\dot{m}^2 RT}{4\pi^2 r_i^5} dz \qquad (9.46)$$

Incorporating the functional relationship for the propellant temperature from Eq. (9.34) which was derived earlier into Eq. (9.46) for the differential pressure drop then yields an expression of the form:

$$p\,dp = f \frac{\dot{m}^2 R}{4\pi^2 r_i^5} \left(T_{\text{in}} + \frac{P_{\text{ave}}}{2} \frac{\pi(r_o^2 - r_i^2)L}{C_p \dot{m}} \left\{1 - \frac{\text{Sin}\left[\alpha\left(\frac{L}{2} - z\right)\right]}{\text{Sin}\left(\frac{\alpha L}{2}\right)}\right\}\right) dz \qquad (9.47)$$

Integrating Eq. (9.47) over the length of the core then gives an expression relating the core pressures to the other reactor parameters such that:

$$\int_{P_{in}}^{P_{out}} p\,dp = f\frac{\dot{m}^2 R}{4\pi^2 r_i^5}\int_0^L \left(T_{in} + \frac{P_{ave}\,\pi(r_o^2 - r_i^2)L}{2\,C_p\dot{m}}\left\{1 - \frac{\operatorname{Sin}\left[\alpha\left(\frac{L}{2} - z\right)\right]}{\operatorname{Sin}\left(\frac{\alpha L}{2}\right)}\right\}\right)dz \Rightarrow \frac{p_{in}^2 - p_{out}^2}{2}$$

$$= f\frac{\dot{m}RL}{8\pi^2 r_i^5 C_p}\left[2\dot{m}C_p T_{in} + P_{ave}\,\pi(r_o^2 - r_i^2)L\right] \qquad (9.48)$$

Incorporating the definition for the specific heat ratio from Eq. (2.12) into Eq. (9.48) and solving for the core outlet pressure then yields:

$$p_{out} = \sqrt{p_{in}^2 - f\frac{\dot{m}L}{4\pi^2 r_i^5}\left(\frac{\gamma - 1}{\gamma}\right)\left[2\dot{m}C_p T_{in} + P_{ave}\,\pi(r_o^2 - r_i^2)L\right]} \qquad (9.49)$$

In Fig. 9.9 the influence of the various parameters from Eq. (9.49) affecting the pressure drop in an equivalent propellant channel flow cell may be examined.

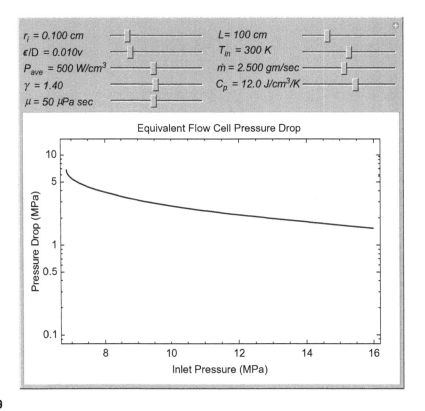

FIGURE 9.9

Pressure drop through an equivalent flow cell.

3. NUCLEAR REACTOR TEMPERATURE AND PRESSURE DISTRIBUTIONS

EXAMPLE

Determine the required mass flow rate and average power density within a single equivalent fuel cell if the maximum allowable fuel centerline temperature can be no higher than 3100K and the pressure drop in the propellant flow channel cannot exceed 0.2 MPa. Plot the propellant temperature and maximum fuel temperature as a function of the axial position in the fuel cell and determine the propellant outlet temperature. Use the design and flow parameters for the fuel cell as given in the following table:

Variable	Value	Units	Description
r_i	0.15	cm	Radius of coolant channel
r_o	0.35	cm	Equivalent radius of fuel cell
L	100	cm	Length of fuel cell
α	0.03	cm^{-1}	Core buckling
T_{in}	300	K	Propellant inlet temperature
C_p	15.2	$\frac{J}{gm\,K}$	Specific heat of propellant
k	0.3	$\frac{W}{cm\,K}$	Thermal conductivity of fuel
h_c	10	$\frac{W}{cm^2\,K}$	Heat transfer coefficient
γ	1.4	–	Propellant specific heat ratio
P_{in}	10	MPa	Propellant inlet pressure
f	0.02	–	Friction factor

Solution

In this problem, there are two unknown variables which must be determined, namely the propellant mass flow rate "$\dot{m}$" and the fuel cell power density "P_{ave}." These variables must be evaluated such that they yield a maximum fuel temperature of 3100K and a pressure drop of 0.2 MPa. Since there are two unknowns to be determined, two equations are required to obtain a solution. One of the required equations is Eq. (9.39) which yields the maximum fuel temperature as a function of axial position:

$$T_f^{max}(z) = T_{in} + \frac{P_{ave}L}{2}\frac{\pi(r_o^2-r_i^2)}{\dot{m}C_p}\left\{1-\frac{\operatorname{Sin}\left[\alpha\left(\frac{L}{2}-z\right)\right]}{\operatorname{Sin}\left(\frac{\alpha L}{2}\right)}\right\} + \frac{\alpha P_{ave}L}{2}\frac{\operatorname{Cos}\left[\alpha\left(\frac{L}{2}-z\right)\right]}{\operatorname{Sin}\left(\frac{\alpha L}{2}\right)} \quad (1)$$

$$\times \left\{\frac{r_o^2-r_i^2}{2h_c r_i} - \frac{1}{2k}\left[\frac{r_o^2-r_i^2}{2} + r_o^2\operatorname{Ln}\left(\frac{r_i}{r_o}\right)\right]\right\}$$

Substituting numerical values into Eq. (1) for the position-dependent maximum fuel temperature then yields:

$$T_f^{max}(z) = 300K + 0.1849\,P_{ave}\operatorname{Cos}[0.03(50-z)] + 1.0334\frac{P_{ave}}{\dot{m}}\{1-1.0025\operatorname{Sin}[0.03(50-z)]\} \quad (2)$$

Since the fuel maximum temperature cannot exceed 3100K, the axial location at which the fuel temperature is a highest must be determined. This location was derived earlier and is expressed by Eq. (9.41) wherein:

$$z_f^{peak} = \frac{L}{2} - \frac{1}{\alpha}\operatorname{Tan}^{-1}\left\{\frac{4\pi k h_c r_i(r_o^2-r_i^2)}{\alpha\dot{m}C_p\left[(h_c r_i - 2k)(r_o^2-r_i^2) + 2h_c r_i r_o^2\operatorname{Ln}\left(\frac{r_i}{r_o}\right)\right]}\right\} \quad (3)$$

Substituting numerical values into Eq. (3) for the location of the highest temperature yields:

$$z_f^{peak} = 50 + 33.33 \,\text{Tan}^{-1}\left(\frac{5.6016}{\dot{m}}\right) = 50 + 33.33 \,\text{Sin}^{-1}\left(\frac{5.6016}{\sqrt{5.6016^2 + \dot{m}^2}}\right)$$

$$= 50 + 33.33 \,\text{Cos}^{-1}\left(\frac{\dot{m}}{\sqrt{5.6016^2 + \dot{m}^2}}\right) \quad (4)$$

Inserting Eq. (4) for the axial location of the peak fuel temperature into Eq. (2) and simplifying the result yields for the peak fuel temperature in a fuel cell an expression of the form:

$$T_f^{peak} = 300\text{K} + \frac{P_{ave}}{\dot{m}}\left(1.0334 + \frac{0.1849\dot{m}^2 + 5.8034}{\sqrt{31.378 + \dot{m}^2}}\right) \quad (5)$$

The other equation required to solve for the average power density and propellant flow rate is an expression for pressure drop in the fuel cell channel. This pressure drop may be determined from Eq. (9.49) such that:

$$\Delta p = p_{in} - p_{out} = p_{in} - \sqrt{p_{in}^2 - f\frac{\dot{m}L}{4\pi^2 r_i^5}\left(\frac{\gamma-1}{\gamma}\right)\left[2\dot{m}C_pT_{in} + P_{ave}\pi(r_o^2 - r_i^2)L\right]} \quad (6)$$

Rearranging Eq. (6) so as to solve for the average fuel power density then yields:

$$P_{ave} = \frac{4\gamma\pi^2 r_i^5 \Delta p(2p_{in} - \Delta p) - 2f(\gamma-1)L\dot{m}^2 C_p T_{in}}{\pi f(\gamma-1)(r_o^2 - r_i^2)L^2 \dot{m}} \quad (7)$$

Substituting numerical values into Eq. (7) yields for the average fuel cell power density an equation of the form:

$$P_{ave} = \frac{6613 - 290.3\dot{m}^2}{\dot{m}} \quad (8)$$

Incorporating Eq. (8) into Eq. (5) for the peak fuel temperature yields an equation which is a function only of the fuel cell propellant mass flow rate such that:

$$T_f^{peak} = 300 + \frac{6613 - 290.3\dot{m}^2}{\dot{m}^2}\left(1.0334 + \frac{0.1849\dot{m}^2 + 5.8034}{\sqrt{31.378 + \dot{m}^2}}\right)\text{K} \quad (9)$$

Plotting the peak fuel temperature as a function of the propellant mass flow rate as illustrated in Fig. 1 indicates that the flow rate required to give a peak fuel temperature of 3100K with a pressure drop of 0.2 MPa is about 2.03 g/s per fuel cell channel.

Using the mass flow rate determined from Fig. 1 into Eq. (8) then yields for the fuel cell average power density a value of about 2.664 kW/cm³. The mass flow rate and average fuel cell power density may now be used with Eq. (2) to determine the maximum fuel temperature as a function of axial position such that:

$$T_f^{max}(z) = 300 + 492.8\,\text{Cos}[0.03(50-z)] + 1355\{1 - 1.0025\,\text{Sin}[0.03(50-z)]\}\text{K} \quad (10)$$

3. NUCLEAR REACTOR TEMPERATURE AND PRESSURE DISTRIBUTIONS

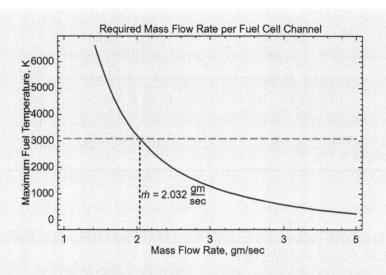

FIGURE 1
Flow rate versus maximum fuel temperature.

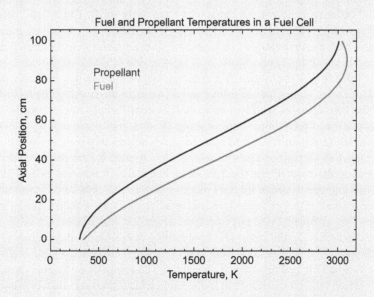

FIGURE 2
Axial temperature distributions.

The mass flow rate and average fuel cell power density may also be used with Eq. (9.34) to determine the propellant temperature as a function of axial position such that:

$$T_p(z) = 300 + 1355\{1 - 1.00251 \sin[0.03(50 - z)]\}\text{K} \tag{11}$$

Using Eq. (11) the propellant exit temperature at the exit of the fuel cell may be determined to be

$$T_p(100) = 300 + 1355\{1 - 1.00251 \sin[0.03(50 - 100)]\}\text{K} = 3010\text{K} \tag{12}$$

A plot of the maximum fuel temperature and the propellant temperature as a function of axial position may now be generated using Eqs. (10) and (11), respectively, to yield the functional relationship illustrated in Fig. 2:

4. NUCLEAR REACTOR FUEL ELEMENT TEMPERATURE DISTRIBUTIONS IN RADIAL FLOW GEOMETRY

Axial flow—type fuel elements similar to those described earlier performed quite successfully in many of the rocket engine tests run during the heyday of the NERVA program; however, the axial flow fuel element design used in the NERVA program has a number of geometric characteristics which limit its performance capabilities. In particular, the long narrow propellant channels running the length of the fuel element yield surface area to volume ratios which are fairly small. These small surface to volume ratios limit the rate at which heat can be transferred from the fuel to the propellant thus restricting the power density at which the fuel element may operate. Lower power densities result in larger, more massive reactor cores which for a given thrust level yield engine thrust to weight ratios which are rather modest. Another limitation of the engine is a consequence of the fact that the axial power profile is cosine shaped. This cosine-shaped power profile causes temperature peaking effects to occur such that the majority of the core operates with fuel temperatures considerably lower than the peak fuel temperature. An illustration of this temperature peaking effect is presented in Fig. 9.8 wherein the fuel centerline and surface temperatures can be seen to vary considerably over the length of the reactor. Since the power level at which the engine may operate is limited by the peak fuel temperature, it is evident that most of the reactor will be forced to operate at temperatures and power densities considerably lower than those which the fuel might otherwise be capable of sustaining. Lower fuel temperatures again result in larger core sizes and smaller thrust to weight ratios. Finally, the long narrow propellant flow channels cause large pressure drops and pumping power losses to occur in the core during operation reducing the thrust level potentially attainable by the engine.

To mitigate some of the performance limitations characteristic of axial flow fuel element geometries, the radial flow particle bed reactor was conceived. The geometry of the fuel particle bed permits high surface to volume ratios to be achieved in the fuel elements and consequently allows the fuel to run at quite high power densities. These high power densities in the fuel enable very compact nuclear rocket engines having very high thrust to weight ratios to be designed. The short path length traversed by the propellant through the particle bed also reduces the pressure drop and pumping power requirements. Unfortunately, as was mentioned earlier, the particle bed design is thermally unstable. During testing, this instability caused local hot spots to appear in the fuel which resulted in localized melting and fuel particle agglomeration in the particle bed. The thermal instability was caused by a thermal hydraulic effect in which the propellant tended to migrate away from locations in the particle bed having slightly higher temperatures. As the propellant migrated away from the hot locations in the particle bed, these locations became even hotter, resulting even less propellant flow into these hot locations and so on until fuel failure occurred.

One solution to the problem of the thermal instability in radial flow fuel element configurations is to use grooved ring fuel elements (GRFEs). This design offers a great deal of flexibility in that it can be configured to have large surface to volume ratios and low pressure drops while maintaining thermal stability. The enhanced thermal stability results from the fact that the groove pattern in the fuel ring constrains the propellant flow to follow prescribed paths through the fuel. In addition, the groove pattern may be optimized so as to maximize fuel performance. If desired, the uranium enrichment in the individual rings may also be varied so as to fairly easily yield an extremely flat axial power distribution. Even if the enrichment is not varied axially, the power peaking resulting from the cosine-shaped power distribution can be accommodated by varying the groove design in the rings so as to force more propellant into those regions where the power is the highest. From a production standpoint, the grooved ring fuel pieces should also be easier to fabricate than NERVA fuel elements, since NERVA fuel elements require fairly complicated materials processing techniques to manufacture. Fig. 9.10 illustrates what a GRFE might look like.

To illustrate how the GRFE may be optimized so as to maximize fuel performance, a derivation is performed to determine a groove wall thickness profile which will yield a constant groove wall centerline temperature radially across the fuel ring. Such an optimization not only enables the majority of the fuel to operate at or near its maximum allowable temperature, but also reduces thermally induced stresses to a minimum by eliminating most of the thermal gradients in the fuel.

The first step in the derivation involves determining the temperature distribution across a grooved wall in the fuel element. In the analysis, it is assumed that the temperature distribution is symmetrical along the centerline of the grooved wall and that all thermal parameters are temperature independent.

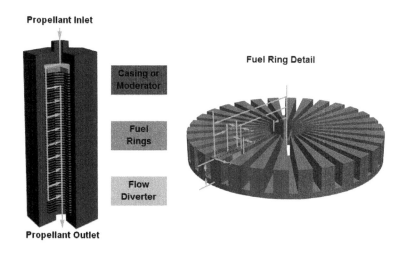

FIGURE 9.10

Radial flow grooved ring fuel element.

It is also assumed that the power density in the fuel ring is constant. While the assumption of constant power density is not strictly true in practice, it should nevertheless be fairly good since, provided the mean free path of neutrons is of the same order as the dimensions of the fuel ring (which is the case for reasonably sized fuel rings), it is unlikely that strong power density gradients would be capable of forming.

Using the one-dimensional form of Poisson's equation as expressed in Eq. (9.10) in Cartesian coordinates yields a governing equation of the form:

$$\frac{d^2 T}{dx^2} + \frac{P}{k} = \frac{d}{dx}\left(\frac{dT}{dx}\right) + \frac{P}{k} = 0 \qquad (9.50)$$

where T = fuel temperature; P = fuel power density; and k = fuel thermal conductivity.

Rearranging Eq. (9.50) and integrating then yields:

$$\frac{d}{dx}\left(\frac{dT}{dx}\right) = -\frac{P}{k} \Rightarrow \frac{dT}{dx} = -\frac{P}{k}\int dx = -\frac{P}{k}x + C_1 \qquad (9.51)$$

Applying the assumption that the temperature gradient is zero at the grooved wall centerline (due to symmetry) then the arbitrary constant "C_1" may be determined such that:

$$0 = -\frac{P}{k}(0) + C_1 \Rightarrow C_1 = 0 \qquad (9.52)$$

Incorporating the expression for the arbitrary constant from Eq. (9.52) into Eq. (9.51) and integrating a second time then yields:

$$T = -\frac{P}{k}\int x\,dx = -\frac{P}{2k}x^2 + C_2 \qquad (9.53)$$

Since the surface temperature of the grooved wall is determined at $x = w$, the arbitrary constant "C_2" may be found from Eq. (9.53), such that:

$$T_s = -\frac{P}{2k}w^2 + C_2 \Rightarrow C_2 = T_s + \frac{P}{2k}w^2 \qquad (9.54)$$

where T_s = fuel temperature at the grooved wall surface and w = half width of the grooved wall.

Incorporating the expression for the arbitrary constant from Eq. (9.54) into Eq. (9.53) then yields an expression for the temperature distribution within the grooved wall of the form:

$$T = -\frac{P}{2k}x^2 + T_s + \frac{P}{2k}w^2 = T_s + \frac{P}{2k}(w^2 - x^2) \qquad (9.55)$$

The grooved wall centerline fuel temperature (eg, at $x = 0$) as a function of the grooved wall width may now be determined from Eq. (9.55), such that:

$$T_f = T_s + \frac{P}{2k}w^2 \qquad (9.56)$$

where T_f = fuel temperature at the grooved wall centerline.

Referencing Fig. 9.10 and performing a heat balance on the differential element, it may be noted that:

$$dQ = PdV = -P(hwdr) = \dot{m}c_p dT_p \Rightarrow \frac{dT_p}{dr} = -\frac{Ph}{\dot{m}c_p}w \tag{9.57}$$

where dQ = amount of power generated in the differential volume "dV"; dT_p = differential change in the propellant temperature; h = height of the grooved wall; $\dot{m}$ = propellant flow rate through an area defined by $h \times t$ in which "t" is the flow channel half width; c_p = specific heat capacity of the propellant; and dr = differential radial length.

The negative sign in Eq. (9.57) indicates that propellant heating is occurring in the negative "r" direction (eg, from the outside of the fuel ring to the inside of the fuel ring). If the heat generated in the differential fuel element volume "dV" is all assumed to transfer into the propellant through the side of the grooved wall, a heat balance may be performed such that:

$$dQ = PdV = P(hwdr) = h_c(T_s - T_p)dA = h_c(T_s - T_p)hdr \Rightarrow T_s = \frac{P}{h_c}w + T_p \tag{9.58}$$

where h_c = heat transfer coefficient from fuel grooved wall to propellant.

Combining Eqs. (9.56) and (9.58), it is possible to arrive at an expression for the fuel grooved wall centerline temperature in terms of the grooved wall half width and the propellant temperature of the form:

$$T_f = \frac{P}{h_c}w + \frac{P}{2k}w^2 + T_p \tag{9.59}$$

To determine an expression describing the width of the grooved wall radially along the fuel ring which yields a constant fuel centerline temperature, the derivative of Eq. (9.59) with respect to "r" is taken to yield:

$$\frac{dT_f}{dr} = \frac{P}{h_c}\frac{dw}{dr} + \frac{P}{k}w\frac{dw}{dr} + \frac{dT_p}{dr} \tag{9.60}$$

A constant fuel centerline temperature requires that the fuel centerline temperature derivative term in Eq. (9.60) equal zero; therefore, setting Eq. (9.60) equal to zero and incorporating the results of Eq. (9.57) it is found that:

$$0 = \frac{P}{h_c}\frac{dw}{dr} + \frac{P}{k}w\frac{dw}{dr} - \frac{Ph}{\dot{m}c_p}w = \frac{1}{h_c}\frac{dw}{dr} + \frac{1}{k}w\frac{dw}{dr} - \frac{h}{\dot{m}c_p}w \Rightarrow \frac{dw}{dr} = \frac{h_c khw}{\dot{m}c_p(k + h_c w)} \tag{9.61}$$

Solving the differential equation expressed by Eq. (9.61) yields a transcendental expression describing the half width of the grooved wall in terms of the radial position such that:

$$0 = h_c hk(r_o - r) - h_c \dot{m}c_p[w_o - w] - k\dot{m}c_p \operatorname{Ln}\left(\frac{w_o}{w}\right) \tag{9.62}$$

where r = radial position along the fuel ring; r_o = outer radius of the fuel ring; and w_o = half width of the grooved wall at the outer radius of the fuel ring.

To determine the propellant temperature as a function of radial position, Eq. (9.59) is first rearranged so as to express the half width of the grooved wall as a function of the propellant temperature, yielding:

$$w = -\frac{k}{h_c} + \frac{k}{h_c}\sqrt{1 + \frac{2h_c^2}{kP}(T_f - T_p)} \quad (9.63)$$

The initial half width of the grooved wall is found from Eq. (9.63) using the temperature of the propellant as it enters the fuel ring giving an initial width value of

$$w_o = -\frac{k}{h_c} + \frac{k}{h_c}\sqrt{1 + \frac{2h_c^2}{kP}(T_f - T_{po})} \quad (9.64)$$

Substituting Eq. (9.63) into Eq. (9.62) so as to eliminate the grooved wall width term yields another transcendental expression relating the propellant temperature to the radial position under the constraint that the fuel grooved wall centerline temperature remains constant yielding:

$$0 = hh_c(r_o - r) - c_p\dot{m}\left(1 - \sqrt{1 + \frac{2h_c^2}{kP}(T_f - T_p)} + \frac{h_c w_o}{k}\right)$$

$$- c_p\dot{m}\mathrm{Ln}\left(\frac{-h_c w_o}{k\left[1 - \sqrt{1 + \frac{2h_c^2}{kP}(T_f - T_p)}\right]}\right) \quad (9.65)$$

To determine the grooved wall surface temperature as a function of radial position, Eq. (9.56) is rearranged so as to express the half width of the grooved wall as a function of the grooved wall surface temperature of the form:

$$w = \sqrt{\frac{2k}{P}(T_f - T_s)} \quad (9.66)$$

Substituting Eq. (9.66) into Eq. (9.62) so as to eliminate the grooved wall width term again yields a transcendental expression, in this case relating the grooved wall surface temperature to the radial position again under the constraint that the fuel grooved wall centerline temperature remains constant yielding:

$$0 = hh_c(r_o - r) - \frac{h_c c_p \dot{m}}{k}\left(w_o - \sqrt{\frac{2k}{P}(T_f - T_s)}\right) - c_p\dot{m}\mathrm{Ln}\left(w_o\sqrt{\frac{P}{2k(T_f - T_s)}}\right) \quad (9.67)$$

In Fig. 9.11, the temperature and channel wall profiles are presented as functions of various thermal and geometric parameters. Note that the flattest overall temperature distribution occurs when the fuel thermal conductivity is high and the heat transfer coefficient to the propellant is low. With this in mind, the goal in designing these fuel rings for maximum possible performance is to maximize the surface to volume ratio of the fuel walls so as to attain the greatest transfer rate from the fuel to the propellant while simultaneously achieving the flattest possible radial temperature profile so as to minimize

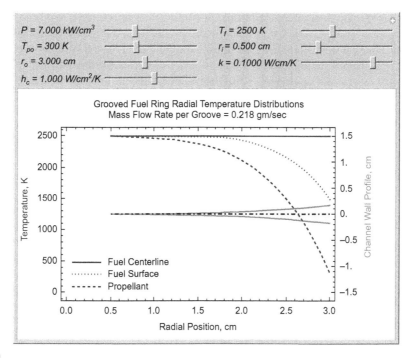

FIGURE 9.11

Temperature and channel wall profiles for a grooved ring fuel element.

thermal stresses in the fuel. This optimization would generally be accomplished by minimizing the widths of the walls so as to maximize the number of channel wall surfaces and by increasing the height of the channel walls within the fuel ring.

5. RADIATORS

Nuclear reactors which operate in space invariably encounter thermal heat rejection issues that must be addressed. For nuclear thermal propulsion systems, even after shutdown, decay heat from fission products and other processes can be significant and must be removed by some means or it is possible that the reactor will suffer damage due to overheating. One method by which this heat may be removed from the engine is to continue to flow propellant through the core for a time after shutdown until sufficient energy has been removed to prevent the engine from overheating. While this is a viable solution, it can waste a considerable amount of precious propellant. Alternatively, it may be possible to use radiators attached to the reactor and configured, such that they efficiently remove at least part of the heat from the engine system and radiate it away to space. Determining how or if radiators are deployed in any given situation depends upon the details of the mission at hand and are necessarily a part of the overall vehicle design process.

For nuclear electric systems where the primary use of the nuclear reactor is to provide power to ion or plasma propulsion systems, the heat rejection issue results from the requirement that in all power cycles, waste heat must at some point be rejected from the power conversion system which converts the thermal power from the reactor into electricity. In this case, it is desirable that the heat be rejected at as low a temperature as possible consistent with other thermodynamic limitations in order to have the most efficient conversion of thermal power to electric power.

Since excess mass is always a concern on spacecraft, radiator design activities always focus on rejecting the maximum amount heat with the minimum possible radiator mass. The amount of waste heat which can be radiated to space is determined from the Stefan–Boltzmann equation and is proportional to the fourth power of the radiator surface temperature. The total amount of heat rejected from the radiator also depends on the thermal emissivity at the radiator surface and total surface area of the radiator.

A possible radiator configuration is shown in Fig. 9.12. In this configuration, heat pipes from the reactor transfer heat to the fin which in turn radiates the heat into space. The problem to be addressed is to determine how to optimize the fin width and thickness as a function of the thermal properties of the fin so as to maximize the amount of heat radiated into space while simultaneously minimizing the total radiator mass [11]. A fin configuration which is thick at the root and tapers to zero at the tip is analyzed since such a design allows more heat to be conducted into the fin where the temperature is high, thus offsetting to some extent the rapid drop in temperature along the fin width.

Using the Stefan–Boltzmann equation to model heat loss due to radiation from the fin and assuming that $W \ll H$, it is possible to write a one-dimensional heat balance equation on the differential volume illustrated in Fig. 9.12, such that:

$$Q_{in} = Q_{out} + Q_{rad} = Q_{in} - \frac{dQ}{dx}\Delta x + q_{rad}dA \Rightarrow \frac{dQ}{dx}\Delta x = q_{rad}Ldx = 2\varepsilon\sigma T^4 L\Delta x \qquad (9.68)$$

where Q_{in} = heat into the differential volume element; Q_{out} = heat out of the differential volume element; $Q_{rad} = q_{rad}dA$ = heat radiated away from fin differential volume element (both sides); T = local temperature of fin; σ = Stefan–Boltzmann constant; and ε = emissivity of fin.

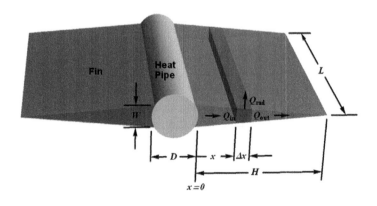

FIGURE 9.12

Space radiator.

Applying the Fourier heat conduction relation from Eq. (9.5) to the expression for the fin heat balance from Eq. (9.68) and assuming constant material properties, it is possible derive a differential equation which represents the temperature distribution in the fin of the form:

$$\frac{dQ}{dx}\Delta x = \frac{d}{dx}\left(kA\frac{dT}{dx}\right)\Delta x = kW\frac{d}{dx}\left[\left(1-\frac{x}{H}\right)\frac{dT}{dx}\right]L\Delta x = 2\varepsilon\sigma T^4 L\Delta x \qquad (9.69)$$

By rearranging the terms in Eq. (9.69), it is possible to put the expression in a dimensionless form such that:

$$H^2\frac{d}{dx}\left[\left(1-\frac{x}{H}\right)\frac{1}{T_0}\frac{dT}{dx}\right] = \underbrace{\frac{2\varepsilon\sigma H^2 T_0^3}{kW}}_{\beta^2}\frac{1}{T_0^4}T^4 \Rightarrow \frac{d}{d\xi}\left[(1-\xi)\frac{d\theta}{d\xi}\right] = \beta^2\theta^4 \qquad (9.70)$$

where $\xi = \frac{x}{H}$ and $\theta = \frac{T}{T_0}$, with boundary conditions: $\theta(0) = 1$ and $\theta'(1) = 0$.

The total amount of heat radiated to space per unit length of the fin assembly is twice the amount of heat conducted into a single fin from the heat pipe plus the amount of heat lost due to radiation from the heat pipe itself:

$$\frac{Q}{L} = -2kW\frac{dT}{dx}\bigg|_{x=0} + c\varepsilon\sigma T_0^4 = -2k\frac{W}{H}T_0\theta'(0) + c\varepsilon\sigma T_0^4 \qquad (9.71)$$

where c = total effective radiating width of the heat pipe.

The mass of the radiator system is simply the sum of the mass contributions from the heat pipe and the two fins such that:

$$m = L\left(\frac{\pi}{4}D^2\rho_{\text{hp}} + 2\frac{HW}{2}\rho_{\text{fin}}\right) \qquad (9.72)$$

where ρ_{hp} = linear density of the heat pipe and ρ_{fin} = density of the fins.

The heat radiated per unit mass can now be calculated using Eqs. (9.71) and (9.72) yielding an equation of the form:

$$\frac{Q}{m} = q = \frac{-2k\frac{W}{H}T_0\theta'(0) + c\varepsilon\sigma T_0^4}{\frac{\pi}{4}D^2\rho_{\text{hp}} + HW\rho_{\text{fin}}} \qquad (9.73)$$

To maximize the heat radiated by the fin assembly, optimum values for the fin width and thickness need to be determined for a given heat pipe mass. It is easier to perform this optimization if a new dimensionless variable is defined such that:

$$R = \frac{8HW\rho_{\text{fin}}}{\pi D^2\rho_{\text{hp}}} \Rightarrow W = \frac{R\pi D^2\rho_{\text{hp}}}{4H\rho_{\text{fin}}} \qquad (9.74)$$

Using the definition for β defined in Eq. (9.70) and the definition for R from Eq. (9.74), the fin width may be determined to be:

$$\beta^2 = \frac{2\varepsilon\sigma H^2 T_0^3}{kW} = \frac{2\varepsilon\sigma H^2 T_0^3}{k}\frac{4H\rho_{\text{fin}}}{R\pi D^2\rho_{\text{hp}}} \Rightarrow H = \beta\left(\frac{R\pi D^2\rho_{\text{hp}}}{4\rho_{\text{fin}}\beta}\right)^{\frac{1}{3}}\left(\frac{k}{2\varepsilon\sigma T_0^3}\right)^{\frac{1}{3}} \qquad (9.75)$$

The fin thickness at its root from Eq. (9.74) may now be recast using the expression for the fin width from Eq. (9.75), such that:

$$W = \frac{R\pi D^2 \rho_{hp}}{4H\rho_{fin}} = \frac{R\pi D^2 \rho_{hp}}{4\rho_{fin}} \frac{1}{\beta} \left(\frac{4\rho_{fin}\beta}{R\pi D^2 \rho_{hp}}\right)^{\frac{1}{3}} \left(\frac{2\varepsilon\sigma T_0^3}{k}\right)^{\frac{1}{3}} = \left(\frac{R\pi D^2 \rho_{hp}}{4\rho_{fin}\beta}\right)^{\frac{2}{3}} \left(\frac{2\varepsilon\sigma T_0^3}{k}\right)^{\frac{1}{3}} \quad (9.76)$$

Using the fin thickness at its root from Eq. (9.74) and the fin width from Eq. (9.76) in the expression for the heat radiated per unit mass from Eq. (9.73), it is possible to obtain a relation of the form:

$$q = \frac{4\varepsilon\sigma T_0^4}{\pi D^2 \rho_{hp}} \left[\frac{c}{1+R} - 2\left(\frac{k\pi D^2 \rho_{hp}}{\varepsilon\sigma T_0^3 \rho_{fin}}\right)^{\frac{1}{3}} \frac{R^{\frac{1}{3}}}{1+R} \frac{\theta'(0)}{\beta^{\frac{4}{3}}}\right] \quad (9.77)$$

Note from Eq. (9.77) that by minimizing $\frac{\theta'(0)}{\beta^{4/3}}$ the heat loss from the radiator may be maximized independent of R. By solving Eq. (9.70) for various values of β, the required minimum value for $\frac{\theta'(0)}{\beta^{4/3}}$ may be determined. This calculation must be done numerically since Eq. (9.70) is a nonlinear differential equation having no analytical solution. The results of this calculation are shown in Fig. 9.13.

By taking the derivative of Eq. (9.74) with respect to R, using the results from Fig. 9.13, and setting the resulting expression equal to zero result in an equation of the form:

$$0 = 2R - 1 - \frac{1}{2} \frac{3\beta^{\frac{4}{3}}}{\theta'(0)} \left(\frac{\varepsilon\sigma T_0^3 \rho_{fin}}{k\pi D^2 \rho_{hp}}\right)^{\frac{1}{3}} cR^{\frac{2}{3}} = 2R - 1 + 3.094c \underbrace{\left(\frac{\varepsilon\sigma T_0^3 \rho_{fin}}{k\pi D^2 \rho_{hp}}\right)^{\frac{1}{3}} R^{\frac{2}{3}}}_{\alpha} \quad (9.78)$$

Solving Eq. (9.78) yields an expression for the optimal value for R such that:

$$R_{opt} = \frac{1}{2} - \frac{\alpha^3}{24} + \frac{24\alpha^3 - \alpha^6 + \left(-216\alpha^3 + 36\alpha^6 - \alpha^9 + 24\sqrt{81\alpha^6 - 3\alpha^9}\right)^{2/3}}{24\left(-216\alpha^3 + 36\alpha^6 - \alpha^9 + 24\sqrt{81\alpha^6 - 3\alpha^9}\right)^{1/3}} \quad (9.79)$$

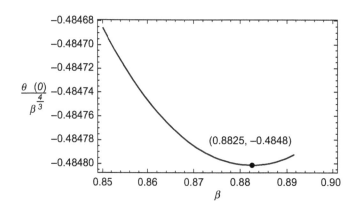

FIGURE 9.13

The maximum heat loss rate per unit mass from the fin assembly may now be determined by using the optimal value for R from Eq. (9.79), the optimal value for $\frac{\theta'(0)}{\beta^{4/3}}$ from Fig. 9.13, and the definition for α from Eq. (9.78) in Eq. (9.77) yielding:

$$q_{max} = \frac{4\varepsilon\sigma T_0^4}{\pi D^2 \rho_{hp}} \left[\frac{c}{1+R_{opt}} + 0.97 \left(\frac{k\pi D^2 \rho_{hp}}{\varepsilon\sigma T_0^3 \rho_{fin}} \right)^{\frac{1}{3}} \frac{R_{opt}^{\frac{1}{3}}}{1+R_{opt}} \right] = \frac{4c\varepsilon\sigma T_0^4}{\pi D^2 \rho_{hp}} \left[\frac{1}{1+R_{opt}} + \frac{3}{\alpha} \left(\frac{R_{opt}^{\frac{1}{3}}}{1+R_{opt}} \right) \right]$$

(9.80)

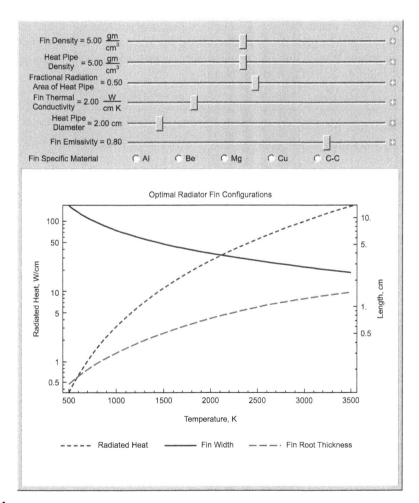

FIGURE 9.14

Space radiator characteristics.

In addition, using the optimal value for β from Fig. 9.13, optimal values for the fin width and thickness at its root may be determined from Eqs. (9.75) and (9.76), respectively, such that:

$$H = 0.92 \left(\frac{R\pi D^2 \rho_{hp}}{4\rho_{fin}}\right)^{\frac{1}{3}} \left(\frac{k}{2\varepsilon\sigma T_0^3}\right)^{\frac{1}{3}} \tag{9.81}$$

and

$$W = 1.087 \left(\frac{R\pi D^2 \rho_{hp}}{4\rho_{fin}}\right)^{\frac{2}{3}} \left(\frac{2\varepsilon\sigma T_0^3}{k}\right)^{\frac{1}{3}} \tag{9.82}$$

In Fig. 9.14, the performance and geometric characteristics of a space radiator at various temperatures are shown. Note that higher heat rejection temperatures result in significant increases in the amount of heat which may be rejected. Also observe that the fins on the radiator become shorter and fatter as the heat rejection temperature goes up, and under certain conditions the fins disappear entirely with all the heat rejection occurring from the heat pipe alone.

REFERENCES

[1] H. Schlichting, Boundary Layer Theory (Translated by J. Kestin), McGraw-Hill Book Company, Inc., New York, 1955.
[2] M.M. El-Wakil, Nuclear Heat Transport, American Nuclear Society, 1981, ISBN 978-0894480140.
[3] F. Kreith, M. Bohn, Principles of Heat Transfer, sixth ed., CL-Engineering, September 2000.
[4] F.W. Dittus, L.M.K. Boelter, Heat Transfer in Automobile Radiators of the Tubular Type, vol. 2 (13), University of California Publications of Engineering, USA, 1930, pp. 443–461.
[5] E.W. Sieder, G.E. Tate, Heat transfer and pressure drop of liquids in tubes, Ind. Eng. Chem. 28 (1936) 1429.
[6] V. Gnielinski, New equations for heat and mass transfer in turbulent pipe and channel flow, Int. Chem. Eng. 16 (2) (1976) 359–368.
[7] C.F. Colebrook, Turbulent flow in pipes, with particular reference to the transition region between smooth and rough pipe laws, J. Inst. Civ. Eng. (February 1939) 133–156.
[8] D.J. Wood, An explicit friction factor relationship, Civ. Eng. ASCE 60 (1966).
[9] L.F. Moody, Friction factors for pipe flow, Trans. ASME 66 (8) (1944) 671–684.
[10] W.M. Kayes, Convective Heat and Mass Transfer, fourth ed., McGraw-Hill, New York, NY, 2005, ISBN 0-07-246876-9.
[11] R.J. Naumann, Optimizing the design of space radiators, Int. J. Thermophys. 25 (6) (November 2004) 1929–1941.

CHAPTER 10

TURBOMACHINERY

Next to the nuclear reactor itself, probably the most important component in a nuclear rocket engine is the turbopump assembly. The job of the turbopump is to transfer the propellant from its storage tanks to the nuclear reactor at the desired flow rate and pressure. A turbopump consists of two major subassemblies which are, as the name implies, a turbine and a pump. The turbine and the pump are coupled to one another through a common shaft. During operation, the turbine portion of the turbopump supplies the power which runs the pump portion of the turbopump. The working fluid required to operate the turbine is the propellant itself which has been preheated and vaporized in various engine components that have themselves absorbed waste heat from the reactor. The exact means by which these components preheat the propellant depends upon the rocket engine cycle chosen for operation.

1. TURBOPUMP OVERVIEW

Besides the reactor, probably the most critical subsystem in a nuclear rocket is the turbopump assembly. The job of the turbopump is to deliver propellant to the nuclear reactor at pressures and flow rates which achieve the desired thrust and specific impulse values from the engine. The turbopump, as the name implies, consists of two major subassemblies, the propellant pump and the turbine. These two subassemblies are coupled to one another either on a common shaft or through a gearbox with the turbine supplying the power needed to run the propellant pump. A representation of a typical turbopump is illustrated in Fig. 10.1. The working fluid for the turbine is the propellant itself which has been raised to a high pressure by the propellant pump and heated by the reactor in some manner. The exact method by which the propellant is heated prior to being introduced into the turbine depends upon the details of the particular rocket engine cycle chosen (eg, hot bleed, cold bleed, and expander).

The ultimate objective in the design of a turbopump is to develop a robust, low-weight assembly which delivers the propellant to the reactor at the desired pressure and flow rate with optimal efficiency and at a reasonable cost. In this regard, the design process is rarely straightforward and of necessity entails a number of compromises between the competing requirements of material strength limits, vibrational effects, performance dictates, and so on involving a number of independent parameters. To more easily analyze the interrelationships between these parameters and the performance of

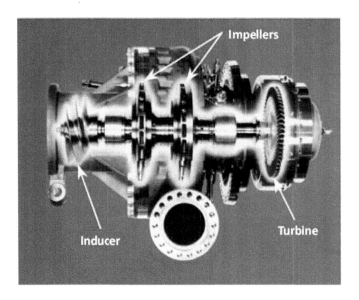

FIGURE 10.1

Typical turbopump layout.

turbopump, the use will again be made of various dimensionless groups. Using these dimensionless groups in conjunction with experiments and various theoretical treatments, empirical relationships can be formulated which relate the dimensionless parameters to various turbopump characteristics. The dimensionless groups most often used in these correlations include:

$$\text{Specific speed} = n_s = \frac{\omega\sqrt{Q}}{(gH)^{3/4}} \qquad \text{Specific diameter} = d_s = \frac{D(gH)^{1/4}}{\sqrt{Q}}$$
$$\text{Reynolds number} = \text{Re} = \frac{DV\rho}{\mu} \qquad \text{Mach number} = M = \frac{V}{c} \tag{10.1}$$

where ω = rotational speed of rotor; V = blade tip speed; D = rotor diameter; Q = total propellant volume flow rate; g = gravitational acceleration; c = sonic velocity; H = total pressure head across the pump or turbine; ρ = propellant density; and μ = propellant viscosity.

Typically, the values for the dimensionless parameters in Eq. (10.1) vary between the pump and turbine portions of the turbopump since the geometry and flow characteristics are generally quite different between the subassemblies. Also, note that in turbomachines, pressure is usually expressed in terms of *head*, which is designated by "H" in the parameter list in Eq. (10.1). Physically, head is given in length units equivalent to the static pressure which would be present at the bottom of a given height of the fluid being pumped as a result of its weight.

In the discussions which follow, only the steady-state operational characteristics of turbopumps are described. Transient operation, while very important to the safe operation of the turbopump, requires a detailed knowledge of the dynamic interactions within the turbopump. These interactions are complex and difficult to analyze and are beyond the scope of this discussion.

2. PUMP CHARACTERISTICS

The purpose of the pump portion of the turbopump is to increase the pressure of the liquid propellant coming from the storage tanks to the desired system pressure and then feed it at the required flow rate into the rest of the engine. There are a number of different pump configurations which can be used for this purpose and the type chosen for any particular turbopump application depends upon a number of factors which are discussed in the following paragraphs. Pump designs generally fall into two broad categories: *centrifugal flow pumps* and *axial flow pumps*.

In centrifugal flow pumps, fluid entering the pump is initially directed into the *impeller* which consists of a disk with curved vanes that rotate on a shaft. The fluid accelerates as it travels radially outward along the impeller vanes until it eventually exits the impeller at high velocity. The high-velocity fluid leaving the impeller is then introduced into the *volute* where it is collected and directed into an expansion section called a *diffuser* where the kinetic energy of the fluid is converted into potential energy which in the case of a pump manifests itself as an increase in pressure. The fluid is then directed either to another impeller assembly if further increases in pressure are required or into the pump discharge port.

In axial flow pumps, the volute section essentially disappears and the impeller section is designed more like a propeller. The increase in flow velocity in an axial flow pump comes as a result of the high-speed impeller rotation. After passing through the impeller, the fluid is introduced immediately into the diffuser section where, as in the case of centrifugal pumps, it may be directed either into another impeller assembly if further increases in head are required or into the pump discharge port.

An illustration of typical centrifugal flow and axial flow pump configurations is presented in Fig. 10.2.

There are other pump configurations in existence such as the *mixed flow pump* and the *Francis-type pump* which embody characteristics of both centrifugal flow and axial flow pumps. One of the most important factors affecting the choice of pump used for a particular application is the specific speed at which the pump is to operate. In general, *centrifugal pumps operate best at low specific speeds* and *axial flow pumps operate best at high specific speeds*. In between, the *mixed flow and Francis-type pumps operate at specific speeds in between those appropriate for use in either centrifugal or axial flow pumps*. The maximum rotational speed that can be tolerated by the impeller is determined primarily by the stress levels at the root of the impeller vanes. These stress levels increase not only with increased impeller rotational speed but also with increases in the length of the impeller vanes. Since centrifugal flow pumps have fairly long impeller vanes it is evident that they are more appropriate for use at lower impeller rotational speeds while axial flow pumps with their short impeller vanes are more appropriate for use at higher impeller rotational speeds. The maximum head which can be achieved by any one impeller stage is determined by the tip speed of the impeller which is proportional to the rotational speed of the impeller times the radius of the impeller.

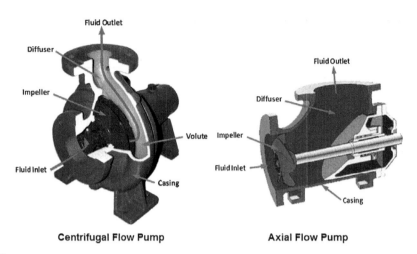

FIGURE 10.2

Pump configurations.

During engine operation, liquid propellant is transferred from its storage tanks into the propellant pump portion of the turbopump. As the propellant approaches the suction side of the pump, fluid dynamic considerations dictate that the propellant will experience a drop in pressure which is proportional to the increase in pressure head at the pump exit. In other words, if the pump speed is increased so as to achieve a greater exit pressure head, the suction pressure at the impeller will drop accordingly. As the pump speed continues to increase, the suction side pressure will continue to decrease until it eventually reaches a point where the pressure at the impeller inlet equals the vapor pressure of the liquid propellant. When this occurs, the propellant will begin to boil in the pump resulting in the formation of vapor bubbles on the impeller blades. This process is called *cavitation,* and the point at which it begins to occur in the pump is called the *point of incipient cavitation*. Pump operation in a regime where cavitation occurs is to be avoided since it results in a significant performance loss in the pump. In addition, as the fluid passes through the impeller to the high pressure side of the pump, the vapor bubbles caused by the cavitation will abruptly collapse, producing shock waves of sufficient force as to often cause damage to the impeller blades. Fig. 10.3 illustrates cavitation on a pump impeller and the damage which can occur to an impeller as a result.

In order to reduce the tendency of the pump to experience cavitation at high pump speeds, the suction side of the pump is generally pressurized. This pressurization can come from static pressure caused by the weight of the liquid propellant situated above the pump or from a dedicated pressurization system. In the NERVA program, pressurized tanks of helium as shown in Fig. 1.1 were used to provide the suction side pressure necessary to prevent pump cavitation. This pressurization is especially necessary for nuclear rocket turbopumps since if the engine is to be started in space, there is no static head resulting from the weight of the propellant until the vehicle starts

2. PUMP CHARACTERISTICS   153

Pump Cavitation   Impeller Damage due to Cavitation

FIGURE 10.3

Pump cavitation effects.

accelerating. As an alternative to externally pressurizing the turbopump inlet feed line, it may be possible to incorporate an *inducer* at the pump inlet ahead of the main impeller. An inducer is a specially designed pump stage designed to tolerate some cavitation. A convenient parameter which may be used to describe the point of incipient cavitation is called the *net-positive suction head* (NPSH) which is defined as:

$$H_{sa} = \text{NPSH}_a = \frac{P_a - P_v}{\rho}$$
$$H_{sr} = \text{NPSH}_r = \frac{P_s - P_v}{\rho} \tag{10.2}$$

where P_a = total inlet pressure available from the propellant feed system; P_v = vapor pressure of the propellant; P_s = minimum propellant feed system pressure required by the pump to prevent cavitation; H_{sa} = NPSH available to the pump from the propellant feed system; and H_{sr} = NPSH required by the pump to prevent cavitation.

Note that in order to avoid cavitation, therefore, the available net-positive suction head supplied to the pump must always exceed the net-positive suction head required by the pump to prevent cavitation. The total head produced by the pump may be given by:

$$H = \frac{P_e - P_a}{\rho} \tag{10.3}$$

where P_e = total exit pressure of the propellant as it leaves the pump.

It was mentioned earlier that fluid dynamic considerations dictate that pressure drop on the pump inlet is proportional to the total head produced by the pump. With this in mind, it is possible to define a parameter called the *Thoma cavitation parameter* such that:

$$\text{Thoma cavitation parameter} = \sigma = \frac{H_{sr}}{H} \tag{10.4}$$

Using the value for H_{sa} defined in Eq. (10.2), it is also possible to define a parameter called the *suction-specific speed* in which:

$$\text{suction-specific speed} = s_s = \frac{\omega\sqrt{Q}}{(gH_{sa})^{3/4}} \quad (10.5)$$

The Thoma parameter and the suction-specific speed have been experimentally correlated with pump-specific speed for a variety of pumps to give an indication as to the conditions likely induce the onset of cavitation [1]. The results of these correlations are illustrated in Fig. 10.4.

Since high pump efficiency is always desirable, it is important in the design of the pump that its geometric and operational characteristics be chosen carefully. From similarity conditions, it has been found that the efficiency of a pump may be completely described by four variables, those being specific speed, specific diameter, the Mach number, and the Reynolds number. Pump efficiency is usually defined as the ratio of the enthalpy change in the flowing propellant resulting from the pressurization process to the input shaft power to the pump rotor such that:

$$\eta_p = \frac{\rho Q H}{W} \quad (10.6)$$

where η_p = pump efficiency and W = input shaft power to pump rotor.

Generally speaking, the Mach number and the Reynolds number are relatively less important with regard to pump performance than are its specific speed and specific diameter. As a consequence, the pump efficiency for a variety of pump configurations can usually be presented adequately on a single contour plot as a function of specific speed and specific diameter. Such a plot is presented in Fig. 10.5 [2].

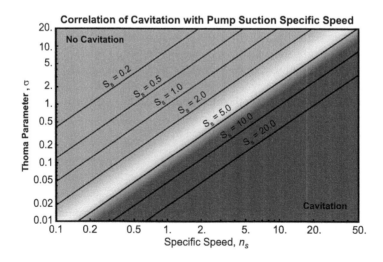

FIGURE 10.4

Cavitation susceptibility of pumps.

2. PUMP CHARACTERISTICS

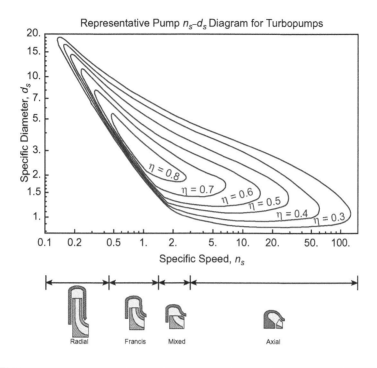

FIGURE 10.5
Typical pump performance characteristics.

The somewhat sophisticated analyses upon which this plot is based includes frictional and leakage effects and various other parasitic losses. It also assumes various geometric parameters and tolerances typical of most pumps.

EXAMPLE
A hydrogen turbopump for a nuclear rocket engine is to be designed in which the pump is to deliver saturated liquid hydrogen to the reactor at 20 K. The inlet pressure to the pump from the propellant feed system will be 0.45 MPa and the required pump outlet pressure will be 6.9 MPa. The liquid hydrogen is to be delivered to the reactor at a flow rate of 21.4 kg/s. Determine a pump impeller diameter and rotational speed which will maximize the pump's efficiency. Also compute the amount of shaft work which must be supplied to the pump by the turbine portion of the turbopump and recommend the most appropriate type of pump to be used in the turbopump assembly.

Solution
The first step in the pump design is to determine the NPSH required by the pump to suppress cavitation. Knowing that the hydrogen to be pumped is saturated at 20 K, the saturation pressure

can be found to be 0.091 MPa and the liquid density to be 71.3 kg/m³. The required NPSH from Eq. (10.2) is then:

$$H_{sr} = \frac{P_s - P_v}{g\rho} = \frac{450{,}000 \text{ Pa} - 91{,}000 \text{ Pa}}{9.8 \frac{m}{s^2} \times 71.3 \frac{kg}{m^3}} = 514 \text{ m} \qquad (1)$$

and the total head produced by the pump from Eq. (10.3) is:

$$H = \frac{P_e - P_a}{g\rho} = \frac{6900000 \text{ Pa} - 450000 \text{ Pa}}{9.8 \frac{m}{s^2} \times 71.3 \frac{kg}{m^3}} = 9230 \text{ m} \qquad (2)$$

The Thoma parameter can now be calculated from Eqs. (1) and (2) such that:

$$\sigma = \frac{H_{sr}}{H} = \frac{514 \text{ m}}{9230 \text{ m}} = 0.0557 \qquad (3)$$

Using Fig. 10.4 and the Thoma parameter calculated from Eq. (3), an upper value for the pump-specific speed may be determined to be about $n_s = 0.5$. Also, the pump must deliver hydrogen at a flow rate of 21.4 kg/s or an equivalent volume flow rate of:

$$Q = \frac{\dot{m}}{\rho} = \frac{21.4}{71.3} = 0.3 \frac{m^3}{s} \qquad (4)$$

From the definition of the pump-specific speed from Eqs. (10.1) and the volume flow rate of hydrogen from Eq. (4), the impeller shaft maximum rotational velocity, which can be tolerated without inducing cavitation, is then found to be:

$$\omega = \frac{n_s(gH)^{3/4}}{\sqrt{Q}} = \frac{0.5\left(9.8 \frac{m}{s^2} \times 9230 \text{ m}\right)^{3/4}}{\sqrt{0.3 \frac{m^3}{s}}} = 4760 \frac{\text{rad}}{\text{s}} = 45{,}500 \text{ rpm} \qquad (5)$$

The pump-specific speed can also be used in conjunction with Fig. 10.5 to determine a pump-specific diameter which will maximize the pump efficiency. An examination of the figure reveals that a pump efficiency of about $n_s = 80\%$ is achievable in a Francis-type pump having a specific diameter of $d_s = 5.2$. Using the definition of specific diameter from Eq. (10.1), it is possible to determine that the pump should have an impeller diameter of:

$$D = \frac{d_s \sqrt{Q}}{(gH)^{1/4}} = \frac{5.2\sqrt{0.3 \frac{m^3}{s}}}{\left(9.8 \frac{m}{s^2} \times 9230 \text{ m}\right)^{1/4}} = 0.164 \text{ m} = 16.4 \text{ cm} \qquad (6)$$

The power required to drive the pump may now be determined by using the definition of efficiency from Eq. (10.6) such that:

$$W = \frac{Q(P_e - P_a)}{\eta_s} = \frac{0.3 \frac{m^3}{s}(6.9 \text{ MPa} - 0.45 \text{ MPa})}{0.8} = 2.42 \text{ MW} \qquad (7)$$

3. TURBINE CHARACTERISTICS

The primary purpose of the turbine portion of the turbopump is to supply the power required to drive the propellant pump. Turbines supply this power by extracting energy from all or part of the propellant flow stream which has previously been vaporized and heated inside special flow circuits within the nuclear reactor. The thermal energy from the propellant gas is converted into kinetic energy in the turbine assembly by passing it through a set of rotating turbine blades. *Most turbopump turbines are of the axial flow type* in which the inlet and outlet propellant flows are parallel to the axis of rotation. Occasionally, especially *if high shaft speeds are involved, radial flow turbines may be used* wherein the inlet flow is introduced into the turbine along the axis of rotation and the outlet flow leaves the turbine through a circumferential outlet. These two types of turbines are illustrated in Fig. 10.6.

In axial flow turbines, the design of the turbine blades varies depending upon the method used to extract the energy from the propellant. Generally, the turbine blades are configured in one of two ways to yield what are called either impulse turbines or reaction turbines.

In impulse turbines, the enthalpy of the propellant is converted into kinetic energy by means of fixed converging/diverging nozzle assemblies which direct the propellant flow onto a set of turbine blades attached to the main turbopump shaft. The turbine blades spin as a consequence of momentum transfer acting to change the magnitude and direction of the propellant flow as it impinges on the turbine blades. The spinning turbine blades drive the main turbopump shaft assembly which is either directly connected to the propellant pump or is connected to the propellant pump through a gearbox. *In impulse turbines, the majority of the pressure drop occurs in the fixed nozzle assemblies, not in the spinning turbine blades.*

In reaction turbines, the enthalpy of the propellant is converted into kinetic energy in spinning converging/diverging nozzle assemblies which are attached to the main turbopump shaft. In this configuration, the turbine shaft spins in reaction to the high-speed discharge of the propellant from the turbine blade nozzle assemblies. As was the case for the impulse turbine, the main shaft assembly is

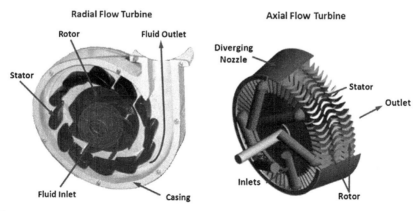

FIGURE 10.6

Turbine design configurations.

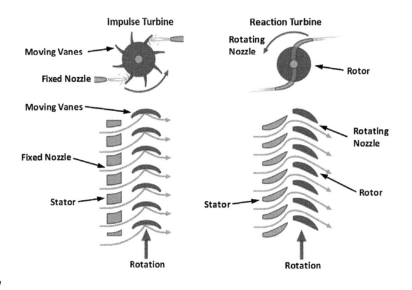

FIGURE 10.7

Turbine blade configurations.

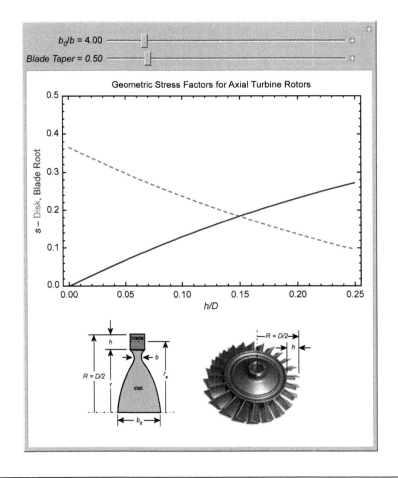

FIGURE 10.8

Axial turbine stress factors.

either directly connected to the propellant pump or is connected to the propellant pump through a gearbox. *In reaction turbines, the majority of the pressure drop occurs in the spinning nozzle assemblies which constitute the turbine blades.* Fig. 10.7 illustrates the difference between impulse and reaction turbines.

In the turbine portion of the turbopump, the propellant is normally in a gaseous form; therefore, cavitation is not an issue as it was in the pump portion of the turbopump. However, because the density of the gaseous propellant is low as compared to propellant in its liquid state, and because it is also desirable to have the turbine operate at the highest efficiencies practical, the volume flow rate of the propellant through the turbine will normally be higher than that in the pump assembly especially if full propellant flow through the turbine is required. High propellant volume flow rates will usually require the turbine operate at high rotor speeds resulting in high rotor assembly stress levels due to centrifugal forces. Because of these high stress levels, under certain circumstances, it may be found necessary to operate the turbine at rotor speeds less than optimal in order to reduce the rotor disk and blade root stresses to acceptable values. These stress values are a function of the rotational speed of the rotor, the density of the rotor, and the geometry of the rotor and may be represented by an equation of the form:

$$\sigma = \xi\, S\, \rho_r\, \omega^2\, R^2 \tag{10.7}$$

where σ = rotor stress; ρ_r = rotor material density; S = rotor geometric factor; ξ = safety factor.

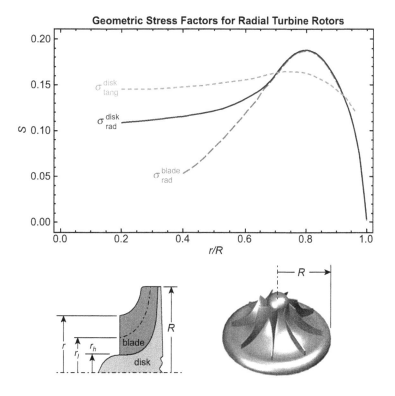

FIGURE 10.9

Radial turbine stress factors.

Typical rotor geometric stress factors may be found in Fig. 10.8 for axial turbines [3] and in Fig. 10.9 for radial turbines [4].

Note that in Fig. 10.8 when h/D ratios are low, large b_0/b ratios are needed to reduce disk stresses while at high h/D ratios, low turbine blade tapers are required to reduce turbine blade root stresses.

In Fig. 10.9 observe that in radial turbines, tangential stresses exceed radial stresses except between r/R ratios of between about 0.7 and 0.9. Between these values radial stresses in both the turbine blades and the disk dominate, reaching a maximum S value of 0.19 at an r/R ratio of about 0.8.

As was the case for pump efficiencies, the turbine efficiency can generally be presented on a single contour plot as a function of specific speed and specific diameter such as is presented in Fig. 10.10 [5];

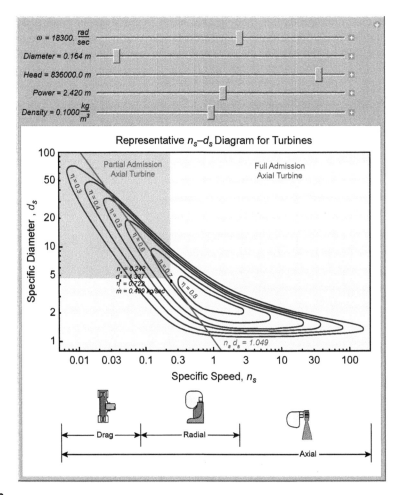

FIGURE 10.10

Typical turbine performance characteristics.

however, since turbopump turbines operate in flow regimes where compressibility effects are important, a selection must be made as to the reference conditions where the specific speed and specific diameter are calculated. Usually the conditions at the turbine exit are better known; therefore, this location is used as the basis for calculating the turbine reference conditions. Since the pressure head defined earlier for incompressible fluids is simply a measure of the fluid energy per unit mass, an equivalent expression for head in compressible fluids may be given by:

$$H = \frac{h_{in} - h_{out}}{g} = c_p \frac{T_{in} - T_{out}}{g} \tag{10.8}$$

where h_{in} = fluid enthalpy entering the turbine and h_{out} = fluid enthalpy leaving the turbine.

Once the turbine efficiency of the turbine has been determined, the shaft power available to run the pump may be calculated. The shaft power produced by the turbine, which is proportional to the efficiency of the turbine, the propellant flow rate through the turbine, and the enthalpy change of the propellant between the inlet and outlet ports of the turbine, may be represented by an equation of the form:

$$W = \eta_t(n_s, d_s)\dot{m}(h_{in} - h_{out}) = \eta_t(n_s, d_s)Q\rho_f(h_{in} - h_{out}) \tag{10.9}$$

where η_t = turbine efficiency and ρ_f = fluid density leaving the turbine.

EXAMPLE

The pump described in the previous example is to be used in a nuclear rocket engine employing a hot bleed cycle. The hot hydrogen bleed from the reactor is available to the drive turbine for the pump at a temperature of 800 K and a pressure of 6.9 MPa. Exhaust from the turbine is routed to the engine nozzle where the pressure is 0.1 MPa. The turbine rotor assembly is to be fabricated out of a material having a yield stress of 1000 MPa and a density of 2000 kg/m³ at design conditions with a disk thickness taper of 4 and a blade area taper of 0.5.

Assuming a safety factor of 1.2, estimate the design parameters for a turbine assembly which will be used to drive the hydrogen pump outlined in the previous example. Determine the optimal turbine rotor h/D ratio, the required hydrogen bleed flow rate, the required gear ratio between the turbine and the pump, and the turbine efficiency.

Solution

The first step in the calculation is to determine the turbine adiabatic head which may be calculated from the turbine inlet and outlet enthalpy values. Since the inlet temperature and pressure are known along with the outlet pressure, the inlet and outlet enthalpies may be determined to be 11,285 kJ/kg and 3094 kJ/kg, respectively, with an outlet fluid density of 0.1 kg/m³. From these enthalpy values, Eq. (10.8) may be used to calculate the turbine adiabatic head yielding:

$$H = \frac{h_{in} - h_{out}}{g} = \frac{11,285,000 \frac{J}{kg} - 3,094,000 \frac{J}{kg}}{9.8 \frac{m}{s^2}} = 836,000 \text{ m} \tag{1}$$

The next step in the calculation involves determining the maximum permissible angular velocity of the rotor consistent with the stress limitations of the material. From Fig. 10.8, it can be seen that for a rotor thickness taper of 4 and a blade taper of 0.5, the minimum geometric stress factor may be determined to be $S = 0.185$ with an $h/D = 0.15$. Rearranging Eq. (10.7) to solve for the angular velocity then yields:

$$\omega = \sqrt{\frac{\sigma}{S\rho\xi r^2}} = \sqrt{\frac{1{,}000{,}000{,}000 \text{ Pa}}{0.185 \times 2000\frac{\text{kg}}{\text{m}^3} \times 1.2 \times \left(\frac{0.164}{2}\right)^2 \text{m}^2}} = 18{,}300\frac{\text{rad}}{\text{s}} = 115{,}000 \text{ rpm} \quad (2)$$

Using the pump angular velocity from Eq. (4) in the pump example described earlier and the turbine angular velocity from Eq. (2), the gear ratio of the turbine to the pump may be found to be:

$$\text{Gear ratio} = \frac{\omega_{\text{turbine}}}{\omega_{\text{pump}}} = \frac{18{,}300}{4760} = 3.84 \quad (3)$$

The specific speed and specific diameter may also be determined by using the results from Eqs. (1) and (2), and Eq. (6) from the pump example such that:

$$n_s = \frac{\omega\sqrt{Q}}{(gH)^{3/4}} = \frac{18{,}300\frac{\text{rad}}{\text{s}}\sqrt{Q\frac{\text{m}^3}{\text{s}}}}{\left(9.8\frac{\text{m}}{\text{s}^2} \times 836{,}000 \text{ m}\right)^{3/4}} = 0.1195\sqrt{Q} \quad (4)$$

and

$$d_s = \frac{D(gH)^{1/4}}{\sqrt{Q}} = \frac{0.164 \text{ m}\left(9.8\frac{\text{m}}{\text{s}^2} \times 836{,}000 \text{ m}\right)^{1/4}}{\sqrt{Q\frac{\text{m}^3}{\text{s}}}} = \frac{8.774}{\sqrt{Q}} \quad (5)$$

The relationship determining the shaft power available from the turbine from Eq. (10.9) may now be used along with the results from Eqs. (4) and (5) to develop an expression which is a function solely of the propellant volumetric flow rate yielding:

$$W = \eta_t[n_s(Q), d_s(Q)]Q\rho(h_{\text{in}} - h_{\text{out}}) = 2{,}420{,}000 \text{ W}$$

$$= \eta_t\left[0.1195\sqrt{Q}, \frac{8.774}{\sqrt{Q}}\right]Q\frac{\text{m}^3}{\text{s}} \times .1015\frac{\text{kg}}{\text{m}^3}\left(11{,}285{,}000\frac{\text{J}}{\text{kg}} - 3{,}094{,}000\frac{\text{J}}{\text{kg}}\right) \quad (6)$$

$$\Rightarrow \eta_t\left[0.1195\sqrt{Q}, \frac{8.774}{\sqrt{Q}}\right]Q = 2.9108$$

Eq. (6) for the turbine power may be iteratively solved to yield a value for the turbine propellant volumetric flow rate of $Q = 5.88 \frac{\text{m}^3}{\text{s}}$. Knowing the turbine mass flow rate, it is possible to calculate values for the turbine-specific speed and diameter, and the turbine efficiency. These calculations may be performed iteratively although they can also be performed automatically through the calculator in Fig. 10.10. Using the volumetric flow rate, the mass flow rate may be calculated as:

$$\dot{m} = \rho_f Q = 0.1\frac{\text{kg}}{\text{m}^3} \times 4.09\frac{\text{m}^3}{\text{s}} = 0.409\frac{\text{kg}}{\text{s}} \quad (7)$$

Using the volumetric flow rate just determined, the specific speed and specific diameter may be calculated from Eqs. (4) and (5) such that:

$$n_s = 0.1195\sqrt{Q} = 0.1195\sqrt{4.09} = 0.242 \quad (8)$$

and

$$d_s = \frac{8.774}{\sqrt{Q}} = \frac{8.774}{\sqrt{4.09}} = 4.337 \tag{9}$$

Using Fig. 8.10 and the specific speed and specific diameter determined from Eqs. (8) and (9), the turbine efficiency may be found to be 72%.

REFERENCES

[1] I.J. Karassik, J.P. Messina, P. Cooper, C.C. Heald, Pump Handbook, fourth ed., McGraw-Hill, 2008.
[2] O.E. Balje, A study on design criteria and matching of turbomachines — Part B, Journal of Engineering for Power, ASME Transactions 84 (1) (January 1962) 83.
[3] O.E. Balje, Turbomachines — A Guide to Design, Selection, and Theory, John Wiley & Sons, Inc, 1981, ISBN 0-471-06036-4, p. 61.
[4] M.J. Schilhansl, Stress analysis of radial flow rotor, Journal of Engineering for Power, ASME Transactions 84 (1) (January 1962) 124.
[5] O.E. Balje, A study on design criteria and matching of turbomachines — Part A, Journal of Engineering for Power, ASME Transactions 84 (1) (January 1962) 83.

CHAPTER 11

NUCLEAR REACTOR KINETICS

Reactor kinetics is the study of those processes which control the time-dependent behavior of a nuclear reactor. Because during fission, the length of time between one neutron generation and the next is extremely short, it might be expected that controlling a nuclear reactor would be quite difficult. Fortunately, however, neutrons are produced not only through the fission process, but also through the radioactive decay of certain nuclides. The neutrons produced through radioactive decay appear over time scales ranging from seconds to several minutes. With proper design, these radioactive decay neutrons can be made to govern the rate of power change in the reactor thus making the control of the reactor much simpler due to the much slower reactor response times. Because these radioactive decay neutrons also persist for a time after the control system brings the reactor subcritical, a considerable amount of reactor power may be produced even after the reactor is shut down. This power must be disposed of in some way and may require wasting significant amounts of fuel to keep the reactor cool until these decay neutrons die away.

1. DERIVATION OF THE POINT KINETICS EQUATIONS

During steady-state nuclear reactor operation the rate of neutron production and the rate of neutron loss are in an exact balance. If the neutron production and loss rates are not in balance, that is if the reactor k_{eff} is not equal to one, the reactor power level will not remain constant, but rather will rise or decay exponentially on a period which is related to the neutron generation rate. Assuming that the length of time between one neutron generation and the next is 0.00005 s and the neutron production rate was to exceed the neutron loss rate by only 0.1%, one would find that the reactor power level would increase by a factor of almost 500,000,000 in only 1 s! Clearly, if the neutron generation rate were to be the only factor controlling the reactor period, one would be presented with a very serious reactor control situation indeed. Fortunately, as it turns out, there are other factors in play which make the reactor control situation much less severe.

As it turns out, the production of neutrons results not only directly from fission (called *prompt neutrons*), but also indirectly from certain fission product nuclides which emit neutrons through radioactive decay. These decay neutrons (called *delayed neutrons*), which are emitted from several seconds to a few minutes after the fission products are produced, are crucial in permitting the practical control of a nuclear reactor. If the reactor is subcritical on prompt neutrons alone, but supercritical when the delayed neutrons are included in the total neutron production rate, it turns out that the delayed neutrons will control the rate of change of the reactor transient. Thus, the delayed neutron will enable the reactor period to be in the order of minutes rather than microseconds—a rate of change which is easily controllable by rather conventional control systems.

To determine just how these delayed neutrons control the rate at which a reactor transient proceeds, we begin by writing the time-dependent form of the one energy group neutron balance equation including a term representing the addition of the delayed neutrons:

$$\frac{dn}{dt} = \underbrace{D\nabla^2\phi}_{\text{Neutron Leakage}} - \underbrace{\Sigma_a\phi}_{\text{Neutron Absorption}} + \underbrace{(1-\beta)\nu\Sigma_f\phi}_{\text{Prompt Fission Neutrons}} + \underbrace{\sum_{i=1}^{6}\lambda_i C_i}_{\text{Delayed Fission Product Neutrons}} \tag{11.1}$$

where β = total yield fraction of fission product nuclides which decay through neutron emission; λ_i = decay constant of neutron-emitting pseudo-nuclide "i"; and C_i = concentration of neutron-emitting pseudo-nuclide "i."

In Eq. (11.1), the individual neutron-emitting precursor nuclides are generally not explicitly specified, but rather are grouped together such that each group, called a pseudo-nuclide, consists of those actual nuclides having similar decay constants. It has been found that six groups of these neutron-emitting pseudo-precursor nuclides are quite adequate to accurately describe the time-dependent behavior of virtually all reactor transients.

Recall that:

$$\phi = nV \quad \text{and} \quad D\nabla^2\phi = -DB^2\phi \tag{11.2}$$

Substituting Eq. (11.2) into Eq. (11.1) and rearranging terms yield:

$$\frac{dn}{dt} = V\nu\Sigma_f\left[-\frac{DB^2 + \Sigma_a}{\nu\Sigma_f} + (1-\beta)\right]n + \sum_{i=1}^{6}\lambda_i C_i = \underbrace{V\nu\Sigma_f}_{1/\Lambda}\left[\underbrace{1 - \frac{1}{k_{\text{eff}}}}_{\rho} - \beta\right]n + \sum_{i=1}^{6}\lambda_i C_i$$

$$= \frac{\rho - \beta}{\Lambda}n + \sum_{i=1}^{6}\lambda_i C_i \tag{11.3}$$

where Λ = prompt neutron lifetime and ρ = reactivity.

The prompt neutron lifetime "Λ" may be understood by recalling that $1/\Sigma_x$ represents the average distance neutrons travel between interactions of type "x." In the present case, $1/(\nu\Sigma_f)$ represents the average distance over which neutrons travel from where they are born in a nucleus undergoing fission to where they are reabsorbed in another nucleus which subsequently also undergoes fission. By dividing this neutron interaction distance by the average neutron velocity "V," one obtains for the *prompt neutron lifetime, a value which represents the time between when a neutron is born in a fission event and when it is reabsorbed in another nucleus subsequently resulting in another fission event.*

The reactivity "ρ" may be thought of as the ratio of the instantaneous net fission neutron production rate (eg, production − loss) to the rate of neutron production (eg, production alone). As such, the reactivity is negative when the reactor is subcritical, zero when the reactor is exactly critical, and positive when the reactor is supercritical.

Note that Eq. (11.3) cannot be solved without the inclusion of other equations which specify the time-dependent behavior of the delayed neutron precursor concentrations. The equations which yield

1. DERIVATION OF THE POINT KINETICS EQUATIONS

the delayed neutron precursor concentrations are determined by writing rate equations which specify the net rate at which the precursor pseudo-nuclides are produced through fission interactions minus the rate at which they are lost through radioactive decay by neutron emission. These equations are

$$\frac{dC_i}{dt} = \beta_i \nu \Sigma_f \phi - \lambda_i C_i = \beta_i \underbrace{V\nu\Sigma_f}_{1/\Lambda} n - \lambda_i C_i = \frac{\beta_i}{\Lambda} n - \lambda_i C_i \quad \text{where:} \quad \beta = \sum_{i=1}^{6} \beta_i \quad (11.4)$$

Eqs. (11.3) and (11.4) are called the *point kinetics equations* because they contain no space dependencies. An essential assumption in their solution is that the neutron flux shape remains constant during power transients. To solve the point kinetics equations, it will prove helpful to assume solutions of the form:

$$n(t) = n_0 e^{\omega t} \Rightarrow \frac{dn}{dt} = \omega n_0 e^{\omega t} \quad (11.5)$$

$$C_i(t) = C_i^0 e^{\omega t} \Rightarrow \frac{dC_i}{dt} = \omega C_i^0 e^{\omega t} \quad (11.6)$$

Substituting Eqs. (11.5) and (11.6) into Eq. (11.4) then yields:

$$\omega C_i^0 e^{\omega t} = \frac{\beta_i}{\Lambda} n_0 e^{\omega t} - \lambda_i C_i^0 e^{\omega t} \Rightarrow C_i^0 = \frac{\beta_i}{\Lambda(\omega + \lambda_i)} n_0 \quad (11.7)$$

where ω = reactor time constant.

Incorporating Eqs. (11.5)–(11.7) into Eq. (11.3), one finds that:

$$\omega n_0 e^{\omega t} = \frac{\rho - \beta}{\Lambda} n_0 e^{\omega t} + \sum_{i=1}^{6} \lambda_i \frac{\beta_i}{\Lambda(\omega + \lambda_i)} n_0 e^{\omega t} \Rightarrow \omega \Lambda = \rho - \sum_{i=1}^{6} \beta_i + \sum_{i=1}^{6} \lambda_i \frac{\beta_i}{(\omega + \lambda_i)} \quad (11.8)$$

Rearranging Eq. (11.8) to solve for the reactivity then yields what is called the *Inhour equation*:

$$\rho = \omega \Lambda + \sum_{i=1}^{6} \frac{\omega \beta_i}{\omega + \lambda_i} \quad (11.9)$$

The Inhour equation got its name because in the early days of nuclear technology the value for ω was quoted in "inverse hours." It turns out that the equation is quite useful in that it provides a simple means by which the material properties of the nuclear reactor may be related to the reactor period. Mathematically, the Inhour equation is an algebraic expression in which the neutron density as a function of time is expressed as a linear combination of seven terms in ω such that:

$$n(t) = A_0 e^{\omega_0 t} + A_1 e^{\omega_1 t} + \cdots + A_6 e^{\omega_6 t} \quad (11.10)$$

In Table 11.1, the delayed neutron data [1] for ^{235}U are presented. The data are moderately neutron energy spectrum dependent with the data in the table representing what would be expected from a thermal neutron energy spectrum.

Using the delayed neutron data for ^{235}U, the Inhour equation may be solved for various values of reactivity. Note that for all cases, six of the time constants are always negative implying that the effects of these transient terms will die out over a period of time. Depending upon the time constant in question these time periods can last from a few milliseconds to a few tens of seconds. The seventh time

CHAPTER 11 NUCLEAR REACTOR KINETICS

Table 11.1 Delayed Neutron Data for ^{235}U

Group	Decay Constant, λ (s^{-1})	Fission Yield
1	0.0124	0.00022
2	0.0305	0.00142
3	0.111	0.00127
4	0.301	0.00257
5	1.14	0.00075
6	3.01	0.00027
Composite	0.0764	0.0065

constant (ω_0), on the other hand, can be positive or negative depending upon whether the reactor is supercritical (eg, $\rho > 0 \Rightarrow \omega_0 > 0$) or subcritical (eg, $\rho < 0 \Rightarrow \omega_0 < 0$). If $\rho=0$, the reactor is exactly critical and $\omega_0 = 0$ indicating that after the effects of the six other transient terms die out, the reactor power will settle to a constant value. The effects of various step reactivity changes on the seven time constants can be seen in the solution of the Inhour equation which is illustrated in Fig. 11.1. Note that the calculated time constants can differ by several orders of magnitude necessitating the highly nonlinear abscissa used in the plot.

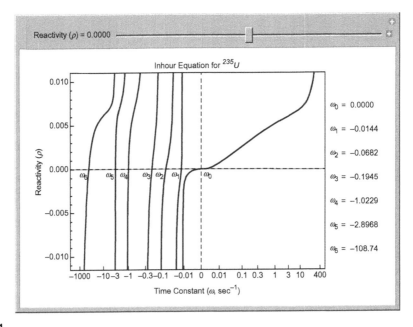

FIGURE 11.1

Inhour equation.

2. SOLUTION OF THE POINT KINETICS EQUATIONS

To gain further insight into the kinetic behavior of a reactor, the point kinetics equations are solved using only one group of delayed neutrons representing a composite value [1] of the six delayed neutrons groups described earlier (which themselves represent a composite of dozens of neutron-emitting nuclides). Using one delayed group of neutrons, the point kinetics equations, Eqs. (11.3) and (11.4) become:

$$\frac{dn}{dt} = \frac{\rho - \beta}{\Lambda} n + \lambda C \tag{11.11}$$

and

$$\frac{dC}{dt} = \frac{\beta}{\Lambda} n - \lambda C \tag{11.12}$$

where $\frac{1}{\lambda} = \frac{1}{\beta} \sum_{i=0}^{6} \frac{\beta_i}{\lambda_i}$.

Making the assumptions that reactivity is introduced in a stepwise fashion (eg, ρ is not a function of time) and that the transient response of the neutron density and the delayed neutron precursors is such that:

$$n(t) = n_0 e^{\omega t} \Rightarrow \frac{dn}{dt} = \omega n_0 e^{\omega t} = \omega n \tag{11.13}$$

and

$$C(t) = C_0 e^{\omega t} \Rightarrow \frac{dC}{dt} = \omega C_0 e^{\omega t} = \omega C \tag{11.14}$$

If these assumptions are now applied to the point kinetics equations represented by Eqs. (11.11) and (11.12), it is found that:

$$\omega n = \frac{\rho - \beta}{\Lambda} n + \lambda C \tag{11.15}$$

and

$$\omega C = \frac{\beta}{\Lambda} n - \lambda C \tag{11.16}$$

Casting Eqs. (11.15) and (11.16) in matrix form to solve for the neutron density and the delayed neutron precursor density will then yield:

$$\begin{bmatrix} \left(\omega - \frac{\rho - \beta}{\Lambda}\right) & -\lambda \\ -\frac{\beta}{\Lambda} & (\omega + \lambda) \end{bmatrix} \begin{bmatrix} n \\ C \end{bmatrix} = 0 \tag{11.17}$$

CHAPTER 11 NUCLEAR REACTOR KINETICS

In order for the neutron density and the neutron precursor density to be nonzero, the determinant of the matrix in Eq. (11.17) must equal zero, therefore,

$$\left(\omega - \frac{\rho - \beta}{\Lambda}\right)(\omega + \lambda) - \frac{\beta\lambda}{\Lambda} = \omega^2 - \left(\frac{\rho - \beta}{\Lambda} - \lambda\right)\omega - \frac{\rho\lambda}{\Lambda} = 0 \qquad (11.18)$$

Eq. (11.18) is a quadratic expression in ω which when solved yields two values for the time constant. These two values are

$$\omega_1 = \frac{1}{2}\left(\frac{\rho - \beta}{\Lambda} - \lambda\right) + \frac{1}{2}\sqrt{\left(\frac{\rho - \beta}{\Lambda} - \lambda\right)^2 + 4\frac{\rho\lambda}{\Lambda}} \qquad (11.19)$$

and

$$\omega_2 = \frac{1}{2}\left(\frac{\rho - \beta}{\Lambda} - \lambda\right) - \frac{1}{2}\sqrt{\left(\frac{\rho - \beta}{\Lambda} - \lambda\right)^2 + 4\frac{\rho\lambda}{\Lambda}} \qquad (11.20)$$

Note that for positive values of ρ, ω_1 will always be positive and ω_2 will always be negative. Also note that for negative values of ρ, both ω_1 and ω_2 will always be negative. By rearranging Eq. (11.16), it may be observed that the neutron precursor density "C" is related to the neutron density "n" such that:

$$C = \frac{\beta}{\Lambda(\omega + \lambda)} n \qquad (11.21)$$

Substituting the time constant values determined from Eqs. (11.19) and (11.20) into the relationships assumed for the neutron density and the neutron precursor density as expressed in Eqs. (11.13) and (11.21), it is found that:

$$n = A_1 e^{\omega_1 t} + A_2 e^{\omega_2 t} \text{ and } C = A_1 \frac{\beta}{\Lambda(\omega_1 + \lambda)} e^{\omega_1 t} + A_2 \frac{\beta}{\Lambda(\omega_2 + \lambda)} e^{\omega_2 t} \qquad (11.22)$$

Except for values of ρ near β, one of the time constants will generally be numerically large and the other time constant will generally be numerically small. For the large time constant, $\omega_1 \gg \frac{\rho\lambda}{\Lambda}$ and $\frac{\rho-\beta}{\Lambda} \gg \lambda$; therefore, one finds that Eq. (11.19) reduces to

$$\omega_1 \approx \frac{1}{2}\left(\frac{\rho - \beta}{\Lambda} - \lambda\right) + \frac{1}{2}\sqrt{\left(\frac{\rho - \beta}{\Lambda} - \lambda\right)^2} \approx \frac{\rho - \beta}{\Lambda} \qquad (11.23)$$

For the small time constant, $\omega_2 \ll \frac{\rho\lambda}{\Lambda}$ and $\omega_2 \ll \lambda$; therefore, it is found from Eq. (11.20) that:

$$\omega_2 = \frac{1}{2}\left(\frac{\rho - \beta}{\Lambda} - \lambda\right) - \frac{1}{2}\sqrt{\left(\frac{\rho - \beta}{\Lambda} - \lambda\right)^2 + 4\frac{\rho\lambda}{\Lambda}} \Rightarrow \left[\omega_2 - \frac{1}{2}\left(\frac{\rho - \beta}{\Lambda} - \lambda\right)\right]^2$$

$$= \frac{1}{4}\left(\frac{\rho - \beta}{\Lambda} - \lambda\right)^2 + \frac{\rho\lambda}{\Lambda} \qquad (11.24)$$

2. SOLUTION OF THE POINT KINETICS EQUATIONS

Expanding Eq. (11.24) and neglecting ω_2^2 and λ then yields:

$$\omega_2^2 - \omega_2\left(\frac{\rho-\beta}{\Lambda} - \lambda\right) + \frac{1}{4}\left(\frac{\rho-\beta}{\Lambda} - \lambda\right)^2 = \frac{1}{4}\left(\frac{\rho-\beta}{\Lambda} - \lambda\right)^2 + \frac{\rho\lambda}{\Lambda}$$

$$\Rightarrow -\omega_2\left(\frac{\rho-\beta}{\Lambda}\right) = \frac{\rho\lambda}{\Lambda} \Rightarrow \omega_2 = \frac{\rho\lambda}{\beta-\rho} \quad (11.25)$$

Substituting the values for ω_1 and ω_2 from Eqs. (11.23) and (11.25) into Eq. (11.22) then yields for the time-dependent neutron density and neutron precursor density:

$$n(t) = A_1 e^{\frac{\rho-\beta}{\Lambda}t} + A_2 e^{\frac{-\rho\lambda}{\rho-\beta}t} \quad (11.26)$$

and

$$C(t) = A_1 \frac{\beta}{\rho-\beta} e^{\frac{\rho-\beta}{\Lambda}t} - A_2 \frac{\rho-\beta}{\Lambda\lambda} e^{\frac{-\rho\lambda}{\rho-\beta}t} \quad (11.27)$$

For $\rho = 0$, the steady-state conditions (eg, $t = \infty$) from Eqs. (11.26) and (11.27) yield a relationship between the neutron density and neutron precursor density at the start of the reactor transient (eg, before the step reactivity change); therefore

$$n(\infty) = A_2 \text{ and } C(\infty) = A_2 \frac{\beta}{\lambda\Lambda} \Rightarrow C_0 = n_0 \frac{\beta}{\lambda\Lambda} \quad (11.28)$$

At the start of the reactor transient (eg, $t = 0$), using the relationship from Eq. (11.28), Eqs. (11.26) and (11.27) yield:

$$n_0 = A_1 + A_2 \quad (11.29)$$

and

$$C_0 = n_0 \frac{\beta}{\Lambda\lambda} = A_1 \frac{\beta}{\rho-\beta} - A_2 \frac{\rho-\beta}{\Lambda\lambda} \quad (11.30)$$

Solving Eqs. (11.29) and (11.30) simultaneously for the unknown constants A_1 and A_2 reveals that:

$$A_1 = \frac{\rho}{(\rho-\beta) + \frac{\Lambda\lambda\beta}{\rho-\beta}} n_0 \approx \frac{\rho}{\rho-\beta} n_0 \quad (11.31)$$

and

$$A_2 = \frac{-\beta + \frac{\Lambda\lambda\beta}{\rho-\beta}}{(\rho-\beta) + \frac{\Lambda\lambda\beta}{\rho-\beta}} n_0 \approx \frac{-\beta}{\rho-\beta} n_0 \quad (11.32)$$

where $\frac{\Lambda\lambda\beta}{\rho-\beta} \ll \rho - \beta$.

Substituting Eqs. (11.31) and (11.32) into the expressions for the neutron density and neutron precursor density as given by Eqs. (11.26) and (11.27) then yields:

$$n(t) \approx n_0 \left[\frac{\rho}{\rho - \beta} e^{\frac{\rho-\beta}{\Lambda}t} - \frac{\beta}{\rho - \beta} e^{\frac{-\rho\lambda}{\rho-\beta}t} \right] \tag{11.33}$$

and

$$C(t) \approx n_0 \left[\frac{\rho\beta}{(\rho - \beta)^2} e^{\frac{\rho-\beta}{\Lambda}t} + \frac{\beta}{\Lambda\lambda} e^{\frac{-\rho\lambda}{\rho-\beta}t} \right] \tag{11.34}$$

Note that the assumptions used to justify the simplifying approximations to ω (eg, ρ is not "close" to β) imply that Eqs. (11.33) and (11.34) will become increasingly inaccurate for reactivity insertions in the near vicinity β. However, since reactivity insertions of this magnitude are seldom encountered in practice, Eqs. (11.33) and (11.34) will, generally speaking, yield reasonably accurate values for the time-dependent point kinetics expressions for the neutron density and neutron precursor density. In Fig. 11.2, the time-dependent behavior of these quantities resulting from various step changes in reactivity is illustrated based upon the ^{235}U composite values for the one group delayed neutron data.

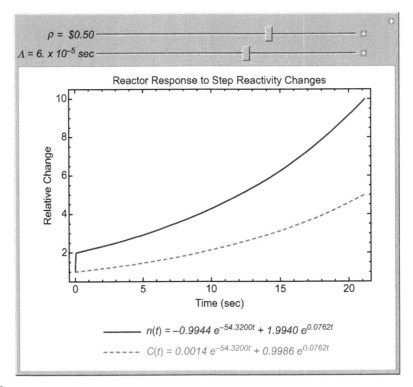

FIGURE 11.2

Reactor response to step reactivity changes.

Note that reactivity is typically measured in dollars ($) where $1.00 represents a change of reactivity equal to β.

Also note in Fig. 11.2 that as the reactivity approaches and then exceeds $1.00, the time required for the neutron density (or neutron flux or power) to reach many times its steady-state value becomes quite short. What is occurring is that as the reactivity insertion exceeds $1.00 (eg, $\rho > \beta$), the reactor goes into what is called a *super prompt critical* state in which the reactor is critical on prompt neutrons alone. This super prompt critical condition is quite dangerous and is to be avoided under normal circumstances since it is extremely difficult to control a reactor in which the power levels are changing so rapidly. As a consequence, reactors are typically limited to reactivity insertions of just few cents (¢). For example, a reactivity insertion of only 6¢ is sufficient to double the reactor power in about 2 min.

Fig. 11.2 also illustrates how for small positive reactivity insertions, the neutron density jumps suddenly and then rises asymptotically (called the *asymptotic rise*) at a much slower rate. This jump, called the *prompt jump*, is the start of nuclear runaway; however, since for small reactivity insertions the reactor is subcritical on prompt neutrons alone, this fast nuclear transient quickly dies away. As the reactivity insertion amount increases, the jump becomes higher and more pronounced until for $\rho \geq \beta$, the slowly rising portion of the curve disappears entirely.

The value of the prompt neutron lifetime "Λ" is typically smaller for fast reactors and larger for thermal reactors. Observe from the plot that as Λ decreases, the rate at which the prompt jump proceeds becomes quicker. The implication for fast reactors is that the reactivity control systems employed must be quite robust since nuclear runaways in these systems occur much more quickly.

3. DECAY HEAT REMOVAL CONSIDERATIONS

From Fig. 11.2, observe that when a nuclear reactor experiences a large negative reactivity insertion such as during a shutdown sequence, the reactor power does not immediately drop to zero but rather decays asymptotically on a negative period the characteristics of which depends upon the amount of negative reactivity inserted. This power decrease eventually approaches the period of the longest lived delayed neutron precursor. Even after the fission processes resulting from the delayed neutrons die down a few minutes after shutdown, large amounts of power can still continue to be produced as a result of the radioactive decay of the fission products. If the reactor operated at high power for a long period of time prior to shutdown such that the reactor contains a large fission product inventory, the amount of power resulting from the decay of these fission products can be quite high for several hours or longer. Fortunately, for the reactor in a nuclear thermal rocket, it is unlikely that the fission product decay energy will be too high due to the relatively short engine firing times during which the fission product inventory may be built up.

Precisely calculating the heat resulting from the decay of fission products requires modeling in detail the spatially dependent production and decay of large numbers of individual nuclides throughout the core of the nuclear rocket. Nevertheless, empirical relations have been derived which can approximate the power generated from the decay of fission products as a function to the operational time of the reactor prior to shutdown and the time interval after shutdown has occurred. Over most of the time interval after reactor shutdown, these empirical relationships are generally accurate to within about 10% and are much easier to apply than the more exact calculations requiring the modeling of the individual nuclides. One of the simplest and also most accurate of these decay heat relationships is one formulated by Todreas and Kazimi [2]. This relationship, which accounts for the decay heat resulting from beta and gamma emissions may be expressed as

$$\frac{P_{\beta\gamma}(t)}{P_{\text{fp}}} = 0.066\left[t^{-0.2} - (t_{\text{fp}} + t)^{-0.2}\right] \quad (11.35)$$

where $P_{\beta\gamma}(t)$ = reactor power level due to beta and gamma decay at time "t" seconds after shutdown; P_{fp} = reactor full power level; t = time in seconds after shutdown; and t_{fp} = time in seconds of engine full power operation prior to shutdown.

By adding the power contribution due to delayed neutrons from Eq. (11.33) with the power contribution from beta and gamma decay from Eq. (11.35), the total decay power as a function of time after shutdown may be determined such that:

$$P_{\text{sd}}(t) = P_{\text{fp}}\left\{\frac{\rho}{\rho-\beta}e^{\frac{\rho-\beta}{\Lambda}t} - \frac{\beta}{\rho-\beta}e^{\frac{-\rho\lambda}{\rho-\beta}t} + 0.066\left[t^{-0.2} - (t_{\text{fp}} + t)^{-0.2}\right]\right\} \quad (11.36)$$

where $P_{\text{sd}}(t)$ = reactor power level at time "t" seconds after shutdown.

Using typical values for the prompt neutron lifetime "Λ," neutron precursor decay constant "λ," and neutron precursor fission yield fraction "β," the characteristics of the transient shutdown power profile from Eq. (11.36) may be examined in Fig. 11.3. Note that while it is true that the reactor power

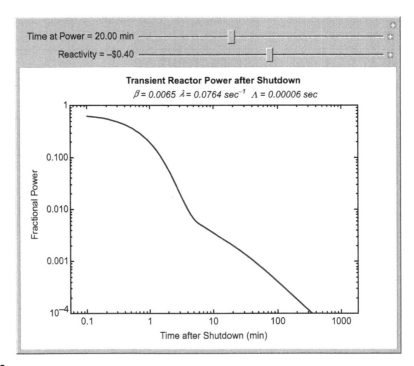

FIGURE 11.3

Reactor power after shutdown.

drops to low levels quite quickly for large negative reactivity insertions, there remains a significant residual power level due to fission product decay whose magnitude depends upon the length of time the reactor at high power.

This continuing power production must be dealt with by reactor systems which are designed to prevent overheating of the reactor core. Several options are potentially available for dealing with the problem decay heating.

The simplest solution for mitigating the unwanted decay heat consists simply of running additional propellant through the core at low flow rates to prevent the core fuel temperatures from rising to unacceptable levels. This warm propellant, after leaving the core, would presumably then be dumped overboard. The main disadvantage to this method is that useful propellant is wasted in the process. Alternatively, it may be possible to design the shutdown system such that the warm propellant is directed through small nozzles to provide some additional thrust to the spacecraft. To estimate the amount of propellant required to provide the necessary reactor shutdown cooling, Eq. (11.36) is first integrated over the shutdown period to yield the total amount of energy which must be dissipated by the cooling system, yielding:

$$Q(t) = P_{fp} \int_0^t \left\{ \frac{\rho}{\rho - \beta} e^{\frac{\rho - \beta}{\lambda} t'} - \frac{\beta}{\rho - \beta} e^{\frac{-\rho \lambda}{\rho - \beta} t'} + 0.066 \left[t'^{-0.2} - (t_{fp} + t')^{-0.2} \right] \right\} dt'$$

$$= P_{fp} \left\{ \frac{\beta}{\lambda \rho} \left[e^{\frac{-\rho \lambda}{\rho - \beta} t} - 1 \right] + \frac{\Lambda \rho}{(\beta - \rho)^2} \left[e^{\frac{\rho - \beta}{\lambda} t} - 1 \right] + 0.0825 \left[t^{0.8} + t_{fp}^{0.8} - (t_{fp} + t)^{0.8} \right] \right\} \quad (11.37)$$

where $Q(t) =$ total energy released "t" seconds after shutdown through delayed neutron fission and fission product beta and gamma decay.

Assuming an infinite shutdown time, it is possible to determine the total energy released from delayed neutron fission and fission product beta and gamma decay by taking the limit of Eq. (11.37) as the shutdown time approaches infinity such that:

$$Q_{sd} = \lim_{t \to \infty} P_{fp} \left\{ \frac{\beta}{\rho \lambda} \left[e^{\frac{\lambda \rho}{\beta - \rho} t} - 1 \right] + \frac{\Lambda \rho}{(\beta - \rho)^2} \left[e^{\frac{(\rho - \beta)}{\lambda} t} - 1 \right] + 0.0825 \left[t^{0.8} + t_{fp}^{0.8} - (t_{fp} + t)^{0.8} \right] \right\}$$

$$= P_{fp} \left[0.0825 t_{fp}^{0.8} - \frac{\beta}{\rho \lambda} - \frac{\Lambda \rho}{(\beta - \rho)^2} \right]$$

(11.38)

Performing a heat balance using the total energy released after shutdown as determined from Eq. (11.38) then yields an equation describing the total mass of propellant required to maintain acceptable fuel temperatures during the course of the shutdown transient:

$$Q_{sd} = P_{fp} \left[0.0825 t_{fp}^{0.8} - \frac{\beta}{\rho \lambda} - \frac{\Lambda \rho}{(\beta - \rho)^2} \right] = m_{dh} C_p \left(T_p^{max} - T_p^{tank} \right)$$

$$\Rightarrow m_{dh} = \frac{P_{fp} \left[0.0825 t_{fp}^{0.8} - \frac{\beta}{\rho \lambda} - \frac{\Lambda \rho}{(\beta - \rho)^2} \right]}{C_p \left(T_p^{max} - T_p^{tank} \right)}$$

(11.39)

where T_p^{max} = maximum propellant temperature yielding acceptable maximum fuel temperatures; T_p^{tank} = temperature of propellant in the propellant tanks; m_{dh} = mass of propellant required to dissipate the reactor decay heat; and C_p = specific heat capacity of the propellant.

To determine the fractional amount of propellant required to remove the reactor decay heat as compared to the total amount of propellant consumed over the entire period of engine operation, it is first necessary to determine the amount of propellant consumed during the course of full power engine firing. The full power propellant mass requirements can be evaluated from:

$$Q_{tot} = P_{fp} t_{fp} = m_{fp} C_p \left(T_p^{max} - T_p^{tank} \right) \Rightarrow m_{fp} = \frac{P_{fp} t_{fp}}{C_p \left(T_p^{max} - T_p^{tank} \right)} \quad (11.40)$$

where m_{fp} = mass of propellant consumed during engine full power operation.

The fractional amount of propellant required to remove the decay heat produced after engine shutdown may then be determined from Eqs. (11.38) and (11.40) such that:

$$f_{sd} = \frac{m_{dh}}{m_{dh} + m_{fp}} = \frac{0.0825 t_{fp}^{0.8} - \frac{\beta}{\rho\lambda} - \frac{\Lambda\rho}{(\beta-\rho)^2}}{t_{fp} + 0.0825 t_{fp}^{0.8} - \frac{\beta}{\rho\lambda} - \frac{\Lambda\rho}{(\beta-\rho)^2}} \quad (11.41)$$

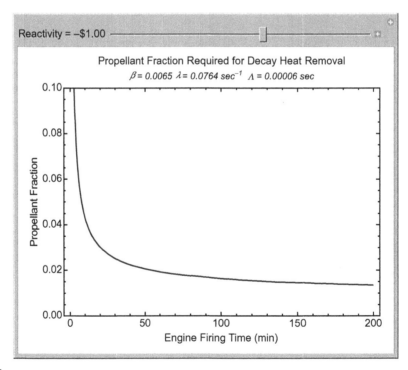

FIGURE 11.4

Propellant required to remove reactor decay heat.

where f_{sd} = fractional amount of propellant required to remove the decay heat produced after engine shutdown.

Again using typical values for the prompt neutron lifetime "Λ," neutron precursor decay constant "λ," and neutron precursor fission yield fraction "β," the propellant fraction required to maintain acceptable core temperatures from Eq. (11.41) may be estimated in Fig. 11.4 as a function of the shutdown core reactivity. Note that large negative reactivity insertions minimize the fractional propellant requirements for decay heat removal, especially for short engine firing times.

Also note that Eq. (11.41) leads to a somewhat conservative result for the propellant mass fraction required to maintain acceptable core temperatures. This is due to the fact that energy loss due to thermal radiation from the hot reactor to the space environment is neglected. In this regard, if space radiators could be incorporated into the design of the nuclear rocket engine, it might be possible to significantly reduce or even eliminate the amount of propellant required to maintain adequate core cooling after shutdown. The advantages of using radiators to reduce the amount of propellant mass required to mitigate the unwanted decay heat must, of course, be balanced with the additional system weight and complexity the use of radiators would entail.

4. NUCLEAR REACTOR TRANSIENT THERMAL RESPONSE

To evaluate the detailed thermal response of a reactor to a change in power level, the general heat conduction equation from Eq. (9.9) must be solved spatially throughout the reactor over the time period during which the thermal response is desired. This calculation is usually quite complicated and often requires considerable computational horsepower to obtain accurate results. To simplify the problem to something which is analytically manageable, the general heat transfer equation is evaluated using what is called the *lumped parameter* technique. This technique can give reasonable estimates of the time-dependent fuel temperature behavior in a reactor assuming that the thermal capacity of the reactor can be treated as a single (or lumped) parameter. In making this assumption, all spatial dependencies in the problem are assumed to be negligible. For this assumption to be valid, the Biot number should be less than about 0.1. Fortunately, for most of the reactor concepts of interest for application in nuclear rockets, the assumption of a low Biot number is fairly good. In the following analysis, it is assumed that the reactor has separate fuel and moderator regions such as might exist in a reactor core where the fuel would be in the form of small fissile particles embedded in a graphite matrix. During a power transient, the temperature of the fissile particles will respond almost immediately to the power change; however, the graphite temperature change is delayed due to the thermal capacity of the graphite and the time required for the heat to travel from the fissile fuel particles into the graphite. With these assumptions in mind, the general heat transfer equation in the fuel region may be represented by

$$\underbrace{\rho_f c_{pf} \frac{dT_f}{dt}}_{\substack{\text{time rate of change} \\ \text{of fuel heat content}}} = \underbrace{k_f \nabla^2 T_f}_{\substack{\text{conduction heat} \\ \text{transfer through fuel}}} + \underbrace{P}_{\substack{\text{power} \\ \text{density}}} - \underbrace{U_{fm} S_{vf}(T_f - T_m)}_{\substack{\text{heat transfer from} \\ \text{fuel to moderator}}} \qquad (11.42)$$

where T_f = fuel temperature; T_m = moderator temperature; ρ_f = density of the fuel; k_f = thermal conductivity of the fuel; c_{pf} = specific heat capacity of the fuel; U_{fm} = average thermal conductance between fuel and moderator; S_{vf} = surface area between fuel and moderator for heat transfer per unit volume; and P = fission power density in the fuel.

Rearranging Eq. (11.42) then yields:

$$\underbrace{\frac{\rho_f c_{pf}}{U_{fm}S_{vf}}}_{\tau_f} \frac{dT_f}{dt} = \underbrace{\frac{k_f}{U_{fm}S_{vf}}}_{Bi_f} \nabla^2 T_f + \frac{P}{U_{fm}S_{vf}} - (T_f - T_m) \quad (11.43)$$

where τ_f = fuel temperature time constant and Bi_f = Biot number for the fuel.

Solving for the conduction heat transfer term in Eq. (11.43) and assuming a small value for the Biot number, all spatial dependencies for thermal conduction in the fuel may be eliminated such that:

$$\nabla^2 T_f = Bi_f \left[\tau_f \frac{dT_f}{dt} + (T_f - T_m) - \frac{P}{U_{fm}S_{vf}} \right] \approx 0 \quad (11.44)$$

Eq. (11.43) may now be simplified to yield:

$$\tau_f \frac{dT_f}{dt} = \frac{P}{U_{fm}S_{vf}} - (T_f - T_m) \quad (11.45)$$

Eq. (11.45) may be further simplified by noting that since virtually all the fission power is produced directly in the fuel, there is little time delay between a change in reactor power and the resulting fuel temperature response. The fuel temperature time constant by implication is, therefore, quite small. Assuming that the fuel temperature time constant is in fact zero, Eq. (11.45) may be rearranged to yield an expression for fuel temperature such that:

$$T_f = \frac{P}{U_{fm}S_{vf}} + T_m \quad (11.46)$$

For the moderator region, generalized heat transfer equation may be represented by

$$\underbrace{\rho_m c_{pm} \frac{dT_m}{dt}}_{\text{time rate of change of moderator heat content}} = \underbrace{k_m \nabla^2 T_m}_{\text{conduction heat transfer through moderator}} + \underbrace{U_{fm}S_{vf}(T_f - T_m)}_{\text{heat transfer rate from fuel to moderator}} - \underbrace{U_{mp}S_{vm}(T_m - T_p)}_{\text{heat transfer rate from moderator to propellant}} \quad (11.47)$$

where T_p = temperature of propellant (assumed constant); k_m = thermal conductivity of the moderator; ρ_m = density of the moderator; c_{pm} = specific heat capacity of the moderator; U_{mp} = average thermal conductance between moderator and the propellant; and S_{vm} = surface area between moderator and propellant for heat transfer per unit volume.

Rearranging Eq. (11.47) then yields:

$$\underbrace{\frac{\rho_m c_{pm}}{U_{mp}S_{vm}}}_{\tau_m} \frac{dT_m}{dt} = \underbrace{\frac{k_m}{U_{mp}S_{vm}}}_{Bi_m} \nabla^2 T_m + \frac{U_{fm}S_{vf}}{U_{mp}S_{vm}}(T_f - T_m) - (T_m - T_p) \quad (11.48)$$

where τ_m = moderator temperature time constant and Bi_m = Biot number for the moderator.

4. NUCLEAR REACTOR TRANSIENT THERMAL RESPONSE

As in the fuel region, Eq. (11.48) for the moderator region is rearranged to solve for the conduction heat transfer term. If again, a small value for the Biot number is assumed, the thermal conduction term in the moderator may be eliminated yielding:

$$\nabla^2 T_m = Bi_m \left[\tau_m \frac{dT_m}{dt} - \frac{U_{fm}S_{vf}}{U_{mp}S_{vm}}(T_f - T_m) + (T_m - T_p) \right] \approx 0 \tag{11.49}$$

Again, since there are no spatial dependencies in the moderator, Eq. (11.48) may be simplified such that:

$$\tau_m \frac{dT_m}{dt} = \frac{U_{fm}S_{vf}}{U_{mp}S_{vm}}(T_f - T_m) - T_m + T_p \tag{11.50}$$

Using Eq. (11.46) to eliminate the fuel temperature variable in Eq. (11.50), a differential equation describing the moderator temperature response to a power transient may be formulated to finally yield:

$$\tau_m \frac{dT_m}{dt} = \frac{P}{U_{mp}S_{vm}} - T_m + T_p \tag{11.51}$$

By now choosing an expression which represents a typical power transient in the reactor, it is possible to derive expressions for the time-dependent behavior of the fuel and moderator temperatures resulting from that power transient. Since many reactor power transients are exponential in nature, an expression of that form is chosen such that:

$$P = P_f - (P_f - P_0)e^{-\frac{t}{\xi}} \tag{11.52}$$

where P_0 = initial power density in the fuel; P_f = final power density in the fuel; and ξ = time constant of the power transient.

Incorporating the power transient represented by Eq. (11.52) into the differential equation describing the moderator temperature behavior from Eq. (11.51), then yields:

$$\tau_m \frac{dT_m}{dt} = \frac{P_f - (P_f - P_0)e^{-\frac{t}{\xi}}}{U_{mp}S_{vm}} - T_m + T_p \tag{11.53}$$

Solving the differential equation for moderator temperature from Eq. (11.53), the temperature response of the moderator due to an exponential power transient in the fuel may be determined such that:

$$T_m = \left[\frac{P_0 - P_f}{U_{mp}S_{vm}\left(1 - \frac{\tau_m}{\xi}\right)} \right] e^{-\frac{t}{\xi}} + \left[\frac{\frac{\tau_m}{\xi}P_f - P_0}{U_{mp}S_{vm}\left(1 - \frac{\tau_m}{\xi}\right)} + T_{m0} - T_p \right] e^{-\frac{t}{\tau_m}} + \frac{P_f}{U_{mp}S_{vm}} + T_p \tag{11.54}$$

where T_{m0} = moderator temperature at the start of the transient.

Assuming steady-state conditions exist at the start of the transient, the time derivative of the moderator temperature at that point in time is zero. From Eq. (11.51), therefore, the moderator temperature may be given by

$$\tau_m \frac{dT_m}{dt} = 0 = \frac{P_0}{U_{mp}S_{vm}} - T_{m0} + T_p \Rightarrow T_{m0} = \frac{P_0}{U_{mp}S_{vm}} + T_p \tag{11.55}$$

Substituting the initial moderator temperature from Eq. (11.55) into the expression for the time-dependent moderator temperature from Eq. (11.54) then yields:

$$T_m = \left[\frac{P_0 - P_f}{U_{mp}S_{vm}\left(1 - \frac{\tau_m}{\xi}\right)}\right]\left(e^{-\frac{t}{\xi}} - \frac{\tau_m}{\xi}e^{-\frac{t}{\tau_m}}\right) + \frac{P_f}{U_{mp}S_{vm}} + T_p \qquad (11.56)$$

The fuel temperature response due to the exponential power transient may be determined by incorporating Eq. (11.56) for the moderator temperature time-dependent behavior and Eq. (11.52)

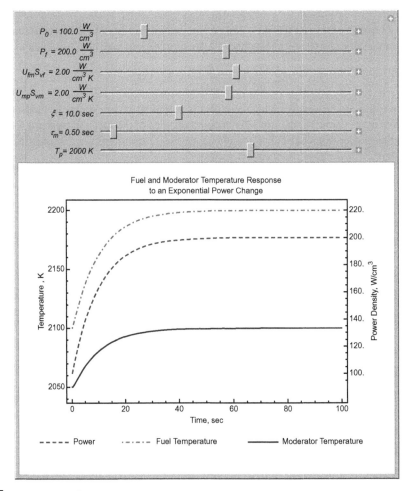

FIGURE 11.5

Fuel and moderator temperature response to an exponential power change.

describing the power transient into Eq. (11.46) describing the fuel transient temperature response yielding:

$$T_\mathrm{f} = \left[\frac{P_0 - P_\mathrm{f}}{U_\mathrm{mp}S_\mathrm{vm}\left(1 - \frac{\tau_\mathrm{m}}{\xi}\right)}\right]\left(\mathrm{e}^{-\frac{t}{\xi}} - \frac{\tau_\mathrm{m}}{\xi}\mathrm{e}^{-\frac{t}{\tau_\mathrm{m}}}\right) + \frac{(P_0 - P_\mathrm{f})\mathrm{e}^{-\frac{t}{\xi}}}{U_\mathrm{fm}S_\mathrm{vm}} + \left(\frac{1}{U_\mathrm{mp}S_\mathrm{vm}} + \frac{1}{U_\mathrm{fm}S_\mathrm{vf}}\right)P_\mathrm{f} + T_\mathrm{p} \quad (11.57)$$

In Fig. 11.5, the temperature response of the fuel and moderator to an exponential power transient is presented. Note that as expected, increasing the thermal conductance between the fuel and moderator minimizes the temperature difference between the two materials, and increasing the thermal conductance between the moderator and the propellant reduces the moderator temperature.

REFERENCES

[1] R.G. Keepin, Physics of Nuclear Kinetics, Addison-Wesley Pub. Co., 1965. Library of Congress QC787.N8 K4.
[2] N.E. Todreas, M.S. Kazimi, Nuclear Systems I, Thermal Hydraulic Fundamentals, vol. 1, Hemisphere Publishing Corporation, 1990, ISBN 0-89116-935-0.

CHAPTER 12

NUCLEAR ROCKET STABILITY

In order to operate in a safe manner, nuclear reactors should operate such that when small naturally occurring random power perturbations occur during operation, the reactor passively responds with corresponding reactivity changes that act to suppress the growth of the perturbations. While it is possible to control reactors, where this is not the case, quite a robust active control system is required to operate safely. Such reactor configurations are generally not desirable and should be avoided if possible. In a nuclear rocket engine, power perturbations manifest themselves through temperature changes in various portions of the reactor. It is possible that some portions of the reactor act to add reactivity when the temperature increases and other portions of the reactor act to decrease the reactivity when the temperature increases. As long as the overall reactivity change is such that the power decreases with increases in temperature, the reactor should be stable during operation.

1. DERIVATION OF THE POINT KINETICS EQUATIONS

At this point, the transient behavior of a reactor system due to changes in reactivity has been modeled by using the point kinetics equations. The question which now needs to be asked is whether a reactor operating in steady state (eg, $\rho = 0$) will remain stable when subjected to small naturally occurring reactivity perturbations.

To answer this question, the point kinetics equations are first written for a critical reactor:

$$\frac{dn}{dt} = \frac{\rho - \beta}{\Lambda} n + \lambda C \approx \frac{\delta k - \beta}{\Lambda} n + \lambda C \tag{12.1}$$

and

$$\frac{dC}{dt} = \frac{\beta}{\Lambda} n - \lambda C \tag{12.2}$$

where $\rho = \frac{k_{eff} - 1}{k_{eff}} \approx k_{eff} - 1 = \delta k$.

Adding Eqs. (12.1) and (12.2) then yields:

$$\frac{dn}{dt} = \frac{\delta k}{\Lambda} n - \frac{dC}{dt} \tag{12.3}$$

If it is assumed that δk undergoes small variations with time, then the neutron density and the neutron precursor density are also induced to undergo variations with time such that:

$$n = n_0 + \delta n \quad \text{and} \quad C = C_0 + \delta C \tag{12.4}$$

where: $\delta n \ll n_0$ and $\delta C \ll C_0$.

Substituting Eqs. (12.4) into Eq. (12.3) then results in:

$$\underbrace{\frac{dn_0}{dt}}_{\approx 0} + \frac{d\delta n}{dt} = \frac{\delta k}{\Lambda}n_0 + \underbrace{\frac{\delta k}{\Lambda}\delta n}_{\approx 0} - \underbrace{\frac{dC_0}{dt}}_{\approx 0} - \frac{d\delta C}{dt} \Rightarrow \frac{d\delta n}{dt} = \frac{\delta k}{\Lambda}n_0 - \frac{d\delta C}{dt} \quad (12.5)$$

Assuming steady-state conditions, it is found from Eq. (12.2) that:

$$\frac{dC}{dt} = \frac{\beta}{\Lambda}n_0 - \lambda C_0 \Rightarrow \frac{\beta}{\Lambda}n_0 = \lambda C_0 \quad (12.6)$$

Substituting Eq. (12.4) into Eq. (12.2) results in:

$$\underbrace{\frac{dC_0}{dt}}_{\approx 0} + \frac{d\delta C}{dt} = \frac{\beta}{\Lambda}n_0 + \frac{\beta}{\Lambda}\delta n - \lambda C_0 - \lambda C \quad (12.7)$$

Incorporating the results from Eq. (12.6) into Eq. (12.7) then yields:

$$\frac{d\delta C}{dt} = \lambda C_0 + \frac{\beta}{\Lambda}\delta n - \lambda C_0 - \lambda \delta C = \frac{\beta}{\Lambda}\delta n - \lambda \delta C \quad (12.8)$$

At this point, the small perturbations in the neutron and neutron precursor densities described in Eq. (12.4) are defined as being sinusoidal in nature such that:

$$\delta n = n_0 e^{i\omega t} \Rightarrow \frac{d\delta n}{dt} = i\omega n_0 e^{i\omega t} = i\omega \delta n \quad (12.9)$$

and

$$\delta C = C_0 e^{i\omega t} \Rightarrow \frac{d\delta C}{dt} = i\omega C_0 e^{i\omega t} = i\omega \delta C \quad (12.10)$$

Substituting Eqs. (12.9) and (12.10) into Eq. (12.5) then yields for the perturbed neutron density:

$$i\omega \delta n = \frac{\delta k}{\Lambda}n_0 - i\omega \delta C \quad (12.11)$$

Also substituting Eqs. (12.9) and (12.10) into Eq. (12.8) gives an expression for the perturbed neutron precursor density of the form:

$$i\omega \delta C = \frac{\beta}{\Lambda}\delta n - \lambda \delta C \quad (12.12)$$

Using Eqs. (12.11) and (12.12) to eliminate the perturbed neutron precursor density and rearranging terms then yields:

$$\frac{\delta n}{\delta k} = \frac{n_0}{i\omega\Lambda + \frac{i\omega\beta}{i\omega+\lambda}} \Rightarrow \Sigma_f v \frac{\delta n}{\delta k} = \frac{\Sigma_f v n_0}{i\omega\Lambda + \frac{i\omega\beta}{i\omega+\lambda}} = \frac{\delta q}{\delta k} = \frac{q_0}{i\omega\Lambda + \frac{i\omega\beta}{i\omega+\lambda}} \quad (12.13)$$

Extending Eq. (12.13) to six groups of delayed neutrons then yields a function of the form:

$$K_R G_R = \frac{\delta q}{\delta k} = \frac{q_0}{i\omega\Lambda + \sum_{i=1}^{6}\frac{i\omega\beta_i}{i\omega+\lambda_i}} \quad (12.14)$$

Eq. (12.14) is what is known as the *reactor kinetics transfer function*. This function gives the change in the power density (output signal) resulting from small perturbations in the reactor k_{eff} (input signal). Typically, the transfer function is broken up into two parts, the gain (K_R) and another factor (G_R) which describes the magnitude and phase relationship of the transfer function. Note from Eq. (12.13) that as the perturbation frequency approaches zero, the reactor transfer function approaches ∞. Such behavior indicates that a nuclear reactor which is exactly critical is inherently unstable because the reactor power will increase without limit as a result of a minor step perturbation in k_{eff}. Since nuclear reactors normally operate in a stable manner, there are obviously other factors at play which serve to stabilize the reactor. It is the objective of the following section to examine these stabilization mechanisms and the manner in which they serve to moderate reactor behavior.

2. REACTOR STABILITY MODEL INCLUDING THERMAL FEEDBACK

Nuclear reactors have several mechanisms which can serve to maintain power stability including control rod manipulation, negative fuel temperature coefficient of reactivity, and negative moderator temperature coefficient of reactivity. Of these mechanisms, the negative fuel temperature coefficient of reactivity is the most important since its effect occurs immediately following a reactor transient. The fuel temperature coefficient of reactivity operates by causing the reactivity to decrease as the fuel temperature increases. This very effective feedback mechanism operates by changing the effective microscopic cross section of the fuel through the Doppler broadening effect described earlier.

Other temperature-related reactivity effects can occur as a result of density changes due to thermal expansion in the fuel or moderator materials or as a result of density changes in the propellant as it flows through the core. These density changes alter the macroscopic cross sections of the materials being heated resulting in changes to the core reactivity. Since it takes some period of time for the heat from the fuel to be transferred to other core materials in the reactor (eg, moderator and structure), the effect of density changes on core reactivity can be somewhat delayed with respect to a power transient.

A block diagram of a closed-loop transfer function of a nuclear rocket reactor system which includes feedback effects due to the fuel, moderator, and propellant temperature coefficients of reactivity is presented in Fig. 12.1. This transfer function is greatly simplified over that required to model the overall stability of a complete nuclear rocket engine system; nevertheless, it serves to demonstrate the dominant thermal feedback mechanisms, and how these feedback mechanisms allow nuclear rocket engines to operate near the upper edge of their operational envelopes.

To determine the form of the reactor transfer function, it should first be noted from Fig. 12.1 that the error signal resulting the fuel temperature feedback may be represented by

$$\epsilon_F = \theta_i - \theta_F = \theta_i - \theta_o K_F G_F \tag{12.15}$$

In like manner, using the results from Eq. (12.15), the error signal resulting from the moderator temperature feedback may be determined to be

$$\epsilon_{FM} = \epsilon_F - \theta_M = \theta_i - \theta_o K_F G_F - \theta_o K_M G_M \tag{12.16}$$

Similarly, using the results from Eq. (12.16), the error signal resulting from the propellant temperature feedback becomes

$$\epsilon_{FMP} = \epsilon_{FM} - \theta_H = \theta_i - \theta_o K_F G_F - \theta_o K_M G_M - \theta_o K_P G_P = \theta_i - (K_F G_F + K_M G_M + K_P G_P)\theta_o \tag{12.17}$$

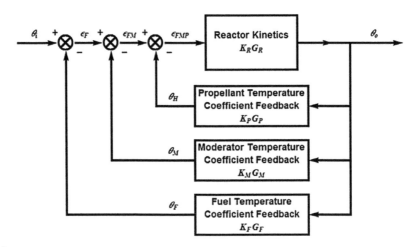

FIGURE 12.1

Simplified reactor transfer function block diagram.
where: θ_i = input signal; θ_o = output signal; θ_F = output signal resulting from fuel temperature feedback; θ_M = output signal resulting from moderator temperature feedback; θ_P = output signal resulting from propellant temperature feedback; ϵ_F = error signal due to fuel temperature feedback; ϵ_{FM} = error signal due to moderator temperature feedback; and ϵ_{FMP} = error signal due to propellant temperature feedback.

Using the results from Eq. (12.17) as the input signal to the reactor kinetics transfer function, it is now possible to derive an overall reactor transfer function such that:

$$\theta_o = \epsilon_{FMP} K_R G_R = [\theta_i - (K_F G_F + K_M G_M + K_P G_P)\theta_o] K_R G_R \Rightarrow \frac{\theta_o}{\theta_i}$$

$$= \frac{K_R G_R}{1 + (K_F G_F + K_M G_M + K_P G_P) K_R G_R} = K_{RT} G_{RT} \qquad (12.18)$$

Eq. (12.18) represents the new reactor transfer function ($K_{RT} G_{RT}$) which incorporates the effects of temperature feedback. It remains now to find expressions for the temperature feedback transfer functions $K_F G_F$, $K_M G_M$, and $K_P G_P$ which properly model the thermal feedback effects. To simplify the analysis, the lumped parameter model is again used to determine the reactor temperatures. In this model, the heat balance equations for the fuel, moderator, and propellant are presented for a system in which heat is first generated in the fuel and subsequently transferred into the moderator after which it is finally transferred into the flowing propellant. These heat balance equations may be expressed as

$$\underbrace{q}_{\substack{\text{power} \\ \text{density}}} = \underbrace{\rho_f c_p^f \frac{dT_f}{dt}}_{\substack{\text{time rate of change} \\ \text{of fuel heat content}}} + \underbrace{U_{fm}(T_f - T_m)}_{\substack{\text{heat transfer rate from fuel} \\ \text{to moderator}}} \qquad (12.19)$$

2. REACTOR STABILITY MODEL INCLUDING THERMAL FEEDBACK

$$\underbrace{U_{\text{fm}}(T_{\text{f}} - T_{\text{m}})}_{\substack{\text{heat transfer rate from}\\\text{fuel to moderator}}} = \underbrace{\rho_{\text{m}} c_{\text{p}}^{\text{m}} \frac{dT_{\text{m}}}{dt}}_{\substack{\text{time rate of change of}\\\text{moderator heat content}}} + \underbrace{U_{\text{mp}}(T_{\text{m}} - T_{\text{p}})}_{\substack{\text{heat transfer rate from}\\\text{moderator to propellant}}} \tag{12.20}$$

$$\underbrace{U_{\text{mp}}(T_{\text{m}} - T_{\text{p}})}_{\substack{\text{heat transfer rate from}\\\text{moderator to propellant}}} = \underbrace{\rho_{\text{p}} c_{\text{p}}^{\text{p}} \frac{dT_{\text{p}}}{dt}}_{\substack{\text{time rate of change of}\\\text{propellant heat content}}} + \underbrace{\dot{m} c_{\text{p}}^{\text{p}}(T_{\text{po}} - T_{\text{pi}})}_{\substack{\text{heat transfer rate out of reactor}\\\text{due to propellant outflow}}} \tag{12.21}$$

where T_{f} = average fuel temperature; T_{m} = average moderator temperature; T_{p} = average propellant temperature; T_{pi} = propellant temperature at reactor inlet; T_{po} = propellant temperature at reactor outlet; ρ_{f} = density of fissionable fuel material; ρ_{m} = density of moderator and associated structural material; ρ_{p} = density of propellant in the reactor at any given instant (assumed constant); $\dot{m}$ = mass of propellant flowing through a unit volume per unit time; c_{p}^{f} = specific heat capacity of the fissionable fuel material; c_{p}^{m} = specific heat capacity of the moderator and associated structural material; c_{p}^{p} = specific heat capacity of the propellant; U_{fm} = thermal conductance per unit volume from fuel to moderator; and U_{mp} = thermal conductance per unit volume from moderator to propellant.

Assuming that the average propellant temperature may be expressed as

$$T_{\text{p}} = \frac{T_{\text{po}} + T_{\text{pi}}}{2} \tag{12.22}$$

Eq. (12.21) may be rewritten as

$$U_{\text{mp}}(T_{\text{m}} - T_{\text{p}}) = \rho_{\text{p}} c_{\text{p}}^{\text{p}} \frac{dT_{\text{p}}}{dt} + 2\dot{m} c_{\text{p}}^{\text{p}}(T_{\text{p}} - T_{\text{pi}}) \tag{12.23}$$

Again assuming small time-dependent fluctuations in the heat generation rate and material temperatures, Eqs. (12.19), (12.20) and (12.23) may be transformed such that:

$$q_0 + \delta q = \rho_{\text{f}} c_{\text{p}}^{\text{f}} \frac{d}{dt}\left(T_{\text{f}}^0 + \delta T_{\text{f}}\right) + U_{\text{fm}}\left(T_{\text{f}}^0 - T_{\text{m}}^0\right) + U_{\text{fm}}(\delta T_{\text{f}} - \delta T_{\text{m}}) \Rightarrow \delta q$$

$$= \rho_{\text{f}} c_{\text{p}}^{\text{f}} \frac{d\delta T_{\text{f}}}{dt} + U_{\text{fm}}(\delta T_{\text{f}} - \delta T_{\text{m}}) \tag{12.24}$$

$$U_{\text{fm}}\left(T_{\text{f}}^0 - T_{\text{m}}^0\right) + U_{\text{fm}}(\delta T_{\text{f}} - \delta T_{\text{m}}) = \rho_{\text{m}} c_{\text{p}}^{\text{m}} \frac{d}{dt}\left(T_{\text{m}}^0 + \delta T_{\text{m}}\right) + U_{\text{mp}}\left(T_{\text{m}}^0 - T_{\text{p}}^0\right)$$
$$+ U_{\text{mp}}(\delta T_{\text{m}} - \delta T_{\text{p}}) \Rightarrow U_{\text{fm}}(\delta T_{\text{f}} - \delta T_{\text{m}})$$
$$= \rho_{\text{m}} c_{\text{p}}^{\text{m}} \frac{d\delta T_{\text{m}}}{dt} + U_{\text{mp}}(\delta T_{\text{m}} - \delta T_{\text{p}}) \tag{12.25}$$

$$U_{\text{mp}}\left(T_{\text{m}}^0 - T_{\text{p}}^0\right) + U_{\text{mp}}(\delta T_{\text{m}} - \delta T_{\text{p}}) = \rho_{\text{p}} c_{\text{p}}^{\text{p}} \frac{d}{dt}\left(T_{\text{p}}^0 + \delta T_{\text{p}}\right) + 2\dot{m} c_{\text{p}}^{\text{p}}\left(T_{\text{p}}^0 - T_{\text{pi}}\right)$$
$$+ 2\dot{m} c_{\text{p}}^{\text{p}}(\delta T_{\text{p}} - T_{\text{pi}}) \Rightarrow U_{\text{mp}}(\delta T_{\text{m}} - \delta T_{\text{p}})$$
$$= \rho_{\text{p}} c_{\text{p}}^{\text{p}} \frac{d\delta T_{\text{p}}}{dt} + 2\dot{m} c_{\text{p}}^{\text{p}} \delta T_{\text{p}} \tag{12.26}$$

CHAPTER 12 NUCLEAR ROCKET STABILITY

Small sinusoidal perturbations in the temperature are again defined as before such that:

$$\delta T_f = T_f^0 e^{i\omega t} \Rightarrow \frac{d\delta T_f}{dt} = i\omega T_f^0 e^{i\omega t} = i\omega \delta T_f \qquad (12.27)$$

$$\delta T_m = T_m^0 e^{i\omega t} \Rightarrow \frac{d\delta T_m}{dt} = i\omega T_m^0 e^{i\omega t} = i\omega \delta T_m \qquad (12.28)$$

$$\delta T_p = T_p^0 e^{i\omega t} \Rightarrow \frac{d\delta T_p}{dt} = i\omega T_p^0 e^{i\omega t} = i\omega \delta T_p \qquad (12.29)$$

where ω = perturbation frequency and t = time.

Substituting the perturbed temperature definitions from Eqs. (12.27) through (12.29) into the perturbed heat balance equations, Eqs. (12.24) through (12.26) then yields:

$$\delta q = i\omega \rho_f c_p^f \delta T_f + U_{fm}(\delta T_f - \delta T_m) \qquad (12.30)$$

$$U_{fm}(\delta T_f - \delta T_m) = i\omega \rho_m c_p^m \delta T_m + U_{mp}(\delta T_m - \delta T_p) \qquad (12.31)$$

$$U_{mp}(\delta T_m - \delta T_p) = i\omega \rho_p c_p^p \delta T_p + 2\dot{m} c_p^p \delta T_p \qquad (12.32)$$

If Eqs. (12.24) through (12.26) are now solved simultaneously for the perturbed temperatures it is found that:

$$\frac{\delta T_f}{\delta q} = \frac{U_{fm}\left(2\dot{m}c_p^p + i\omega\rho_p c_p^p + U_{mp}\right)\left[\frac{i\omega\rho_m c_p^m + U_{fm} + U_{mp}}{U_{fm}} - \frac{U_{mp}^2}{U_{fm}\left(2\dot{m}c_p^p + i\omega\rho_p c_p^p + U_{mp}\right)}\right]}{U_{fm}^2 + \left(i\omega\rho_f c_p^f + U_{fm}\right)\left(i\omega\rho_m c_p^m + U_{fm} + U_{mp}\right)\left(2\dot{m}c_p^p + i\omega\rho_p c_p^p + U_{mp}\right) - \left(i\omega\rho_f c_p^f + U_{fm}\right)U_{mp}^2} \qquad (12.33)$$

$$\frac{\delta T_m}{\delta q} = \frac{U_{fm}\left(2\dot{m}c_p^p + i\omega\rho_p c_p^p + U_{mp}\right)}{U_{fm}^2 + \left(i\omega\rho_f c_p^f + U_{fm}\right)\left(i\omega\rho_m c_p^m + U_{fm} + U_{mp}\right)\left(2\dot{m}c_p^p + i\omega\rho_p c_p^p + U_{mp}\right) - \left(i\omega\rho_f c_p^f + U_{fm}\right)U_{mp}^2} \qquad (12.34)$$

$$\frac{\delta T_p}{\delta q} = \frac{U_{fm} U_{mp}}{U_{fm}^2 + \left(i\omega\rho_f c_p^f + U_{fm}\right)\left(i\omega\rho_m c_p^m + U_{fm} + U_{mp}\right)\left(2\dot{m}c_p^p + i\omega\rho_p c_p^p + U_{mp}\right) - \left(i\omega\rho_f c_p^f + U_{fm}\right)U_{mp}^2} \qquad (12.35)$$

In the perturbed temperature equations (12.33)–(12.35) the grayed out terms will be neglected since they are either of zeroth or first order in frequency and will be small for higher frequencies as compared to the other (third-order) term. To further simplify Eqs. (12.33)–(12.35), the following definitions are now made:

$$\tau_f = \frac{\rho_f c_p^f}{U_{fm}}, \quad \tau_m = \frac{\rho_m c_p^m}{U_{fm} + U_{mp}}, \quad \tau_p = \frac{\rho_p c_p^p}{U_{mp} + 2\dot{m}c_p^p},$$

$$A_f = \frac{1}{U_{fm}}, \quad A_m = \frac{1}{U_{fm} + U_{mp}}, \quad A_p = \frac{U_{mp}}{(U_{fm} + U_{mp})(U_{mp} + 2\dot{m}c_p^p)}$$

Note in the above equations, that τ_f may be thought of as a time constant related to the rate of change of the fuel temperature, τ_m may be thought of as a time constant related to the rate of change of the moderator temperature, and τ_p may be thought of a time constant related to the rate of change of the propellant temperature. The parameters A_f, A_m, and A_p may be thought of as thermal resistances which relate fuel, moderator, and propellant temperature changes to changes in reactor power. With these thoughts in mind, Eqs. (12.33) through (12.35) may now be rewritten to yield:

$$\frac{\delta T_f}{\delta q} = \frac{A_f(1 + i\omega\tau_p)(1 + i\omega\tau_m)}{(1 + i\omega\tau_f)(1 + i\omega\tau_m)(1 + i\omega\tau_p)} = \frac{A_f}{1 + i\omega\tau_f} \quad (12.36)$$

$$\frac{\delta T_m}{\delta q} = \frac{A_m(1 + i\omega\tau_p)}{(1 + i\omega\tau_f)(1 + i\omega\tau_m)(1 + i\omega\tau_p)} = \frac{A_m}{(1 + i\omega\tau_f)(1 + i\omega\tau_m)} \quad (12.37)$$

$$\frac{\delta T_p}{\delta q} = \frac{A_p}{(1 + i\omega\tau_f)(1 + i\omega\tau_m)(1 + i\omega\tau_p)} \quad (12.38)$$

In order to develop feedback transfer functions appropriate for use in Eq. (12.18), Eqs. (12.36) through (12.38) must be modified such that they yield changes in k_{eff} rather than changes in temperature as a result of changes in power. To accomplish this transformation, use is made of the temperature coefficients of reactivity which were described earlier. Temperature coefficient of reactivity are, in fact, themselves functions of temperature; however, for the present analysis, they are treated as constants. Restricting these temperature coefficients of reactivity to constant values should not introduce large errors into the results since typically these coefficients are only mild functions of temperature.

The temperature coefficients of reactivity for the fuel and moderator can be positive or negative depending upon the design of the reactor. In low enriched water moderated reactors, for example, the fuel temperature coefficients of reactivity are almost always negative due to the large amount of ^{238}U present. Increases in temperature shift the neutron flux such that a greater percentage of the neutron are captured in the large absorption resonance at 6.67 eV and lost. Such behavior implies that an increase in the reactor temperature will result in a decrease in the reactor k_{eff}. The decrease in k_{eff} results in a decrease in the reactor power, eventually leading to a decrease in the reactor temperature. Such a situation is desirable since it implies that the reactor will operate in a naturally stable manner. Virtually all nuclear reactors in the United States are of this type. Moderator temperature coefficients of reactivity are also typically negative since increases in temperature leads to density decreases in the moderator. This density reduction in the moderator decreases its ability to slow-down neutrons which again leads to decreases in the reactivity of the reactor.

In highly enriched reactors with graphite moderation, as opposed to low enriched water reactors, the fuel temperature coefficient of reactivity can sometimes be positive since the Doppler effect described earlier will primarily be broadening low-lying fission resonances in ^{235}U (and especially ^{239}Pu) rather than the neutron absorption resonances in ^{238}U. Doppler broadening the fission resonances in preference to the capture resonances results in the reactor k_{eff} going up with increasing reactor temperature, leading to an increase in reactor power which in turn leads to further increases in reactor temperature and so on until the destruction of the reactor occurs or some other process comes into play which serves to terminate the instability transient. Positive values of the fuel temperature coefficient of reactivity are thus generally to be avoided in reactor designs since they can lead to destructive power instabilities in the reactor. The Chernobyl reactor disaster in Russia was a direct result of its having a positive temperature coefficient of reactivity. In this case, however, it was the reactor's positive coolant temperature coefficient of reactivity which resulted in the power instability. It turns out that Chernobyl-type reactors are fueled with natural uranium and moderated by graphite. They are cooled with light water. What is interesting (and dangerous) in this particular design is that the water actually acts more as an absorber of neutrons than a moderator of neutrons due to the relatively higher absorption cross section of the hydrogen as compared to the graphite. A power increase in a Chernobyl-type reactor results in a decrease in the density of the water thereby reducing the water's ability to absorb excess neutrons. With more neutrons thus available for fission, an increase in water temperature leads to an increase in reactor power (eg, therefore the positive coolant reactivity coefficient). In the Chernobyl disaster, the reactor control system, which was designed to actively control the power, was disengaged to perform some testing. During the test, a small power perturbation occurred, which due to the positive coolant temperature reactivity coefficient led to a transient in which the power increased rapidly, ultimately leading to the destruction of the reactor.

Since many nuclear rocket engine concepts presently under consideration are moderated with graphite and "cooled" with the hydrogen propellant, it appears that they have certain similarities to the Russian Chernobyl reactor. As a consequence, care must be taken in the design of future nuclear rocket engines such that they incorporate materials and design features which result in the reactor having an overall negative temperature coefficient of reactivity.

Incorporating these temperature coefficients of reactivity into Eqs. (12.36) and (12.37) then yields fuel and moderator temperature feedback transfer functions of the form:

$$\frac{\delta k_{\text{eff}}^f}{\delta q} = \alpha_f \frac{\delta T_f}{\delta q} = \frac{\alpha_f A_f}{1 + i\omega \tau_f} = K_F G_F \tag{12.39}$$

$$\frac{\delta k_{\text{eff}}^m}{\delta q} = \alpha_m \frac{\delta T_m}{\delta q} = \frac{\alpha_m A_m}{(1 + i\omega \tau_f)(1 + i\omega \tau_m)} = K_M G_M \tag{12.40}$$

$$\frac{\delta k_{\text{eff}}^p}{\delta q} = \beta_p \frac{\delta \rho_p}{\delta T_m} \frac{\delta T_p}{\delta q} = \frac{\beta_p A_p}{(1 + i\omega \tau_f)(1 + i\omega \tau_m)(1 + i\omega \tau_p)} \frac{\delta \rho_p}{\delta T_p} = K_P G_P \tag{12.41}$$

where $\alpha_f = \frac{\delta k_{\text{eff}}^f}{\delta T_f}$ = fuel temperature coefficient of reactivity; $\alpha_m = \frac{\delta k_{\text{eff}}^m}{\delta T_m}$ = moderator temperature coefficient of reactivity; and $\beta_p = \frac{\delta k_{\text{eff}}^p}{\delta \rho_p}$ = propellant density coefficient of reactivity.

In order to determine a final expression for the propellant temperature feedback transfer function expressed in Eq. (12.41), additional effort must be put forth since the propellant is a gas and the reactivity effects are due mainly to changes in the propellant density rather than to changes in temperature. Assuming that the propellant obeys the ideal gas law, it is possible to write:

$$\rho_p = \frac{P_p}{R_p T_p} \tag{12.42}$$

where P_p = propellant pressure and R_p = propellant gas constant.

From the definition of the propellant time constant "τ_p" presented earlier, the use of Eq. (12.42) then yields:

$$\tau_p = \frac{\rho_p c_p^p}{U_{mp} + 2\dot{m}c_p^p} = \frac{P_p c_p^p}{R_p T_p (U_{mp} + 2\dot{m}c_p^p)} \tag{12.43}$$

If the system pressure remains essentially constant, a perturbation in the propellant temperature will result in a perturbation in the propellant density such that:

$$\rho_{p0} + \delta\rho_p = \frac{P_{p0}}{R_p(T_{p0} + \delta T_p)} = \frac{P_{p0}(T_{p0} - \delta T_p)}{R_p(T_{p0} + \delta T_p)(T_{p0} - \delta T_p)} = \frac{P_{p0}(T_{p0} - \delta T_p)}{R_p(T_{p0}^2 - \delta T_p^2)} = \frac{P_{p0}}{R_p T_{p0}^2} - \frac{P_{p0}\delta T_p}{R_p T_{p0}^2} \tag{12.44}$$

Canceling out terms in Eq. (12.44) and noting that the second-order term δT_p^2 is small, it is found that:

$$\delta\rho_p = -\frac{P_{p0}\delta T_p}{R_p T_{p0}^2} \Rightarrow \frac{\delta\rho_p}{\delta T_p} = -\frac{P_{p0}}{R_p T_{p0}^2} \tag{12.45}$$

Substituting Eq. (12.45) into Eq. (12.41) then yields for the propellant temperature feedback transfer function:

$$\frac{\delta k_{eff}^p}{\delta q} = \frac{-\beta_p A_p}{(1 + i\omega\tau_f)(1 + i\omega\tau_m)(1 + i\omega\tau_p)} \frac{P_{p0}}{R_p T_{p0}^2} = \frac{-\beta_p B_p}{(1 + i\omega\tau_f)(1 + i\omega\tau_m)(1 + i\omega\tau_p)} = K_P G_P \tag{12.46}$$

where $B_p = \frac{A_p P_{p0}}{R_p T_{p0}^2} = \frac{U_{mp} P_{p0}}{R_p T_{p0}^2 (U_{fm} + U_{mp})(U_{mp} + 2\dot{m}c_p^p)}$.

Unlike the fuel and moderator temperature coefficients discussed earlier, one usually finds that the propellant density coefficient is almost always positive. Nevertheless, noting the minus sign in Eq. (12.46), it will be observed that propellant density fluctuations normally act so as to decrease the reactivity during a transient.

At this point, it is possible to present the overall system stability model for a nuclear rocket wherein the design is such that heat generated in the fuel is transferred through a moderator and then into the propellant flow stream. By substituting the temperature feedback transfer function expressions as given by Eqs. (12.39), (12.40), and (12.46), and the reactor nuclear kinetics transfer function from Eq. (12.13) into the overall reactor system transfer model as given by Eq. (12.18), the nuclear rocket transfer function may be determined to be

$$K_{RT} G_{RT} = \frac{1}{\frac{i\omega\Lambda}{q_0} + \frac{1}{q_0}\sum_{i=1}^{6}\frac{i\omega\beta_i}{i\omega + \lambda_i} + \frac{\alpha_f A_f}{1 + i\omega\tau_f} + \frac{\alpha_m A_m}{(1 + i\omega\tau_f)(1 + i\omega\tau_m)} - \frac{\beta_p B_p}{(1 + i\omega\tau_f)(1 + i\omega\tau_m)(1 + i\omega\tau_p)}} \tag{12.47}$$

The stability of the rocket reactor system may now be examined by presenting Eq. (12.47) in the form of a *Bode plot* as shown in Fig. 12.2. In a Bode plot, the gain (in decibels) and the phase shift of

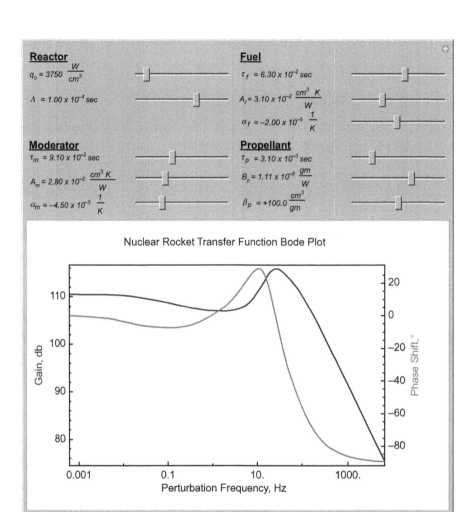

FIGURE 12.2

Bode plot of a nuclear rocket transfer function with thermal feedback.

the response function are plotted against the perturbation frequency. The gain and phase shift are normally defined in the following manner:

$$\text{Gain} = 20\,\text{Re}[\log(K_{RT}G_{RT})] \quad \text{and} \quad \text{Phase shift} = \tan^{-1}\left[\frac{\text{Im}(K_{RT}G_{RT})}{\text{Re}(K_{RT}G_{RT})}\right] \quad (12.48)$$

2. REACTOR STABILITY MODEL INCLUDING THERMAL FEEDBACK

EXAMPLE

Use a Bode plot to evaluate the stability of the nuclear reactor part of a NERVA nuclear rocket. Assume that the design is such that the fuel element consists of small fissile particles embedded in a graphite matrix with hydrogen propellant. Use the parameter values presented in the following table to determine the stability characteristics of the reactor:

Variable	Value	Units	Description
q_0	3750	$\frac{W}{cm^3}$	Core average power density
Λ	0.0001	s	Prompt neutron lifetime
S_{fm}	3.5	cm^{-1}	Surface area of fuel per unit volume of fuel element
c_p^f	0.15	$\frac{W\,s}{g\,K}$	Specific heat of fuel
k^f	0.23	$\frac{W}{cm\,K}$	Thermal conductivity of fuel
ρ^f	13.5	$\frac{g}{cm^3}$	Density of fuel
α^f	−0.000020	K^{-1}	Fuel temperature coefficient of reactivity
S_{mp}	8.3	cm^{-1}	Surface area of coolant holes per unit volume of fuel element
c_p^m	1.9	$\frac{W\,s}{g\,K}$	Specific heat of graphite moderator
k^m	0.31	$\frac{W}{cm\,K}$	Thermal conductivity of graphite moderator
ρ^m	1.7	$\frac{g}{cm^3}$	Density of graphite moderator
α^m	−0.000045	K^{-1}	Moderator temperature coefficient of reactivity
c_p^p	16.8	$\frac{W\,s}{g\,K}$	Specific heat of hydrogen propellant
h_c	0.5	$\frac{W}{cm^2 K}$	Heat transfer coefficient of hydrogen
P	7	MPa	Average hydrogen propellant pressure
T	1500	K	Average hydrogen propellant temperature
R	4.2	$\frac{MPa\,cm^3}{g\,K}$	Gas constant for hydrogen
$\dot{m}$	0.083	$\frac{g}{s\,cm^3}$	Hydrogen flow rate per unit volume of fuel element
β^p	100	$\frac{cm^3}{g}$	Hydrogen density coefficient of reactivity
r^f	0.025	cm	Characteristic distance for heat transfer through fuel
r^m	0.2	cm	Characteristic distance for heat transfer through graphite matrix

Solution

To begin the analysis, it is necessary to first calculate effective values for the thermal conductance in the fuel and moderator. For the fuel, the thermal conductance of the fuel particles per unit fuel element volume is primarily a function of the fuel thermal conductivity and various geometric factors. From basic heat transfer considerations, a reasonable approximation for the thermal conductance may be described by

$$U_{fm} = \frac{k^f S_{fm}}{r^f} = 32.3 \frac{W}{cm^3 K}$$

For the moderator, the thermal conductance of the graphite per unit fuel element volume is not only a function of the fuel thermal conductivity and various geometric factors, but also of the heat transfer coefficient between the moderator and the propellant. For this case, a reasonable approximation for the thermal conductance may be described by

$$U_{mp} = \frac{S_{mp}}{\frac{r^m}{k^m} + \frac{1}{h_c}} = 3.14 \frac{W}{cm^3 K}$$

Knowing the thermal conductance's of the fuel and moderator, it is now possible to calculate values for the thermal resistance constants and the time constants such that:

$$\tau_f = \frac{\rho_f c_p^f}{U_{fm}} = 0.063 \text{ s}, \quad \tau_m = \frac{\rho_m c_p^m}{U_{fm} + U_{mp}} = 0.091 \text{ s},$$

$$\tau_p = \frac{P c_p^p}{RT(U_{mp} + 2\dot{m}c_p^p)} = 0.0031 \text{ s}, \quad A_m = \frac{1}{U_{fm} + U_{mp}} = 0.028 \frac{cm^3 K}{W},$$

$$A_f = \frac{1}{U_{fm}} = 0.031 \frac{cm^3 K}{W}, \quad B_p = \frac{U_{mp} P}{RT^2 (U_{fm} + U_{mp})(U_{mp} + 2\dot{m}c_p^p)} = 1.11 \times 10^{-8} \frac{g}{W}$$

Using the time constants and thermal resistance constants just calculated plus the other specified parameters, it is now possible to create a Bode plot describing the nuclear rocket system stability. Note that the Bode plot described earlier in Fig. 12.2 uses these parameter values by default. It is observed in the plot that the gain is finite for all frequencies and that the phase shift is always less than 180 degree. Such behavior indicates that the nuclear rocket engine is stable. If the design were modified such that the propellant density coefficient of reactivity was reduced to $-170 \frac{cm^3}{g}$ or less as shown in Fig. 1, a 180 degree phase shift would occur at low frequencies, indicating the occurrence of unstable engine operation in which the reactor would experience a constantly increasing power.

Also observe that stable engine operation is possible even when some parts of the reactor have positive thermal reactivity coefficients. As long as the overall thermal reactivity coefficient is negative, stable reactor operation is possible. In addition, note that if all the time and thermal resistance constants are set to zero so as to turn off all thermal reactivity feedback, the Bode plot confirms the statement made earlier that an exactly critical reactor is unstable since the gain becomes infinite as the frequency approaches zero.

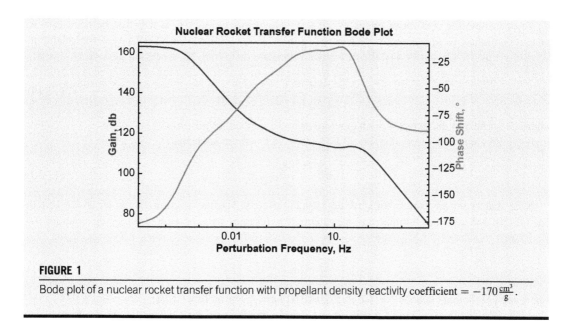

FIGURE 1

Bode plot of a nuclear rocket transfer function with propellant density reactivity coefficient $= -170 \frac{cm^3}{g}$.

3. THERMAL FLUID INSTABILITIES

It has been postulated by Bussard [1] that because the viscosity of most gases increases as the temperature increases, the possibility exists that a flow instability may manifest itself in heated flow channels such as are present in NERVA-type fuel elements. The instability could occur if, for instance, there was an increase in the heating rate in the fuel surrounding an individual flow channel. Physically, the additional heating would cause an increase in the gas temperature resulting in an increase in the gas viscosity. This increase in the gas viscosity would tend to cause the pressure drop in that flow channel to also increase. Assuming that the gas flow is unconstrained, that is, the gas is free to travel down any of the available flow channels in the fuel element and also assuming that the total pressure drop across the core as a whole remains constant, the gas flow rate in the channel experiencing the additional heating should decrease to compensate for the increase in the gas viscosity. The flow instability results from the fact that the gas flow rate continues to decrease and the gas temperature increase until the pressure drop equalizes across all the fuel element flow channels. Under certain conditions, the gas flow rate in the channel experiencing the additional heating decreases to such an extent that the resulting increase in the gas temperature causes the temperature in the fuel adjacent to the channel experiencing the flow reduction to exceed acceptable limits.

To determine the conditions under which this instability could exist, it will first be necessary to write an equation for the pressure drop across a differential section of a NERVA-type fuel element. Recalling Eqs. (9.42) and (9.44), the differential pressure drop may be expressed as

$$dP = f \frac{dz}{D} \frac{\rho}{2} \left(\frac{\dot{m}}{\rho A}\right)^2 = f \frac{\dot{m}^2}{2\rho A^2 D} dz \tag{12.49}$$

where D = channel diameter; A = channel area; and z = channel position.

If the flow channel is circular, Eq. (12.49) may be rewritten such that:

$$dP = f \frac{8\dot{m}^2}{\rho \pi^2 D^5} dz \tag{12.50}$$

Note that from an examination of Eqs. (9.23) and (9.24), the friction factor (f) expressed in Eq. (12.50) for both the laminar and turbulent flow regimes may be expressed by an equation of the form:

$$f = \alpha + \frac{\beta}{Re^\xi} \tag{12.51}$$

where the coefficients needed for use in Eq. (12.51) are given in Table 12.1:

Also note that the viscosity for most gases may be represented by a power law of the form:

$$\mu = \mu_0 T^n \tag{12.52}$$

Using Eq. (12.52), the Reynolds number may now be expressed as

$$Re = \frac{4\dot{m}}{\pi D \mu} = \frac{4\dot{m}}{\pi D \mu_0 T^n} \tag{12.53}$$

In Eq. (12.51) for the friction factor, the Reynolds number may now be replaced by the relationship given in Eq. (12.53) such that:

$$f = \alpha + \frac{\beta}{Re^\xi} = \alpha + \beta \left(\frac{\pi D \mu_0 T^n}{4\dot{m}}\right)^\xi \tag{12.54}$$

Finally, recall that in the ideal gas law, the gas density may be expressed as a function of temperature and pressure such that:

$$\rho = \frac{P}{RT} \tag{12.55}$$

Table 12.1 Friction Factor Coefficients for Tubes

Parameter	Laminar [2]	Turbulent [3]
α	0	$0.094\left(\frac{\epsilon}{D}\right)^{0.225} + 0.53\left(\frac{\epsilon}{D}\right)$
β	64	$88\left(\frac{\epsilon}{D}\right)^{0.44}$
ξ	1	$1.62\left(\frac{\epsilon}{D}\right)^{0.134}$

3. THERMAL FLUID INSTABILITIES

Incorporating the relationships expressed in Eqs. (12.54) and (12.55) into Eq. (12.50), the differential pressure drop may be given by an expression of the form:

$$dP = f\frac{8\dot{m}^2}{\rho\pi^2 D^5}dz = \left[\alpha + \beta\left(\frac{\pi D\mu_0 T^n}{4\dot{m}}\right)^{\xi}\right]\frac{RT}{P}\frac{8\dot{m}^2}{\pi^2 D^5}dz \qquad (12.56)$$

In order to integrate Eq. (12.56) to determine the pressure drop over the entire length of the channel, it is necessary to determine the fluid temperature as a function of position. This relationship may be determined from the first law of thermodynamics such that:

$$qz = \dot{m}c_p(T - T_{in}) \Rightarrow T = \frac{qz}{\dot{m}c_p} + T_{in} \qquad (12.57)$$

where q = heating rate per unit length of channel; T = temperature of the fluid at position "z" in the channel; and T_{in} = temperature of the fluid at the channel inlet.

In the current analysis, the heating rate "q" specified in Eq. (12.57) will be assumed to be a constant even though the parameter is normally a function of position (eg, chopped cosine). This assumption of a constant heating rate will greatly simplify the analysis and will not affect the results too much in most cases. Therefore, incorporating Eq. (12.57) into Eq. (12.56), rearranging terms and integrating then yields:

$$\int_{P_{out}}^{P_{in}} P\,dP = \int_0^L \frac{8\dot{m}^2 R}{\pi^2 D^5}\left[\alpha + \beta\left(\frac{\pi D\mu_0}{4\dot{m}}\right)^{\xi}\left(\frac{qz}{\dot{m}c_p} + T_{in}\right)^{n\xi}\right]\left(\frac{qz}{\dot{m}c_p} + T_{in}\right)dz \qquad (12.58)$$

Carrying out the integrations presented in Eq. (12.58) then yields an equation of the form:

$$P_{in}^2 - P_{out}^2 = \frac{4\dot{m}RL}{\pi^2 D^5}\left(\frac{qL}{c_p} + 2\dot{m}T_{in}\right)\alpha + \frac{2^{3-2\xi}c_p\dot{m}^{3-\xi}R}{q(2+n\xi)\pi^{2-\xi}D^{5-\xi}}\left[\left(\frac{qL}{\dot{m}c_p} + T_{in}\right)^{2+n\xi} - T_{in}^{2+n\xi}\right]\beta\mu_0^{\xi} \qquad (12.59)$$

It is interesting to note that in the laminar flow regime, Eq. (12.59) reduces to a form of the Hagen–Poiseuille law given by

$$P_{in}^2 - P_{out}^2 = \frac{128 c_p\dot{m}^2 R}{q(2+n)\pi D^4}\left[\left(\frac{qL}{\dot{m}c_p} + T_{in}\right)^{2+n} - T_{in}^{2+n}\right]\mu_0 \qquad (12.60)$$

Fig. 12.3 illustrates the results of applying Eq. (12.59) to a NERVA-type fuel element operating at full power. Assuming that the desired outlet temperature from the reactor is 3000 K, the required full power flow rate of 1.12 g/s results in the flow being well into the turbulent regime. The graph indicates that the turbulent flow is stable since a decrease in the flow rate results in a lower pressure drop. In order to maintain a constant pressure drop across all channels, the channel experiencing the flow reduction will respond by increasing the flow rate to its original value thus bringing the channel back into pressure equilibrium with the other flow channels. Similar results can be seen for grooved ring-type fuel elements though, obviously, the characteristic geometric parameters and the heating rate are quite different.

Problems may arise, however, if low power operation is desired such as what might occur during engine shutdown or throttling operations. The graph shows that if the heating rate in the flow channel is reduced to 45 W/cm, reactor operation with an outlet temperature of 3000 K requires a channel flow rate of only 0.131 g/s. At this flow rate the channel flow is just barely turbulent. Should a flow perturbation in the channel occur wherein the flow is reduced even further, laminar flow in the channel

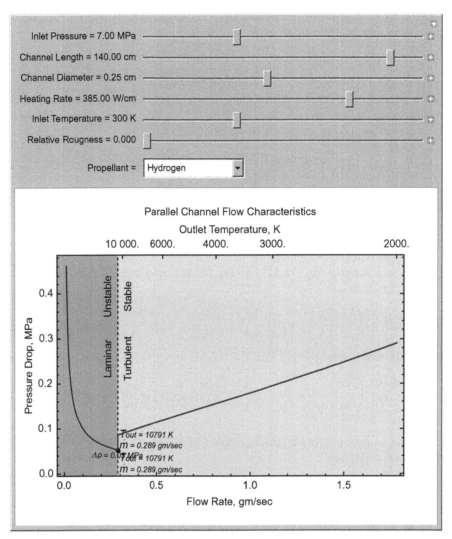

FIGURE 12.3

Thermal fluid stability in fuel elements having parallel flow channels.

could start to develop. In this unfortunate situation, the pressure drop in the channel would suddenly fall precipitously. In this case, however, to equalize the pressure drop with the other channels, the flow rate would continue to decrease until the pressure drops in all channels finally came into equilibrium. This flow rate reduction necessarily results in an increase in the flow channel outlet temperature along with a consequent increase in the fuel temperature. The increase in the channel flow outlet temperature can be quite dramatic with the temperature increasing to almost 7100 K before the flow stabilizes at 0.0521 g/s in this particular case.

3. THERMAL FLUID INSTABILITIES

Table 12.2 Friction Factor Coefficients for Particle Beds

Parameter	Particle Bed [4]
α	$\frac{3.5(1-\epsilon)}{\epsilon^3}$
β	$\frac{300(1-\epsilon)^2}{\epsilon^3}$
ξ	1

One might suppose from the preceding discussion that all the flow instabilities occur at the laminar to turbulent flow boundary: however, this is not the case. If the heating rate is reduced even further to values below 7.3 W/cm, it will be observed that portions of the laminar flow regime will be stable and other portions will be unstable. At power levels this low, however, the temperature differences between the high and low flow channels will generally not be nearly as dramatic as with the turbulent to laminar flow transition. Much higher flow perturbations would be normally required to initiate instabilities where large temperature differences between the high and low flow channels would result.

A similar analysis to that given above may be used to investigate thermal fluid instabilities in particle bed reactors. The only difference in the analyses is that the coefficients used in the friction factor equation employ a relationship [4] derived from the Ergun correlation [5] which describes the pressure drop experienced by fluids flowing through packed beds. The form of the friction factor equation is identical to that presented in Eq. (12.51); however, in this case, the coefficients used would be as in Table 12.2.where ϵ = porosity of the particle bed.

The correlation used to calculate the Reynolds number of the fluid flow in the bed is also somewhat different from that presented in Eq. (12.53). In this case, the Reynolds number is represented by an equation of the form:

$$Re = \frac{\dot{m}D_p}{(1-\epsilon)\mu} \tag{12.61}$$

where D_p = diameter of the fuel particle and $\dot{m}$ = mass flow rate of the fluid per unit surface area of the particle bed.

Incorporating this new definition for the Reynolds number into Eq. (12.54) for the friction factor then yields:

$$f = \alpha + \frac{\beta}{Re^\xi} = \alpha + \beta \left[\frac{(1-\epsilon)\mu}{\dot{m}D_p}\right]^\xi \tag{12.62}$$

Inserting the relationships expressed in Eqs. (12.55) and (12.62) into Eq. (9.42), the differential pressure drop for a particle bed fuel element may be given by an expression of the form:

$$dP = f\frac{dz}{D_p}\frac{\rho V^2}{2} = f\frac{\dot{m}^2}{2\rho D_p}dz = \left\{\alpha + \beta\left[\frac{(1-\epsilon)\mu_0 T^n}{\dot{m}D_p}\right]^\xi\right\}\frac{RT}{P}\frac{\dot{m}^2}{2D_p}dz \tag{12.63}$$

Incorporating Eq. (12.57) into Eq. (12.63), rearranging terms and integrating then yields:

$$\int_{P_{out}}^{P_{in}} P\, dP = \int_0^L \frac{\dot{m}^2 R}{2D_p} \left[\alpha + \beta\mu_0^\xi \left(\frac{1-\epsilon}{\dot{m}D_p}\right)^\xi \left(\frac{qz}{\dot{m}c_p} + T_{in}\right)^{n\xi}\right] \left(\frac{qz}{\dot{m}c_p} + T_{in}\right) dz \quad (12.64)$$

where q = power density in the fuel particle bed.

Carrying out the integrations presented in Eq. (12.64) then yields an equation of the form:

$$P_{in}^2 - P_{out}^2 = \frac{\dot{m}RL}{4D_p}\left(\frac{qL}{c_p} + 2\dot{m}T_{in}\right)\alpha + \frac{c_p R \dot{m}^{3-\xi}(1-\epsilon)^\xi}{2(2+n\xi)qD_p^{1+\xi}}\left[\left(\frac{qL}{\dot{m}c_p} + T_{in}\right)^{2+n\xi} - T_{in}^{2+n\xi}\right]\beta\mu_0^\xi \quad (12.65)$$

Fig. 12.4 illustrates the results of applying Eq. (12.65) to a particle bed-type fuel element operating at full power. An examination of the plot reveals that at the high power densities likely to be typical of

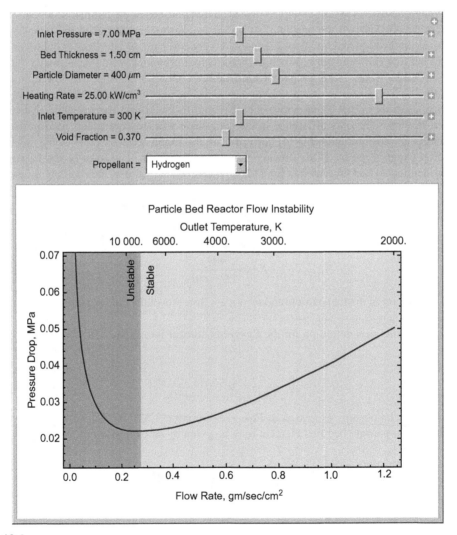

FIGURE 12.4

Thermal fluid stability in particle bed fuel elements.

particle bed reactor operation, stable operation should be possible. At lower power densities, however, (eg, roughly 3 kW/cm^3) thermal instabilities will likely begin to manifest themselves at the outlet temperatures usually desired for nuclear rocket operation. Note also that the Reynolds number for the flow in the bed is typically quite low implying that laminar flow conditions usually exist within the particle bed. Just as in the parallel flow case described earlier, stable operating conditions in the particle bed reactor depend upon where in the laminar flow regime the reactor normally operates. One caveat in this analysis which should be mentioned is that it is tacitly assumed that fluid which enters the particle bed at a particular location always follows the same path to the exit of the bed. This assumption is not necessarily correct, however, since the fluid is free to follow whatever three-dimensional path through the bed it chooses as a result of pressure and flow fluctuations or varying bed conditions. Studies [6] have shown that the three-dimensional flow effects coupled with the particle bed thermal conductivity can have a significant effect on flow stability within the fuel element. Nonetheless, Fig. 12.4 should give qualitatively correct results, especially if the particle bed thermal conductivity is not too high.

REFERENCES

[1] R.W. Bussard, R.D. DeLauer, Nuclear Rocket Propulsion, McGraw-Hill, New York, 1958.
[2] C.F. Colebrook, Turbulent flow in pipes, with particular reference to the transition region between smooth and rough pipe laws, Journal of the Institution of Civil Engineers (London) 11 (4) (February 1939).
[3] D.J. Wood, An explicit friction factor relationship, Civil Engineering, ASCE 60 (1966).
[4] G. Maise, Flow Stability in the Particle Bed Reactor, Brookhaven National Laboratory, 1991. Informal Report BNL/RSD-91-002.
[5] S. Ergun, Fluid flow through packed columns, Chemical Engineering Progress 48 (2) (1952) 89-94.
[6] J. Kalamas, A Three-dimensional Flow Stability Analysis of the Particle Bed Reactor (Masters thesis), Massachusetts Institute of Technology, 1993.

CHAPTER 13

FUEL BURNUP AND TRANSMUTATION

During nuclear reactor operation, the fission products created build up over time and gradually poison the reactor. Eventually, these fission products along with the depletion of the fissionable material can reduce the reactor k_{eff} to the point where criticality can no longer be maintained. For nuclear thermal rockets, fuel burnup generally does not pose much of a problem since the time of operation is so short that there is not sufficient time for fission products to accumulate to any great extent. For power-producing reactors, which operate for long periods of time at relatively high neutron fluxes, however, fuel burnup can be significant and has to be accounted for in the design of the reactor system. Such long-term operation would be required if, for instance, a nuclear reactor were being used to power some type of electric propulsion system. While there are many fission product nuclides created during the fission process, there are two nuclides in particular which, because of their extremely high neutron absorption cross sections and high fission yield probabilities, should be accounted for in the design of even nuclear thermal rocket engines. These nuclides are ^{135}Xe and ^{149}Sm. Of the two, ^{135}Xe is considerably more important than ^{149}Sm due to the fact that ^{135}Xe has a significantly higher neutron absorption cross section.

1. FISSION PRODUCT BUILDUP AND TRANSMUTATION

In nuclear reactors designed to produce power over long periods of time, fuel depletion or burnup can have a significant effect on reactor operation. Such power reactors would be required for systems used to drive electric or ion propulsion systems in deep space where solar energy would not be an option. Under these conditions, reactor systems would have to be designed so as to account for fuel-depletion effects. Nuclear thermal rockets, because they operate for such short periods of time will normally be affected only to a small extent by fuel burnup because there is such a limited amount of time for fission products to build up.

Fissile nuclides, such as ^{235}U, that generally fission upon the absorption of a neutron, yield a spectrum of fission products which slowly poison the reactor core by increasing the rate of parasitic neutron absorption. Two nuclides, in particular, have such high neutron absorption cross sections and fission yield probabilities that they will be discussed in detail. These two nuclides are ^{135}Xe and ^{149}Sm. The details of the cross sections of these two nuclides are shown in Fig. 13.1 and their effect on reactor operations are discussed later.

Besides the increase in the rate of neutron absorption due to the buildup of fission products, it is also the case that the rate of neutron production gradually decreases due to the fission and consequent loss of fissile ^{235}U. This reduction in the rate of neutron production due to the loss of ^{235}U is usually

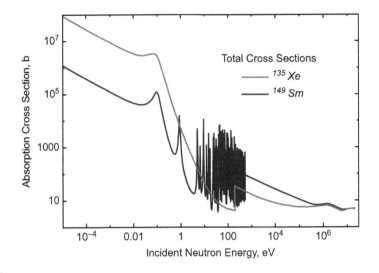

FIGURE 13.1

Total cross sections of ^{135}Xe and ^{149}Sm.

offset to a greater or lesser extent by the fact that the reactor fuel also contains what are called *fertile nuclides* such as ^{238}U. Fertile nuclides are nuclides which upon neutron capture and subsequent beta decay transmute into fissile nuclides. Under certain conditions, it is even possible to design reactors which transmute fertile nuclides into fissile nuclides at a rate greater than the rate at which the fissile nuclides deplete. Such reactors are called *breeder reactors* because they breed fissile nuclides which can then be used to fuel other reactors.

The ratio of the initial fissile atom density to the total of the fissile and fertile atom densities is known as the *fuel enrichment*. Typical commercial reactors use enrichments of roughly 3–6%. Bomb-grade nuclear fuel is 93% enriched. The fuel used in the nuclear engine for rocket vehicle application (NERVA) reactors was 93% enriched and was thus bomb grade. In between, there are isotope production and test reactors which use fuel that is 20% enriched. Fuel of 20% enrichment has been determined to be too low to make any kind of practical nuclear weapon but high enough to construct fairly small and compact reactors which, no doubt, could also be used for nuclear rockets.

As was the case with the burnup of fissile nuclides, the transmutation of fertile nuclides into fissile nuclides is normally of little importance in nuclear rocket engine operation; however, for power reactors the transmutation of fertile nuclides into fissile nuclides can have a significant effect on reactor operation. This is especially true for reactors designed to operate at high power levels for long periods of time. In other applications, some of the fissile nuclides created in breeder reactors could prove to be excellent fuels for nuclear rocket reactors due to their high fission cross sections. Such nuclides include ^{239}Pu, ^{241}Pu, and ^{242m}Am. The neutron absorption transmutation chains starting from ^{238}U which yield the various fissile nuclides is given in Fig. 13.2.

The transmutation chains illustrated in Fig. 13.2 can be represented by a series of coupled linear differential equations which can be solved using a variety of standard techniques. A very abbreviated

1. FISSION PRODUCT BUILDUP AND TRANSMUTATION

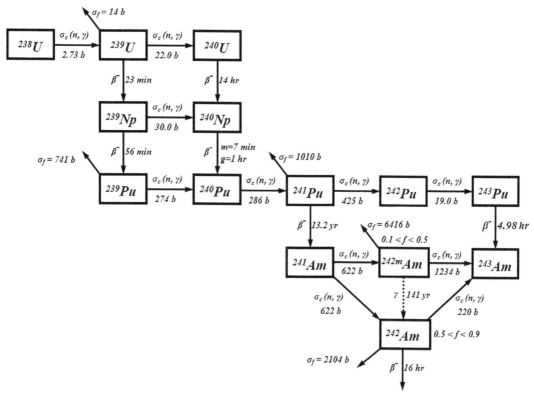

FIGURE 13.2

Transmutation chain for ^{238}U.

set of those differential equations is given by Eqs. (13.1)–(13.4), which illustrates the transmutation of fertile ^{238}U into fissile ^{239}Pu. The full set of differential equations for the entire transmutation chain could be given, but the analytical solution of those equations representing the time-dependent nuclide atom densities would be quite complicated:

$$\frac{dN_{U238}}{dt} = -\sigma_a^{U238}\phi N_{U238} \tag{13.1}$$

$$\frac{dN_{U239}}{dt} = \sigma_c^{U238}\phi N_{U238} - \left(\sigma_a^{U239}\phi + \lambda^{U239}\right)N_{U239} \tag{13.2}$$

$$\frac{dN_{Np239}}{dt} = \lambda^{U239} N_{U239} - \left(\sigma_a^{Np239}\phi + \lambda^{Np239}\right)N_{Np239} \tag{13.3}$$

$$\frac{dN_{Pu239}}{dt} = \lambda^{Np239} N_{Np239} - \sigma_a^{Pu239}\phi N_{Pu239} \tag{13.4}$$

Technically, these differential equations are nonlinear since the reaction cross sections are functions of energy, and the energy spectrum of the neutron flux varies somewhat over time due to density changes in the reactor nuclide distribution. In addition, at constant reactor power, the absolute neutron flux level generally tends to rise with time as a consequence of the decrease in the macroscopic fission cross section resulting from the depletion of the fissile nuclides. These cross section and neutron flux level changes normally occur quite slowly, however, and for calculational purposes can be treated as constant over fairly long periods of time (generally 10s of days).

Rearranging Eq. (13.1) and integrating yields for the time-dependent ^{238}U atom density:

$$\int \frac{dN_{U238}}{N_{U238}} = -\sigma_a^{U238} \phi \int dt \quad \Rightarrow \quad \ln(N_{U238}) = -\sigma_a^{U238} \phi t + C \tag{13.5}$$

Assuming that the initial concentration of ^{238}U is N^0_{U238} it is possible to use Eq. (13.5) to determine a value for the constant of integration such that:

$$\ln(N^0_{U238}) = -\sigma_a^{U238} \phi(0) + C \quad \Rightarrow \quad C = \ln(N^0_{U238}) \tag{13.6}$$

Substituting the constant of integration determined in Eq. (13.6) into Eq. (13.5), it is found that:

$$\ln(N_{U238}) = -\sigma_a^{U238} \phi t + \ln(N^0_{U238}) \quad \Rightarrow \quad \ln\left(\frac{N_{U238}}{N^0_{U238}}\right) = -\sigma_a^{U238} \phi t \tag{13.7}$$

Rearranging Eq. (13.7) to solve for the time-dependent concentration of ^{238}U then yields:

$$N_{U238}(t) = N^0_{U238} e^{-\sigma_a^{U238} \phi t} \tag{13.8}$$

When solving for the transmutation of ^{238}U into ^{239}Pu, Eqs. (13.2) and (13.3) are typically ignored because the decay half-life of ^{239}U (23 min) and ^{239}Np (56 min) are so short. As a consequence, the assumption is made that neutron absorption in ^{238}U immediately results in the production of ^{239}Pu. This assumption results in little error in the calculated nuclide densities and considerably simplifies the nuclide density derivations. The differential equation for ^{239}Pu as expressed by Eq. (13.4) then becomes:

$$\frac{dN_{Pu239}}{dt} = \sigma_c^{U238} \phi N_{U238} - \sigma_a^{Pu239} \phi N_{Pu239} \tag{13.9}$$

Substituting Eq. (13.8) into Eq. (13.9) and rearranging then yields:

$$\frac{dN_{Pu239}}{dt} + \underbrace{\sigma_a^{Pu239} \phi}_{P} N_{Pu239} = \underbrace{\sigma_c^{U238} \phi N^0_{U238} e^{-\sigma_a^{U238} \phi t}}_{Q} \tag{13.10}$$

The differential equation for the time dependence of ^{239}Pu as expressed by Eq. (13.10) may be solved through the use of an *integrating factor* (μ) wherein:

$$\mu = e^{\int P\, dt} = e^{\int \sigma_a^{Pu239} \phi\, dt} = e^{\sigma_a^{Pu239} \phi t} \tag{13.11}$$

Using the integrating factor found in Eq. (13.11), the solution to the differential equation as expressed by Eq. (13.10) is then:

$$N_{Pu239}(t)\mu = \int \mu Q\, dt + C = \int \mu \sigma_c^{U238} \phi N^0_{U238} e^{-\sigma_a^{U238} \phi t}\, dt + C = N_{Pu239}(t) e^{\sigma_a^{Pu239} \phi t}$$

$$= \sigma_c^{U238} \phi N^0_{U238} \int e^{\sigma_a^{Pu239} \phi t} e^{-\sigma_a^{U238} \phi t}\, dt + C \tag{13.12}$$

Rearranging Eq. (13.12) and integrating then yields:

$$N_{Pu239}(t) = e^{-\sigma_a^{Pu239}\phi t}\frac{\sigma_c^{U238}\phi N_{U238}^0}{\phi(\sigma_a^{Pu239}-\sigma_a^{U238})}e^{\sigma_a^{Pu239}\phi t}e^{-\sigma_a^{U238}\phi t} + Ce^{-\sigma_a^{Pu239}\phi t} \quad (13.13)$$

Assuming that initially there is no ^{239}Pu present, Eq. (13.13) may be used to determine a value for the arbitrary constant "C" such that at $t = 0$:

$$N_{Pu239}(0) = 0 = \frac{\sigma_c^{U238}N_{U238}^0}{\sigma_a^{Pu239}-\sigma_a^{U238}} + C \Rightarrow C = -\frac{N_{U238}^0 \sigma_c^{U238}}{\sigma_a^{Pu239}-\sigma_a^{U238}} \quad (13.14)$$

Substituting Eq. (13.14) into Eq. (13.13) then yields for the time-dependent concentration of ^{239}Pu:

$$\begin{aligned}N_{Pu239}(t) &= \frac{N_{U238}^0 \sigma_c^{U238}}{\sigma_a^{Pu239}-\sigma_a^{U238}}e^{-\sigma_a^{U238}\phi t} - \frac{N_{U238}^0 \sigma_c^{U238}}{\sigma_a^{Pu239}-\sigma_a^{U238}}e^{-\sigma_a^{Pu239}\phi t} \\ &= N_{U238}^0 \frac{\sigma_c^{U238}\left(e^{-\sigma_a^{U238}\phi t}-e^{-\sigma_a^{Pu239}\phi t}\right)}{\sigma_a^{Pu239}-\sigma_a^{U238}}\end{aligned} \quad (13.15)$$

In a typical depletion calculation, a multigroup diffusion calculation is initially performed to determine the space-dependent multigroup neutron fluxes. The diffusion calculation is then followed by a depletion calculation to determine the space-dependent nuclide density distribution. This nuclide distribution then forms the basis for a subsequent multigroup diffusion calculation which is used to renormalize the space-dependent multigroup neutron fluxes. These neutron fluxes are then used as the basis for another depletion calculation which yields a new space-dependent nuclide density distribution. This calculational sequence is continued until the desired degree of fuel burnup is attained. Results describing the time-dependent burnup and transmutation of the most important fertile and fissile nuclides in the ^{238}U chain are given in Fig. 13.3 under the assumptions of constant neutron flux and reaction cross sections.

Note in Fig. 13.3 that the equilibrium atom densities for plutonium and americium are much higher when the ratio of the thermal flux to the fast flux is low, in other words, when the neutron flux is mostly fast. Reactors in which most of the flux is in the fast neutron energy groups are called "fast" reactors, and it has been found that a fast neutron energy spectrum is required for the practical construction of breeder reactors. The reason that fast reactors are required to breed significant amounts of fissile material is that typically the fission cross sections are quite high at thermal energies and as a consequence the bred nuclides tend to fission about as fast as they get created when subjected to significant thermal neutron flux levels.

2. XENON 135 POISONING

Probably the most significant fission product produced during reactor operation is xenon 135. This nuclide has an extremely high thermal neutron capture cross section of 2.7×10^6 b due to a large absorption resonance at 0.082 eV plus it has a high fission yield probability. Actually, the majority of the ^{135}Xe produced during fission does not come directly from the fission process ($\gamma^{Xe} = 0.003$) but, rather, is produced as a result of a series of β^- decays starting with ^{135}Te which has a much higher fission yield fraction ($\gamma^{Te} = 0.064$). From the decay chain for ^{135}Xe outlined in Fig. 13.4, it can be seen that the decay of ^{135}Te is quite rapid (~ 43 s).

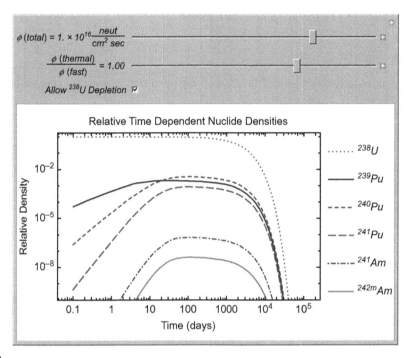

FIGURE 13.3

Nuclide production resulting from ^{238}U transmutation.

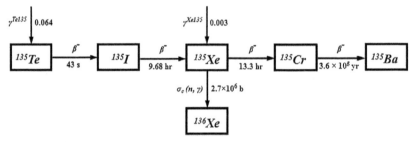

FIGURE 13.4

Depletion chain of ^{135}Xe.

As a result, little error is introduced into the decay calculations if ^{135}Te is neglected and the assumption is made that ^{135}I is produced directly from fission. With this assumption, the ^{135}Xe decay chain equations become

$$\frac{dN_{I135}}{dt} = \gamma^{Te135}\Sigma_f\phi - \lambda^{I135}N_{I135} \tag{13.16}$$

$$\frac{dN_{Xe135}}{dt} = \gamma^{Xe135}\Sigma_f\phi + \lambda^{I135}N_{I135} - \left(\sigma_a^{Xe135}\phi + \lambda^{Xe135}\right)N_{Xe135} \tag{13.17}$$

Solving Eqs. (13.16) and (13.17) yields the time-dependent concentrations of ^{135}I and ^{135}Xe:

$$N_{I135}(t) = \frac{\gamma^{Te135}\Sigma_f\phi}{\lambda^{I135}}\left(1 - e^{-\lambda^{I135}t}\right) + N_{I135}(0)e^{-\lambda^{I135}t} \tag{13.18}$$

$$N_{Xe135}(t) = \frac{\left(\gamma^{Te135} + \gamma^{Xe135}\right)\Sigma_f\phi}{\sigma_a^{Xe135}\phi + \lambda^{Xe135}}\left[1 - e^{-\left(\sigma_a^{Xe135}\phi + \lambda^{Xe135}\right)t}\right]$$

$$- \frac{\gamma^{Te135}\Sigma_f\phi - \lambda^{I135}N_{I135}(0)}{\lambda^{I135} - \lambda^{Xe135} - \sigma_a^{Xe135}\phi}\left[e^{-\left(\sigma_a^{Xe135}\phi + \lambda^{Xe135}\right)t} - e^{-\lambda^{I135}t}\right]$$

$$+ N_{Xe135}(0)e^{-\left(\sigma_a^{Xe135}\phi + \lambda^{Xe135}\right)t}$$

$$\tag{13.19}$$

Following startup, there is no ^{135}I or ^{135}Xe in the reactor; however, after a sufficiently long period of time the nuclides reach equilibrium conditions where from Eqs. (13.18) and (13.19) one finds that:

$$N_{I135}(\infty) = \frac{\gamma^{Te135}\Sigma_f\phi}{\lambda^{I135}} \tag{13.20}$$

$$N_{Xe135}(\infty) = \frac{\left(\gamma^{Te135} + \gamma^{Xe135}\right)\Sigma_f\phi}{\sigma_a^{Xe135}\phi + \lambda^{Xe135}} \tag{13.21}$$

Note from Eq. (13.21) that the equilibrium value for Σ_a^{Xe135} depends upon the neutron flux level in the reactor, wherein:

$$\Sigma_a^{Xe135} = \sigma_a^{Xe135}N_{Xe135}(\infty) = \frac{\left(\gamma^{Te135} + \gamma^{Xe135}\right)\Sigma_f}{1 + \frac{\lambda^{Xe135}}{\sigma_a^{Xe135}\phi}} \tag{13.22}$$

If a reactor shutdown occurs such that the neutron flux level goes to zero after a period of time of powered operation, Eqs. (13.18) and (13.19) also show that ^{135}I and ^{135}Xe will behave according to

$$N_{I135}(t)|_{t>t_{sd}} = N_{I135}(t_{sd})e^{-\lambda^{I135}(t-t_{sd})} \stackrel{t_{sd}\to\infty}{=} \frac{\gamma^{Te135}\Sigma_f\phi}{\lambda^{I135}}e^{-\lambda^{I135}(t-t_{sd})} \tag{13.23}$$

$$N_{I135}(t)|_{t>t_{sd}} = \frac{\lambda^{I135}N_{I135}(t_{sd})}{\lambda^{I135} - \lambda^{Xe135}}\left[e^{-\lambda^{Xe135}(t-t_{sd})} - e^{-\lambda^{I135}(t-t_{sd})}\right]$$

$$+ N_{Xe135}(t_{sd})e^{-\lambda^{Xe135}(t-t_{sd})} \stackrel{t_{sd}\to\infty}{=} \frac{\gamma^{Te135}\Sigma_f\phi}{\lambda^{I135} - \lambda^{Xe135}}\left[e^{-\lambda^{Xe135}(t-t_{sd})} - e^{-\lambda^{I135}(t-t_{sd})}\right]$$

$$+ \frac{\left(\gamma^{Te135} + \gamma^{Xe135}\right)\Sigma_f\phi}{\sigma_a^{Xe135}\phi + \lambda^{Xe135}}e^{-\lambda^{Xe135}(t-t_{sd})}$$

$$\tag{13.24}$$

where t_{sd} = time at which reactor shutdown occurs (eg, time when the neutron flux goes to zero).

By now taking the derivative of Eq. (13.24) at $t = t_{sd}$, it is possible to gain some insight into the behavior of the time rate of change of the ^{135}Xe concentration after shutdown occurs. The time derivative of the ^{135}Xe concentration is found to be

$$\frac{dN_{Xe135}}{dt} = \Sigma_f\phi\left(\frac{\phi\sigma_a^{Xe135} - \gamma^{Xe135}\lambda^{Xe135}}{\phi\sigma_a^{Xe135} + \lambda^{Xe135}}\right) \tag{13.25}$$

Note from Eq. (13.25) that if $\sigma_a^{Xe135}\phi > \gamma^{Xe135}\lambda^{Xe135}$ the time derivative of the ^{135}Xe concentration at shutdown will be positive and the ^{135}Xe concentration will begin to increase with time. This situation occurs at neutron flux levels of about 3×10^{11} neut/cm²/s. The maximum ^{135}Xe concentration generally occurs about 10 h after shutdown, but it can take 40–50 h or even longer at very high neutron flux level to return to its equilibrium value.

The implication of this ^{135}Xe concentration increase is that if the reactor has been operating at high power and there is little excess reactivity in the fuel, it is possible that the reactor will be unable to restart for a period of time after shutdown due to the high parasitic neutron absorption rate of the excess xenon. Xenon oscillations can also occur in a reactor. In regions of the reactor core which operate at high neutron flux levels, the ^{135}Xe concentration will buildup and eventually begin to suppress the neutron flux (and hence power) in that region. The ^{135}Xe concentration in that region will then begin to fall causing the neutron flux (and hence power) to again increase in that region. The xenon oscillation time is typically about 10–15 h. Results describing the time-dependent behavior of ^{135}Xe are given in Fig. 13.5 as a function of the characteristics of the neutron flux and the operation time at power.

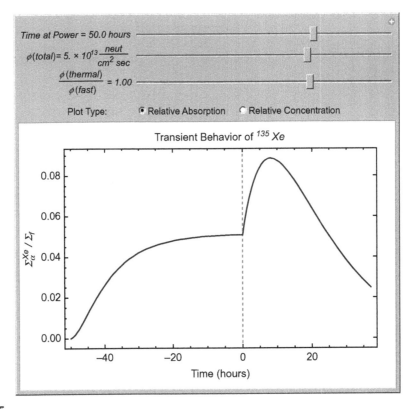

FIGURE 13.5

Transient behavior of ^{135}Xe during reactor operation and shutdown.

3. SAMARIUM 149 POISONING

As was stated previously, during reactor operation the fission process yields a variety of fission products which over a period of time gradually poison the reactor. One of the most important of these fission products is samarium 149. It has a large thermal neutron absorption cross section of about 40,800 b at 0.025 eV plus there is quite a bit of resonance neutron absorption at intermediate neutron energies. In addition, its precursor nuclide ^{149}Nd, has a fairly high fission yield probability of 0.0113. The decay chain for ^{149}Sm is given in Fig. 13.6.

Since ^{149}Nd decays so quickly in relation to ^{149}Pm, it is possible to neglect the effects of ^{149}Nd without introducing serious errors into the decay calculations. Under this assumption, ^{149}Pm appears directly from fission with a yield fraction of 0.0113. With this assumption, the decay equations for ^{149}Sm become

$$\frac{dN_{\text{Pm}149}}{dt} = \gamma^{\text{Nd}149}\Sigma_f\phi - N_{\text{Pm}149}\lambda^{\text{Pm}149} \tag{13.26}$$

$$\frac{dN_{\text{Sm}149}}{dt} = N_{\text{Pm}149}\lambda^{\text{Pm}149} - \phi N_{\text{Sm}149}\sigma_a^{\text{Sm}149} \tag{13.27}$$

Solving differential Eqs. (13.26) and (13.27) yields the time-dependent concentrations for ^{149}Pm and ^{149}Sm:

$$N_{\text{Pm}149}(t) = \frac{\gamma^{\text{Nd}149}\Sigma_f\phi}{\lambda^{\text{Pm}149}}\left(1 - e^{-\lambda^{\text{Pm}149}t}\right) + N_{\text{Pm}149}(0)e^{-\lambda^{\text{Pm}149}t} \tag{13.28}$$

$$N_{\text{Sm}149}(t) = \frac{\gamma^{\text{Nd}149}\Sigma_f}{\sigma_a^{\text{Sm}149}}\left(1 - e^{-\sigma_a^{\text{Sm}149}\phi t}\right) - \frac{\gamma^{\text{Nd}149}\Sigma_f\phi - \lambda^{\text{Pm}149}N_{\text{Pm}149}(0)}{\lambda^{\text{Pm}149} - \sigma_a^{\text{Sm}149}\phi}\left(e^{-\sigma_a^{\text{Sm}149}\phi t} - e^{-\lambda^{\text{Pm}149}t}\right)$$
$$+ N_{\text{Sm}149}(0)e^{-\sigma_a^{\text{Sm}149}\phi t} \tag{13.29}$$

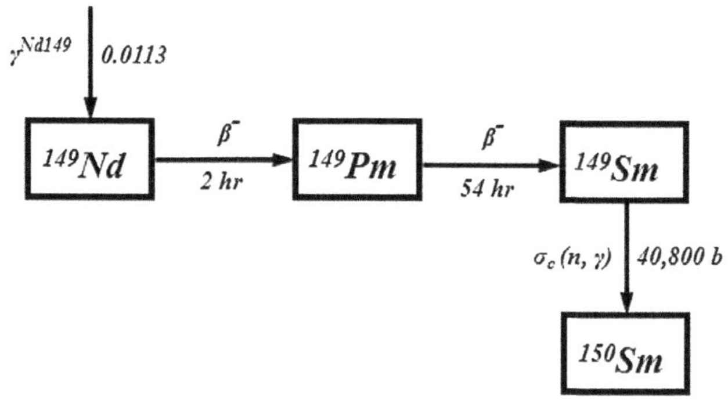

FIGURE 13.6
Depletion chain for ^{149}Sm.

Initially, upon startup, there is no ^{149}Pm or ^{149}Sm in the reactor; however, after a sufficiently long period of time the nuclides reach equilibrium conditions where from Eqs. (13.28) and (13.29), one finds that:

$$N_{\text{Pm149}}(\infty) = \frac{\gamma^{\text{Nd149}} \Sigma_f \phi}{\lambda^{\text{Pm149}}} \tag{13.30}$$

$$N_{\text{Sm149}}(\infty) = \frac{\gamma^{\text{Nd149}} \Sigma_f}{\sigma_a^{\text{Sm149}}} \tag{13.31}$$

At low power densities, these equilibrium concentration values for ^{149}Sm can take a very long time (eg, years) to achieve. This may be illustrated by noting that upon initial startup when there is no ^{149}Pm or ^{149}Sm present, Eq. (13.29) reduces to

$$N_{\text{Sm149}}(t) = \frac{\gamma^{\text{Nd149}} \Sigma_f}{\sigma_a^{\text{Sm149}}} \left(1 - e^{-\sigma_a^{\text{Sm149}} \phi t}\right) \tag{13.32}$$

If one assumes that equilibrium is reached when the concentration of ^{149}Sm reaches 99% of its full equilibrium value, it is found from Eq. (13.32) that for thermal reactors.

$$t \approx \frac{-\ln(0.01)}{\sigma_a^{\text{Sm149}} \phi} \approx \frac{3 \times 10^{16}}{\phi} \text{ (hours)} \tag{13.33}$$

From Eq. (13.33), it will be observed that for a neutron flux level of 10^{12} neut/cm^2/s equilibrium levels of ^{149}Sm will not be reached for about 3.4 years. At high power densities such as is typical in nuclear rocket engines, the buildup of the ^{149}Sm would be considerably faster. If the reactor is shutdown (eg, $\phi = 0$) after a time of high powered operation, Eqs. (13.28) and (13.29) also show that ^{149}Pm and ^{149}Sm will behave according to

$$N_{\text{Pm149}}(t)|_{t>t_{\text{sd}}} = N_{\text{Pm149}}(t_{\text{sd}}) e^{-\lambda^{\text{Pm149}}(t-t_{\text{sd}})} \overset{t_{\text{sd}} \to \infty}{=} \frac{\gamma^{\text{Nd149}} \Sigma_f \phi}{\lambda^{\text{Pm149}}} e^{-\lambda^{\text{Pm149}}(t-t_{\text{sd}})} \tag{13.34}$$

$$N_{\text{Sm149}}(t)|_{t>t_{\text{sd}}} = N_{\text{Pm149}}(t_{\text{sd}}) \left(1 - e^{-\lambda^{\text{Pm149}}(t-t_{\text{sd}})}\right) + N_{\text{Sm149}}(t_{\text{sd}})$$

$$\overset{t_{\text{sd}} \to \infty}{=} \frac{\gamma^{\text{Nd149}} \Sigma_f \phi}{\lambda^{\text{Pm149}}} \left(1 - e^{-\lambda^{\text{Pm149}}(t-t_{\text{sd}})}\right) + \frac{\gamma^{\text{Nd149}} \Sigma_f}{\sigma_a^{\text{Sm149}}} \tag{13.35}$$

where t_{sd} = time at which reactor shutdown occurs (eg, time when the neutron flux goes to zero).

Since ^{149}Sm is stable and its precursor ^{149}Pm is not, it is apparent that eventually all the ^{149}Pm in the core at the time of shutdown will decay into ^{149}Sm causing the concentration of ^{149}Sm at the time of shutdown to increase by that amount. Thus, from Eq. (13.35):

$$\lim_{t-t_{\text{sd}} \to \infty} N_{\text{Sm149}}(t) = N_{\text{Pm149}}(t_{\text{sd}}) + N_{\text{Sm149}}(t_{\text{sd}}) \overset{t_{\text{sd}} \to \infty}{=} \frac{\gamma^{\text{Nd149}} \Sigma_f \phi}{\lambda^{\text{Pm149}}} + \frac{\gamma^{\text{Nd149}} \Sigma_f}{\sigma_a^{\text{Sm149}}} \tag{13.36}$$

From Eq. (13.35), it can be noted that the amount of ^{149}Sm present in the core, a long time after shutdown from an initial state in which the concentrations of ^{149}Pm and ^{149}Sm were in equilibrium, is a function of the average neutron flux level present for a time preceding the shutdown of the reactor.

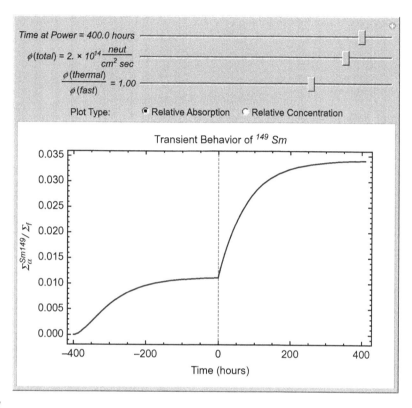

FIGURE 13.7

Transient behavior of ^{149}Sm during reactor operation and shutdown.

To determine a neutron flux level which will result in more than doubling of the ^{149}Sm after shutdown under equilibrium conditions requires that:

$$\frac{\gamma^{Nd149}\Sigma_f \phi}{\lambda^{Pm149}} > \frac{\gamma^{Nd149}\Sigma_f}{\sigma_a^{Sm149}} \quad \Rightarrow \quad \phi > \frac{\lambda^{Pm149}}{\sigma_a^{Sm149}} \approx 10^{14} \frac{\text{neut}}{\text{cm}^2 \text{ s}} \quad (13.37)$$

A description of the time-dependent behavior of ^{149}Sm is given in Fig. 13.7 as a function of the characteristics of the neutron flux and the operation time at power.

4. FUEL BURNUP EFFECTS ON REACTOR OPERATION

Besides the increase in the rate of neutron absorption due to the buildup of fission products, it is also the case that the rate of neutron production gradually decreases due to the fission and consequent loss of the primary fissile nuclide, which is typically ^{235}U. The net result of the reduction in the rate of neutron production and the increase in the rate of neutron absorption is to cause the reactor k_{eff}, which is always somewhat greater than one upon initial reactor startup, to decrease with time. In a typical

214 CHAPTER 13 FUEL BURNUP AND TRANSMUTATION

reactor, exact criticality is maintained through the use of a control system employing highly neutron-absorbing control rods or control drums (typically composed of compounds of boron) which are slowly removed or rotated as the reactor operates to compensate for the continual decrease in k_{eff} due to fuel burnup such that:

$$k_{eff} \equiv 1 = \frac{\text{Neutron production due to fission}}{\text{Neutron leakage} + \text{Neutron absorption due to fission products, structure, etc.} + \text{Neutron absorption due to control system} + \text{Burnable poison absorption}} \quad (13.38)$$

Once the control rods are fully removed or the control drums fully rotated, further reactor operation results in the k_{eff} falling below one and the reactor becoming subcritical. At this point the neutron chain reactions cease and the reactor shuts down. In nuclear rockets, fuel burnup is typically of minor significance since little fuel depletion can occur during the relatively short time the rocket engine is firing even though the reactor power levels are quite high. For power-producing reactors, however, such as would be used to drive ion propulsion systems, fuel-depletion effects would need to be accounted for during the long operational period of the reactor.

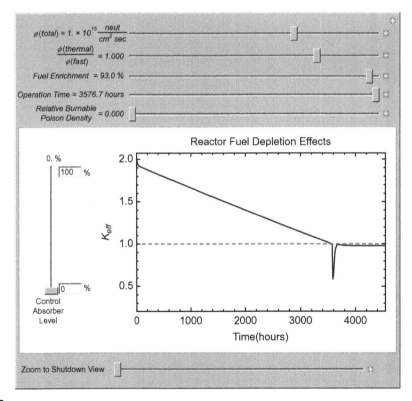

FIGURE 13.8

Reactor operation with depletion effects.

4. FUEL BURNUP EFFECTS ON REACTOR OPERATION

In order to minimize the Δk_{eff} for which the control system must compensate, especially in reactors which are designed to operate for extremely long periods of time, a common practice is to add a *burnable poison* (typically a boron or gadolinium compound) to the nuclear fuel in the reactor. The burnable poison reduces the k_{eff} at the beginning of life of the reactor and progressively burns up over time, compensating to a large extent for the loss of fissile material and the buildup of fission products. Usually, the burnable poison is designed to be almost completely consumed by the end of life of the reactor core. The burnable poison also minimizes the amount of control drive movement required to maintain reactor criticality during the period of reactor operation.

Given in Fig. 13.8 is an interactive plot which simulates the effect on reactor k_{eff} of fuel burnup and transmutation, xenon and samarium buildup, control poison operation, and burnable poison effects. While only a limited number of nuclides are modeled in the plot, the results are qualitatively correct and can simulate a wide range of reactor conditions.

CHAPTER 14

RADIATION SHIELDING FOR NUCLEAR ROCKETS

During nuclear rocket operation, the fissioning reactor emits a tremendous amount of radiation which, if not properly shielded against, could prove lethal to the crew. Generally speaking, only gamma radiation and neutron radiation are important in shield design since beta radiation (high-energy electrons) and alpha radiation (helium nuclei) are easily attenuated by fairly thin layers of material. Since radiation shields are basically just dead weight in the spacecraft, care must be taken in the shield design to achieve the greatest possible radiation protection for the crew for the least possible weight. One method by which the shield weight can be reduced significantly is to use what are called a "shadow shields" wherein shielding is implemented only in the direction of the crew habitat. Other way in which the shield weight can be reduced is to incorporate multiple layers of different materials into the shield design where each layer serves to attenuate a different type of radiation.

1. DERIVATION OF SHIELDING FORMULAS

Shielding designs for nuclear rocket engines present unique challenges which are generally not faced on earth-based nuclear reactors. Since effective radiation shielding generally requires a considerable thickness of high-density material, it is imperative that shielding materials be used as judiciously as possible to minimize the weight of the overall system. Commercial power reactors often use a combination of concrete and iron for shielding. These shields, while quite effective and cheap, are nevertheless quite heavy and as a consequence, generally are unsuitable for use in spacecraft.

Shield design for spacecraft is also complicated by the fact that, typically, there will be a considerable amount of equipment present in the vicinity of the reactor which scatters radiation emanating from the core in many (often unforeseen) directions. If care is not taken to properly account for this equipment and structure around the reactor during the shield design process, even small seams or open areas around this equipment can lead to radiation backscattering that results in dose levels in various parts of the spacecraft that is much higher than what would be predicted. In addition, the gamma radiation emanating from the reactor is multifaceted in the sense that it results from a wide variety of complicated nuclear processes (eg, prompt fission, fission product decay, and neutron capture) as illustrated in Fig. 14.1.

In the analyses which follow, only gamma radiation resulting directly from the fission process, that is, those gammas which originate from within the core (prompt fission gammas) and gammas resulting from slow neutron capture within the shield (capture gammas) are analyzed. These gammas are generally the most important from a shield design standpoint plus they lend themselves more easily to analysis.

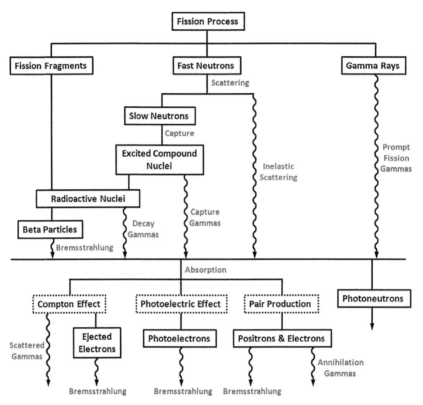

FIGURE 14.1

Nuclear processes resulting from fission.

It should be noted that radiation shields which completely surround a nuclear rocket engine are quite impractical from a design standpoint since at the very least the nozzle area of the engine must be open to allow for the rocket exhaust. It has also been found that such shields are extremely heavy. A strategy which would almost certainly be employed to reduce the amount of shielding required on a spacecraft would be to use what are called "shadow" shields to protect the crew and other vital parts of the spacecraft. In shadow shields, shielding material would only be placed between the nuclear engine and the crew compartment plus those other locations found to be susceptible to radiation backscattering. In general, it is unnecessary to provide shielding in directions exposed only to empty space. Radiation shields could also reduce weight by using several layers of different materials to better shield against the different types of radiation and to complement the shielding properties of each other. An illustration of such a multilayer shield is presented in Fig. 14.2.

Because of the complications which result from trying to account for geometric effects in shields with regard to radiation behavior, only one-dimensional semi-infinite shield configurations are considered in the derivations which follow. These derivations, while being somewhat crude, nevertheless give insight into the manner in which various types of radiation that are attenuated within shields.

1. DERIVATION OF SHIELDING FORMULAS

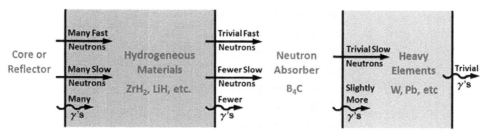

FIGURE 14.2

Typical multilayer shield configuration.

1.1 NEUTRON ATTENUATION

During operation, a nuclear thermal rocket (NTR) engine will generate very high neutron fluxes which will have to be greatly attenuated to protect the crew and sensitive equipment near the reactor. Since fast neutron cross sections are nearly always quite low, it is advantageous to slow these neutrons down to thermal energies by using scattering interactions of some type of hydrogen-bearing material and then use a strong thermal neutron absorber such as boron to attenuate the resulting slow thermal neutrons. To simplify the analysis of this process, the fast neutron flux will be assumed to decay in an exponential manner such that:

$$\phi^1(z) = \phi_0^1 e^{-\left(\Sigma_s^{1 \to 2} + \Sigma_c^1\right)z} \tag{14.1}$$

The thermal flux, on the other hand, will be assumed to behave according to diffusion theory as discussed earlier with neutron scattering from the fast energy group as expressed by Eq. (14.1) acting as a source term for the thermal energy group such that:

$$D^2 \frac{d^2}{dz^2}\phi^2 - \Sigma_c^2 \phi^2 + \phi_0^1 e^{-\Sigma_s^{1 \to 2} z} = 0 \tag{14.2}$$

Solving the thermal flux differential equation as expressed by Eq. (14.2) then yields:

$$\phi^2(z) = \left(\phi_0^2 - \frac{\Sigma_s^{1 \to 2} \phi_0^1}{D^2 \left(\Sigma_s^{1 \to 2}\right)^2 - \Sigma_c^2}\right) e^{-z\sqrt{\frac{\Sigma_c^2}{D^2}}} - \frac{\Sigma_s^{1 \to 2} \phi_0^1}{D^2 \left(\Sigma_s^{1 \to 2}\right)^2 - \Sigma_c^2} e^{-\Sigma_s^{1 \to 2} z} \tag{14.3}$$

where ϕ_0^1 = fast flux at reactor/shield interface (eg, $z = 0$); $\phi_0^2 = \phi_0^2$ = thermal flux at reactor/shield interface (eg, $z = 0$).

As the neutron flux transitions between material regions, the neutron fluxes must be continuous, therefore, from Eqs. (14.1) and (14.3):

$$\phi^1 \sum_{j=1}^{i} h_j = \phi_{0,i}^1 e^{-\left(\Sigma_c^1 + \Sigma_s^{1 \to 2}\right)h_i} = \phi_{0,i+1}^1 \tag{14.4}$$

and

$$\phi^2 \sum_{j=1}^{i} h_j = \left(\phi_{0,i}^2 - \frac{\Sigma_s^{1 \to 2} \phi_{0,i}^1}{D^2 \left(\Sigma_s^{1 \to 2}\right)^2 - \Sigma_c^2} \right) e^{-h_i \sqrt{\frac{\Sigma_c^2}{D^2}}} - \frac{\Sigma_s^{1 \to 2} \phi_{0,i}^1}{D^2 \left(\Sigma_s^{1 \to 2}\right)^2 - \Sigma_c^2} e^{-\Sigma_s^{1 \to 2} h_i} = \phi_{0,i+1}^2 \quad (14.5)$$

where $\phi_{0,i}^1$ = fast flux at the beginning of material region "i"; $\phi_{0,i}^2$ = thermal flux at the beginning of material region "i."

Eqs. (14.4) and (14.5), while not rigorously correct due to the diffusion theory approximation employed in regions, where the neutron flux gradients are high, are nevertheless probably sufficient to give rough values of the neutron attenuation. Generally speaking, transport theory is usually required to obtain numerically accurate solutions.

1.2 PROMPT FISSION GAMMA ATTENUATION

Prompt fission gamma rays are those gamma rays which result directly from the fission process. These gamma rays are quite penetrating and although some will be lost within the reactor itself due to self-shielding effects (eg, the constituents of the reactor absorb the radiation), many of the gamma rays will escape. Those gamma rays which do escape reactor must be attenuated by the radiation shield to manageable levels. The gamma rays produced during fission are not of a single energy, but rather follow a distribution function [1] which is illustrated in Fig. 14.3.

In the analysis which follows, a nuclear reactor will be assumed to be the source of a uniform gamma-ray flux falling on the face of a radiation shield which borders the outside edge of the reactor. As the gamma rays pass through the shield they interact with its material constitutions, undergoing various types of scattering and absorption interactions that serve to attenuate the radiation prior to emerging from the shield. In Fig. 14.4, the gamma radiation from the reactor is assumed to penetrate a cylindrical shield centered at point "O." An expression for the attenuation characteristics of gamma rays will be derived as the radiation passes through the shield, finally emerging at point "P."

FIGURE 14.3

Prompt fission gamma-ray energy distribution.

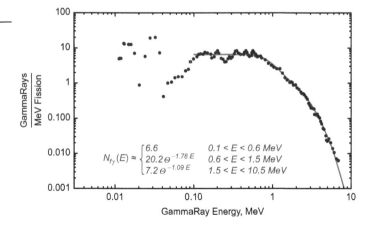

1. DERIVATION OF SHIELDING FORMULAS

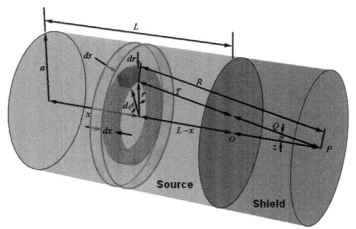

FIGURE 14.4

Gamma-ray flux from a cylindrical volume source.

The analysis begins by noting that the gamma radiation emanating from a small differential volume in the source region may be expressed by

$$ds = S_v N_{\gamma f}(E) dV = S_v N_{\gamma f}(E) r d\phi dr dx \tag{14.6}$$

where S_v = gamma-ray point source strength $\frac{\gamma' s}{cm^3 \, s}$; $N_{\gamma f}(E)$ = energy distribution of fission produced gamma rays.

Also recall that previously it was shown that the attenuation of a collimated narrow beam of neutrons (or gamma rays) may be represented by a simple exponential function. Assuming that gamma rays emanating from the differential source located at "ds" in the Fig. (14.4) travel along a straight line path defined by the collinear line segments "T" in the source region and "Q" in the shield region to a point "P," the differential attenuation of the gamma rays along this path can be defined by an expression of the form:

$$d\phi_{\gamma f}(R) = e^{-\mu_r T} e^{-\mu_s Q} ds \tag{14.7}$$

where μ_r and μ_s = gamma-ray attenuation coefficients for the source and shield regions, respectively.

The specific gamma-ray attenuation coefficients in Eq. (14.7) are dependent upon the material through which the radiation passes and the energy of the radiation. A plot of the specific gamma-ray attenuation coefficients for several materials is illustrated in Fig. 14.5. The data for these curves were taken from the National Institute of Standards and Technology, Tables of X-ray Mass Attenuation Coefficients and Mass Energy-Absorption Coefficients. Note how close the curves are for water and polyethylene. This high degree of correlation is fairly typical for many plastic materials and water.

In addition, assuming that the gamma rays radiate isotropically from the differential source volume "ds" rather than from a collimated beam, Eq. (14.7) must be modified to account for the geometric spreading of the radiation such that:

$$d\phi_{\gamma f}(R) = e^{-\mu_r T} \frac{e^{-\mu_s Q}}{4\pi R^2} ds \tag{14.8}$$

FIGURE 14.5

Gamma-ray mass attenuation coefficients.

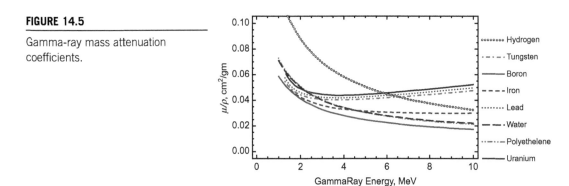

It should also be noted that the tacit assumption in the attenuation expression of Eq. (14.8) is that all particles are either absorbed on their first collision or scattered out of the picture. The term $d\phi_{\gamma f}(R)$, therefore, represents uncollided gamma rays emanating from a point source. This assumption is appropriate for thin shields where secondary collisions are negligible. When the shield is thick, however, and multiple collisions are common, the attenuation calculated will be too low as illustrated in Fig. 14.6.

To account for the attenuation error resulting from these multiple collisions, an empirical term called a buildup factor or "$B(\mu R)$" is included in the point attenuation expression for gamma rays. The buildup factor term is generally added ad hoc into Eq. (14.8) such that:

$$d\phi_{\gamma f}(R) = e^{-\mu_t T} B(\mu_s R) \frac{e^{-\mu_s Q}}{4\pi R^2} ds \qquad (14.9)$$

Buildup factors are generally only used for gamma radiation. In theory, buildup factors could also be used for neutrons; however, this is almost never done. There are several formulas for the buildup factor in the literature; however, for the analyses which follow, the Taylor formulation [2] will be used since it is easy to apply and fairly accurate. The buildup factor for the Taylor formulation is expressed as the sum of two exponential terms such that:

$$B(\mu Y) = Ae^{-\alpha\mu Y} + (1-A)e^{-\beta\mu Y} \qquad (14.10)$$

The coefficients for the Taylor buildup formulation [3,4], for several materials are given in Table 14.1. These coefficients give buildup factors which are generally accurate to within 5%.

Using the Taylor form of the buildup factor from Eq. (14.10) in the differential form of the gamma-ray attenuation expression from Eq. (14.9) then yields:

$$d\phi_{\gamma f}(R) = e^{-\mu_t T}\left[Ae^{-\alpha\mu_s Q} + (1-A)e^{-\beta\mu_s Q}\right]\frac{e^{-\mu_s Q}}{4\pi R^2}ds = e^{-\mu_t T}G_\gamma(\mu_s Q)ds \qquad (14.11)$$

FIGURE 14.6

Scattering differences between thin and thick shields.

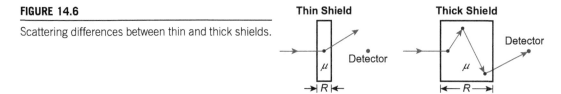

1. DERIVATION OF SHIELDING FORMULAS

Table 14.1 Coefficients for Taylor Gamma-Ray Buildup Formulation for Several Materials

Material	Parameter	Gamma-Ray Energy (MeV)						
		1.0	2.0	3.0	4.0	6.0	8.0	10.0
Water	A	11	6.4	5.2	4.5	3.55	3.05	2.7
	α	−0.104	−0.076	−0.062	−0.055	−0.050	−0.045	−0.042
	β	0.030	0.092	0.110	0.117	0.124	0.128	0.13
Liquid hydrogen	A	3.22	34.8	22.9	20.2	6.82	–	–
	α	−0.165	0.021	0.031	0.025	−0.016	–	–
	β	0.078	0.042	0.058	0.050	0.043	–	–
Iron	A	8.0	5.5	4.4	3.75	2.9	2.35	2.0
	α	−0.089	−0.079	−0.077	−0.075	−0.082	−0.083	−0.095
	β	0.04	0.07	0.075	0.082	0.075	0.055	0.012
Lead	A	6.0	11.6	10.9	6.29	3.5	3.5	2.39
	α	−0.009	−0.021	−0.036	−0.060	−0.108	−0.157	−0.214
	β	0.053	0.015	0.001	−0.007	−0.030	−0.100	−0.092
Tungsten	A	3.3	2.9	2.7	2.05	1.2	0.7	0.6
	α	−0.043	−0.069	−0.086	−0.118	−0.171	−0.205	−0.212
	β	0.148	0.188	0.134	0.070	0.00	0.052	0.144
Uranium	A	2.081	3.550	4.883	2.800	0.975	0.602	0.399
	α	−0.0386	−0.0344	−0.0495	−0.0824	−0.1589	−0.1919	−0.2131
	β	0.2264	0.0881	0.0098	0.0037	0.2110	0.0277	0.0208

In Eq. (14.11), "$G_\gamma(\mu_s Q)$" is known as the gamma-ray point attenuation kernel where

$$G_\gamma(\mu_s Q) = \left[Ae^{-\alpha \mu_s Q} + (1-A)e^{-\beta \mu_s Q}\right]\frac{e^{-\mu_s Q}}{4\pi R^2} \qquad (14.12)$$

The point attenuation kernel is essentially a relationship which yields the degree to which gamma radiation emanating from a point source attenuates as it travels a given distance through a material. The total gamma-ray flux at point "P" may be determined by integrating Eq. (14.9) over the entire volume described in Fig. 14.4 such that:

$$\phi_{\gamma f}(z) = \int_{\text{Volume}} e^{-\mu_t T} G_\gamma(\mu_s Q) ds \qquad (14.13)$$

Substituting Eqs. (14.6), (14.9), and (14.10) into Eq. (14.13) then yields:

$$\phi_{\gamma f}(z) = S v N_{\gamma f}(E) \int_0^L \int_0^a \int_0^{2\pi} r\left[Ae^{-\alpha \mu_s Q} + (1-A)e^{-\beta \mu_s Q}\right]\frac{e^{-\mu_t T}e^{-\mu_s Q}}{4\pi R^2} d\phi dr dx \qquad (14.14)$$

From Fig. 14.4, the following geometric relationships may be noted:

$$T = \frac{L-x}{\text{Cos}(\theta)} \quad \text{and} \quad Q = \frac{z}{\text{Cos}(\theta)} \qquad (14.15)$$

where $\text{Cos}(\theta) = \frac{L-x+z}{R} \Rightarrow T = \frac{(L-x)R}{L-x+z}$ and $Q = \frac{zR}{L-x+z}$

Also observe from Fig. 14.4 that:

$$R^2 = (L - x + z)^2 + r^2 \tag{14.16}$$

Taking the derivative of Eq. (14.7) with respect to "r" then yields:

$$RdR = rdr \tag{14.17}$$

Substituting Eqs. (14.15) and (14.17) into Eq. (14.14) and changing the integration variable from "r" to "R" yields after integrating over "ϕ" an expression of the form:

$$\phi_{\gamma f}(z) = SvN_{\gamma f}(E) \int_0^L \int_{L-x+z}^{\sqrt{(L-x+z)^2+a^2}} 2\pi R \left[Ae^{-\alpha\mu_s\frac{z}{L-x+z}R} \right.$$

$$\left. + (1-A)e^{-\beta\mu_s\frac{z}{L-x+z}R} \right] \frac{e^{-\mu_r\frac{L-x}{L-x+z}R} e^{-\mu_s\frac{z}{L-x+z}R}}{4\pi R^2} dRdx \tag{14.18}$$

By solving the integral expressed in Eq. (14.18), an expression for the gamma-ray attenuation through the centerline of a disk shield of radius "a" may be obtained as a function of the thickness of the gamma-ray source region. To simplify the resulting expression and to make the results more general, the integration is taken as $a \to \infty$ and as $L \to \infty$ with the result being a description of the gamma-ray attenuation through a semi-infinite slab shield resulting from an infinitely large gamma-ray source.

$$\phi_{\gamma f}(z) = \frac{SvN_{\gamma f}(E)}{2} \lim_{L\to\infty} \int_0^L \lim_{a\to\infty} \left\{ \int_{L-x+z}^{\sqrt{(L-x+z)^2+a^2}} \frac{1}{R} \left[Ae^{-\alpha\mu_s\frac{z}{L-x+z}R} + (1-A)e^{-\beta\mu_s\frac{z}{L-x+z}R} \right] \right.$$

$$\left. \times \left(e^{-\mu_r\frac{L-x}{L-x+z}R} e^{-\mu_s\frac{z}{L-x+z}R} \right) dR \right\} dx$$

$$= \frac{SvN_{\gamma f}(E)}{2\mu_r} \{AE_2[\mu_s(1+\alpha)z] + (1-A)E_2[\mu_s(1+\beta)z]\} \tag{14.19}$$

where ExponentialIntegralfunction $= E_n(z) = \int_1^\infty \frac{e^{-zt}}{t^n} dt$.

At the reactor/shield interface (eg, at $z = 0$), it is found that Eq. (14.19), reduces to

$$\phi_{\gamma f}(z) = \frac{SvN_{\gamma f}(E)}{2\mu_r} = S_a(E) \tag{14.20}$$

where $S_a(E)$ = gamma-ray source intensity at the shield surface.

By evaluating the gamma-ray intensity at the surface of the shield, it is thus now possible to calculate the attenuation of the gamma radiation through an arbitrary shield configuration without needing to explicitly know the details of the gamma-ray production within the reactor itself. Using Eq. (14.20) in Eq. (14.19), the gamma-ray distribution within the shield now becomes:

$$\phi_{\gamma f}(z) = S_a(E)\{AE_2[\mu_s(1+\alpha)z] + (1-A)E_2[\mu_s(1+\beta)z]\} \tag{14.21}$$

1. DERIVATION OF SHIELDING FORMULAS

As the gamma flux transitions between material regions, the gamma fluxes must be continuous, therefore, from Eq. (14.21):

$$S_a^i(E)\{A^i E_2[\mu_s^i(1+\alpha^i)h^i] + (1-A^i)E_2[\mu_s^i(1+\beta^i)h^i]\} = S_a^{i+1}(E) \quad (14.22)$$

where i = region index; h^i = thickness of region "i."

1.3 CAPTURE GAMMA ATTENUATION

Capture gamma rays are those gamma rays, which result from neutron capture events within a material. Since the neutron capture cross section for most materials is usually much larger for thermal neutrons than for fast neutrons, the rate of gamma-ray production in the shield will typically follow the thermal neutron flux distribution in the shield. Capture gamma rays differ somewhat from the prompt fission gamma rays described in the previous section in that these gamma rays are emitted at fairly discrete energy levels which depend upon the type of material within which the neutron is captured. Compilations of gamma-ray emission lines resulting from neutron capture events can be found at the National Nuclear Data Center. Because most materials have many gamma-ray emission lines, the individual lines are usually combined together into several broad energy groups so as to ease the calculational requirements in evaluating the energy-dependent gamma-ray fluxes. To effectively attenuate these capture gamma rays, the radiation shield must be designed so as to reduce the neutron flux level to a low level only short distance after the neutrons first penetrate the shield. Once the neutron capture rate (and hence the gamma-ray production rate) has thus been reduced, the balance of the shield can be designed to attenuate the capture gamma rays created near the inside face of the shield where the neutron flux level was high.

In the analysis which follows, the capture gamma rays created within the shield will be assumed to be proportional to the thermal neutron flux distribution derived earlier. The neutrons from the core are assumed to penetrate a cylindrical shield centered at point "O" as shown in Fig. 14.7. The neutrons will

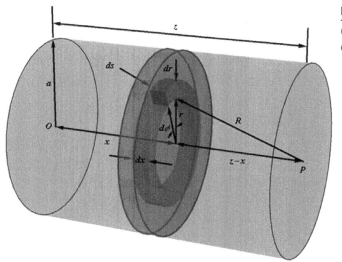

FIGURE 14.7

Gamma-ray flux resulting from neutron capture.

then be progressively absorbed in the shield producing gamma rays at the point where they are captured. The gamma rays thus produced will then be progressively attenuated in the remainder of the shield as they pass through it until they finally exit the shield through its exterior surface. The following analysis calculates the gamma-ray intensity at the exterior surface of the shield at point "P."

The analysis begins by noting that the gamma radiation emanating from a small differential volume as a result of neutron capture within that differential volume is

$$ds = S_p(x)dV = S_p(x)rdrd\phi dx \tag{14.23}$$

The gamma-ray source "$S_p(x)$" will be assumed to be proportional to the thermal neutron flux distribution described in Eq. (14.3) such that:

$$Sp(x) = f_\gamma(E) \sum_c \phi^2(x) = f_\gamma(E) \sum_c \left(Be^{-\xi x} - Ce^{-\psi x}\right)^2 \tag{14.24}$$

where

$$B = \phi_0^2 - \frac{\Sigma_s^{1\to2}\phi_0^1}{D^2\left(\Sigma_s^{1\to2}\right)^2 - \Sigma_c^2}; \quad \xi = \sqrt{\frac{\Sigma_c^2}{D^2}}; \quad C = \frac{\Sigma_s^{1\to2}\phi_0^1}{D^2\left(\Sigma_s^{1\to2}\right)^2 - \Sigma_c^2}; \quad \psi = \Sigma_s^{1\to2}$$

with $f_\gamma(E)$ = capture gamma-ray emission line distribution function.

The capture gamma radiation flux at the point "P" is obtained by integrating the point gamma-ray source over the entire shield volume centered along the line "$\overline{OP}$." If it is assumed that the gamma-ray source is independent of "r," one finds that:

$$\phi_{\gamma c} = \int_0^z \int_0^a \int_0^{2\pi} S_p(x) r G_\gamma(R) d\phi dr dx \tag{14.25}$$

Noting from Fig. 14.7 that:

$$R^2 = r^2 + (z-x)^2 \tag{14.26}$$

Taking the derivative of Eq. (14.26) with respect to "r" then yields:

$$RdR = rdr \tag{14.27}$$

Incorporating Eqs. (14.26) and (14.27) into Eq. (14.25) so as to change the integration variable from "r" to "R" then yields after integrating over "ϕ" an expression of the form:

$$\phi_{\gamma c}(z) = 2\pi \int_0^z \int_{z-x}^{\sqrt{(z-x)^2+a^2}} S_p(x) R G_\gamma(R) dR dx \tag{14.28}$$

Substituting the neutron capture induced gamma-ray source from Eq. (14.24) and a gamma-ray point attenuation kernel similar to that found Eq. (14.12) into the neutron capture shielding relationship of Eq. (14.28) then yields:

$$\phi_{\gamma c}(z) = 2\pi \int_0^z \int_{z-x}^{\sqrt{(z-x)^2+a^2}} f_\gamma(E) \sum_c \left(Be^{-\xi x} - Ce^{-\psi x}\right)^2 \left[Ae^{-\alpha\mu_s R} + (1-A)e^{-\beta\mu_s R}\right] R \frac{e^{-\mu_s R}}{4\pi R^2} dR dx \tag{14.29}$$

1. DERIVATION OF SHIELDING FORMULAS

By solving the integral expressed in Eq. (14.29), an expression for the gamma-ray attenuation through the centerline of a disk shield of radius "a" may be obtained as a function of the thermal neutron capture distribution in the shield. To simplify the resulting expression and to make the results more general, the integration is taken as $a \to \infty$ with the result being a description of the gamma-ray attenuation through a semi-infinite slab shield as a function of gamma-ray producing thermal neutron capture events within the shield:

$$\phi_{\gamma c}(z) = \frac{f_\gamma(E)\Sigma_c^2}{2} \lim_{a \to \infty} \left\{ \int_0^z \int_{z-x}^{\sqrt{(z-x)^2+a^2}} \left(Be^{-\xi x} - Ce^{-\psi x}\right)\left[Ae^{-\alpha\mu_s R} + (1-A)e^{-\beta\mu_s R}\right]\frac{e^{-\mu_s R}}{4\pi R} dR dx \right\}$$

$$= \frac{f_\gamma(E)\Sigma_c^2}{2} \frac{Ce^{-\psi z}}{\psi} \left\{ A\left(E_1[z(\mu_s + \alpha\mu_s - \psi)] + \operatorname{Ln}\left[\frac{\mu_s + \alpha\mu_s - \psi}{\mu_s + \alpha\mu_s}\right]\right)\right.$$

$$\left. + (1-A)\left(E_1[z(\mu_s + \beta\mu_s - \psi)] + \operatorname{Ln}\left[\frac{\mu_s + \beta\mu_s - \psi}{\mu_s + \beta\mu_s}\right]\right)\right\}$$

$$- \frac{f_\gamma(E)\Sigma_c^2}{2} \frac{Be^{-\xi z}}{\xi} \left\{ A\left(E_1[z(\mu_s + \alpha\mu_s - \xi)] + \operatorname{Ln}\left[\frac{\mu_s + \alpha\mu_s - \xi}{\mu_s + \alpha\mu_s}\right]\right)\right.$$

$$\left. + (1-A)\left(E_1[z(\mu_s + \beta\mu_s - \xi)] + \operatorname{Ln}\left[\frac{\mu_s + \beta\mu_s - \xi}{\mu_s + \beta\mu_s}\right]\right)\right\}$$

$$- \frac{f_\gamma(E)\Sigma_c^2}{2} \frac{C\xi - B\psi}{\xi\psi} \{AE_1[z(1+\alpha)\mu_s] + (1-A)E_1[z(1+\beta)\mu_s]\}$$

(14.30)

As the capture gamma flux transitions between material regions, the capture gamma fluxes must be continuous; however, at the beginning boundary between material interfaces, it is necessary for $\phi_{\gamma c}(0) = 0$. The total capture gamma flux for an interior material region, therefore, becomes

$$\text{Total } \phi_{\gamma c}^{i+1}(z) = \phi_{\gamma c}^i(h^i) + \phi_{\gamma c}^{i+1}(z) \tag{14.31}$$

where i = material index and h^i = thickness of material region "i."

1.4 RADIATION ATTENUATION IN A MULTILAYER SHIELD

Based upon the radiation shielding relationships previously derived, a multilayer shield configuration similar to that described in Fig. 14.2 are analyzed. Note, in particular, how in Fig. 14.8, the radiation levels can change dramatically as a function of energy due to the neutron resonance capture/gamma-ray emission characteristics of the shield materials. The areal mass value presented in the plot represents the mass of a 1 cm^2 core cut perpendicularly through the thickness of the shield. If this shield was being designed to provide protection from radiation emanating from an NTR engine, a good deal of effort would be spent optimizing the shield configuration so as to minimize the areal mass required to achieve a specified level of neutron and gamma-ray attenuation.

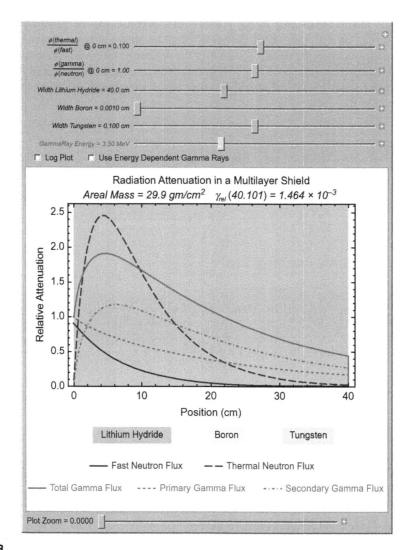

FIGURE 14.8

Radiation attenuation in a multilayer shield.

2. RADIATION PROTECTION AND HEALTH PHYSICS

In order to determine the amount of shielding necessary to protect a crew from radiation originating from an operating NTR engine, it is necessary to discuss briefly the effects different types radiation have on the human body. It should also be pointed out at this point that deep space missions to the moon or Mars will entail crews being exposed to radiation from sources other than that from the operation of the NTR. These other sources include radiation from solar proton events (SPEs), which normally occur as a result of solar flares and galactic cosmic radiation (GCR) which originates from points outside the solar system as a result of supernovae, etc., and consists primarily of high charge and energy (HZE) atomic nuclei. These sources of space (or cosmic) radiation are quite important and if not mitigated properly could easily dwarf the radiation dose received from NTR operation. Unfortunately, the mitigation of cosmic radiation is a fairly complex topic and is outside the scope of this present discourse. The topic of cosmic radiation is, therefore, not further discussed. Those interested in pursuing the topic of cosmic radiation, may want to refer to a very thorough study by NASA [5] which covers the topic of cosmic radiation in considerable detail.

The quantity of radiation received by a person (or anything else) is termed the dose. Radiation doses are generally categorized by the length of time over which the exposure to the radiation occurs. A dose of high-intensity radiation which is received over a short period of time is called an acute exposure, while a dose of low-intensity radiation received over a long period of time is called a chronic exposure. The human body reacts much differently to a radiation dose received over a brief period of time as compared to an equivalent dose received over a much longer period of time. Acute doses are always more detrimental to the health of an individual. An analogy would be similar to that of a person who normally easily survives the minor cuts and bruises they receive over a lifetime. On the other hand, those same injuries, if received over a very short period of time, could very well prove to be deadly. Therefore, when calculating the level of harm expected to result from a given radiation dose, both the dose level and the dose rate are important.

The location on the body where the radiation dose is received is also important when assessing radiation health effects. As a consequence, radiation doses are often distinguished as being either partial or whole body doses. Generally speaking, one finds with regard to radiation sensitivity that:

Greatest sensitivity—lymph nodes, bone marrow, gastrointestinal, reproductive, and eyes.
Medium sensitivity—skin, lungs, and liver.
Least sensitivity—muscles and bones.

The biological effects caused by radiation are the result of ionization interactions in tissues. Ionizing radiation which is absorbed in tissues dislodges electrons from the atoms comprising the molecules of the tissue. When a shared electron in a molecular bond is ejected as a result of absorbing the radiation, the chemical bond is broken causing the atoms forming the molecule to separate from one another resulting in the molecule splitting apart. When ionizing radiation interacts with cells, it sometimes strikes a critical molecule of the cell such as the chromosome. Since the chromosome contains the genetic information required for the cell to perform its function and to reproduce, such damage often, though not always, results in the destruction of the cell. Sometimes the repair mechanisms of the cell can fix the cellular damage, even damage to the chromosome; however, if the cellular repairs are performed incorrectly but not to the extent that they result in the death of the cell, it is possible that a cancer can result. Those types of radiation which are the most ionizing and thus the most damaging to cellular tissue are:

α particles—These particles are highly ionizing but have a very short range in tissues. This radiation can be stopped at the surface of the skin. The greatest tissue damage occurs when the dose is absorbed internally.
β particles—These particles are moderately ionizing and also have a short range in tissues. This radiation can be stopped after only slightly penetrating the skin. The greatest tissue damage occurs when the dose is absorbed internally.

Gamma rays—This radiation is moderately ionizing and quite penetrating. Because this radiation can penetrate deeply into tissues, the external dose is important.

Neutrons—These particles are moderately ionizing and are also quite penetrating. Because the particles can penetrate deeply into tissues, the external dose is important.

Neutrinos—This radiation causes almost no ionization although it is extremely penetrating. Because the radiation interacts with tissues to such an infinitesimal extent, it is completely unimportant with regard to dose.

Radiation doses are typically measured in *RADs* where a *RAD* is the amount of energy absorbed from radiation per mass of material and is defined such that one $RAD = 100$ erg/g. The SI unit for absorbed dose is the *Gray (Gr)* where $1\ Gy = 100\ RAD$. The problem with using such units as *RADs* or *Grays* as a measure of biological damage is that equivalent amounts of various types of radiation affect the physiology of the body differently. For instance, the biological damage resulting from one *RAD* of α particles is different from that one would receive from one *RAD* of γ rays. In order to put all of the various radiation types on an equivalent basis, biologically speaking, another factor called the relative biological effectiveness (RBE) must be applied to the different types of radiation. The RBE basically normalizes the various types of radiation to a single reference radiation so as to provide a consistent measure of biological damage per unit of absorbed dose of any radiation. The RBE is thus defined as follows:

$$\text{RBE} = \frac{\text{Physical dose of 200 KeV } \gamma \text{ rays}}{\text{Physical dose of another type of radiation producing the same biological effect}}$$

The International Commission on Radiological Protection (ICRP) has described the effectiveness of different types of radiation by a series of these RBE factors [6]. The Commission chose a value of one for all radiations having low energy transfer and gamma radiations of all energies. The other values were selected as being broadly representative of the results observed in biological studies, particularly those dealing with cancer and hereditary effects. Table 14.2 presents the recommended ICRP RBE values for several different types of radiation.

Using RBE factors, it is possible to define new units of radiation dose called the Roentgen Equivalent Man (REM) and the Sievert (Sv). These units are biologically more meaningful than the *RAD* or *Gray* and are represented by equations of the form:

$$\text{REM} = \sum_{i=1}^{n} \text{RBE}(i) \times \text{RAD}(i) \quad \text{and} \quad Sv = \sum_{i=1}^{n} \text{RBE}(i) \times Gr(i) = 100\ \text{REM} \qquad (14.32)$$

Table 14.2 Relative Biological Effectiveness Factors

Radiation Type	RBE
Gamma rays	1
Fast neutrons	20
Slow neutrons	5
α particles	20
β particles	1
Protons	5

where i = index representing a radiation type for which a dose is to be calculated and n = total number of radiation types for which a dose is to be calculated.

For neutrons, if it is more convenient to measure the radiation exposure in terms of a neutron fluence (eg, neutron flux × time), then some alternative conversion factors are required. These conversion factors have been determined by the Occupational Safety and Health Administration (OSHA) and compiled into Table 14.3 [7]. If the energy distribution of the neutron flux is unknown, OSHA recommends a conversion factor such that 1 REM = 1.4×10^7 neutron/cm^2.

The acute dose an individual receives in REM can be directly correlated with various physiological effects which would be expected to occur as a result of the radiation exposure. In Table 14.4, these physiological effects are described. The information in this table was excerpted from a report by the Institute of Medicine and the National Research Council [8] from data compiled largely from studies

Table 14.3 Neutron Flux Dose Equivalents

Neutron Energy (MeV)	Neutron Fluence Equivalent to 1 REM (neutron/cm^2)	Neutron Flux Equivalent to 100 mREM in 40 h (neutron/cm^2/s)
Thermal	9.7×10^8	670
0.0001	7.2×10^8	500
0.005	8.2×10^8	570
0.02	4.0×10^8	280
0.1	1.2×10^8	80
0.5	4.3×10^7	30
1.0	2.6×10^7	18
2.5	2.9×10^7	20
5.0	2.6×10^7	18
7.5	2.4×10^7	17
10	2.4×10^7	17
10–30	1.4×10^7	10

Table 14.4 Biological Effects of Acute Radiation Exposure

REM	Health Effect	Lethality (Without Treatment)
0–25	None observable	0%
25–100	Slight blood changes, nausea	0%
100–200	Nausea and vomiting, moderate blood changes	<5%
200–300	Nausea and vomiting, hair loss, severe blood changes	<50%
300–600	Nausea and vomiting, severe blood changes, gastrointestinal damage, hemorrhaging	50–99%
600–1000	Nausea and vomiting, severe gastrointestinal damage, severe hemorrhaging	99–100%

of nuclear weapons survivors from the atomic bomb attacks on the Japanese cities of Hiroshima and Nagasaki at the end of World War II and the Chernobyl accident in Russia.

The effects of low-level chronic exposures to radiation with regard to the formation of cancers and other diseases is much less well understood than for high-level acute exposures to radiation. Nevertheless, many studies have been performed over the years to statistically study the effects of chronic radiation exposures. From these studies, a number of different models have been proposed to estimate the health effects (specifically, cancer) of low doses of radiation based on data extrapolated from observations of the health effects resulting from high doses of radiation. Fig. 14.9 illustrates the trends of these various radiation health models. Although the details of these models have been hotly debated over the years, none of the models have conclusively been shown to be correct.

Briefly, the linear model suggests that cancer risk is directly proportional to the radiation dose received from high-level radiation exposures and that there is no dose below which there is no increased cancer risk (eg, no threshold dose). The supralinear model is similar to the linear model except that it predicts that the cancer risk due to radiation exposure will be somewhat higher than that which would be predicted by the linear model. Some of the available health data, however, seems to indicate that the linear and supralinear models overpredict the cancer risk. As a result, other risk models have been proposed which yield reduced cancer risks at low levels of radiation exposure. These models include the linear quadratic model which reduces the cancer risk as a function of dose quadratically rather than linearly and the linear dose-rate effectiveness factor model (linear DREF) which assumes that the cancer risk varies linearly with dose but does not extrapolate directly from the cancer risks at high radiation exposure levels. Both the Linear Quadratic and the Linear DREF models retain the assumption that there is no threshold dose below which the cancer risk is zero. Other models which do not include the "no threshold" dose risk assumption have also been proposed. These models include the *Linear Threshold* model in which the extrapolated cancer risk goes to zero a some small, but nonzero dose level and the Hormesis model which postulates that there may actually be some health benefits to be derived from exposures to very low levels of radiation.

The Nuclear Regulatory Commission (NRC) has issued regulations for the general public [9] and for radiation workers [10] controlling the maximum permissible radiation dose an individual may receive. These regulations are such that the maximum permitted radiation exposure to either group is well below that which would be expected to cause observable health effects. Because the dose levels

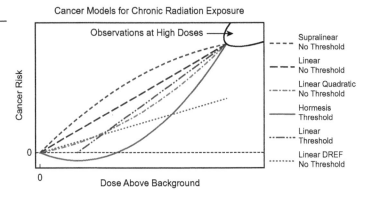

FIGURE 14.9

Cancer model for chronic radiation exposure.

Table 14.5 Regulations on Individual Dose Limits for Adults

Group	Annual Whole Body Dose Limit (REM)
General public	0.1
Radiation worker	5
Astronaut	50

astronauts are expected to receive during deep space missions is well above the NRC regulations, NASA has adopted recommendations from the National Council on Radiation Protection and Measurements (NCRP) as the basis for a supplementary standard for astronaut radiation exposures [11]. These doses are significantly higher than those allowed for radiation workers or the general public although even so, they would not be expected to increase an astronaut's lifetime cancer risk by more than 3% above normal. In Table 14.5, a summary of the applicable regulations is presented.

To minimize the dose received by individuals engaged in activities involving exposure to ionizing radiation, the health physics community has formulated a concept with the acronym *ALARA* which stands for "*A*s *L*ow *A*s *R*easonably *A*chievable." ALARA is a radiation safety principle in which radiation doses and releases of radioactive are minimized by all *reasonable* means. The three major ALARA principles with regard to minimizing the amount of radiation received by individuals consist of:

Time—Minimizing the time during which the exposure to radiation occurs
Distance—Maximizing the distance from the radiation source, minimizes exposure since the radiation intensity falls off as the square of the separation distance
Shielding—Blocking the radiation using materials which are effective radiation absorbers

Space vehicles employing nuclear propulsion would be expected to lend themselves well to the implementation of the ALARA principle, at least with regard to the nuclear engine system, since the engines generally operate for only short periods of time, the vehicle itself would probably be quite long, and due to the vehicle configuration could also probably incorporate relatively small shadow shields.

REFERENCES

[1] R.W. Peele, F.C. Maienschein, The Absolute Spectrum of Photons Emitted in Coincidence with Thermal Neutron Fission of Uranium 235, ORNL-4457, Oak Ridge National Laboratory, Oak Ridge, TN, 1970.
[2] J.J. Taylor, Application of Gamma-Ray Buildup Data to Shield Design, Westinghouse Electric Corporation, Atomic Power Division, 1954. U.S. AEC Report WAPD-RM-217.
[3] O.J. Wallace, Gamma-Ray Dose and Energy Absorption Build-up Factor Data for Use in Reactor Shield Calculations, Report WAPD-TM-1012, June 1974.
[4] M.O. Burrell, Nuclear Radiation Transfer and Heat Deposition Rates in Liquid Hydrogen, NASA Technical Note TN D-1115, August 1962.
[5] J.W. Wilson, et al., Transport Methods and Interactions for Space Radiation, NASA Reference Publication 1257 (NASA-RP-1257), December 1991.

[6] ICRP, Recommendations of the International Commission on Radiological Protection, Ann. ICRP 1, No. 3, 1991.
[7] OSHA Regulation 29CFR1910.1096(b)(1).
[8] Institute of Medicine and National Research Council, Exposure of the American People to Iodine-131 from Nevada Nuclear-bomb Tests. Review of the National Cancer Institute Report and Public Health Implications, National Academy Press, Washington, DC, 1999.
[9] NRC Regulation 10CFR20.1301.
[10] NRC Regulation 10CFR20.1201.
[11] National Council on Radiation Protection, Guidance on Radiation Received in Space Activities, Report 98, July 1989.

CHAPTER 15

MATERIALS FOR NUCLEAR THERMAL ROCKETS

Due to the harsh environment present during nuclear engine operation, the materials used for the fuel, moderator, and other components must be quite resilient in order to survive these severe conditions for any extended length of time. Temperatures within the fuel, for example, may reach 3000 K or even higher. Besides the extremely high temperatures, the corrosive effects of hydrogen or other propellant gases along with the extremely high-radiation environment must be considered during the materials selection process. As a result of these factors, the number of options available with regard to the fabrication of the fuel elements is, not surprisingly, quite limited. The moderator and other structural components must also survive a similarly harsh environment, although, generally not quite as severe as that encountered by the fuel. Since these are material limitations, which constrain the ultimate performance of the nuclear engine, considerable effort will be required in the future if the operational envelope over which the nuclear engine can function is to be significantly expanded over that which is currently envisioned.

1. FUELS

The fissionable nuclide usually proposed for use in nuclear rocket fuel elements is ^{235}U. This isotope of uranium occurs naturally and exists in natural uranium at a concentration of about 0.7% by weight. The other 99.3% of natural uranium consists of nonfissionable ^{235}U. At these concentrations, it is quite difficult (although not impossible) to construct a critical reactor configuration. Natural uranium is fairly common in the earth's crust, existing at a concentration of about 4 ppm. High-grade uranium ore as mined usually contains about 1–4% uranium in a mineral called pitchblende. These high-grade uranium ores are found in Canada and parts of Africa. Medium grades ores containing 0.1–0.5% uranium are found in many other places throughout the world including the United States.

Typically, the concentration of ^{235}U used in fuel for commercial power reactors is increased (or enriched) to about 5% to enable the design of reasonably sized reactors using commonly available materials. This uranium is almost always in the form of an alloy or compound having some desirable set of mechanical and thermal properties. Often the uranium compound used is uranium dioxide (UO_2). While UO_2 is adequate for use in power reactors, it does have a number of drawbacks which would likely preclude its use in nuclear rocket engines. These drawbacks include a fairly low thermal conductivity leading to high fuel temperatures relative to the propellant and only a moderately high melting point of 2865°C.

As was seen earlier, the specific impulse of nuclear rocket engines increases with increasing propellant exhaust temperature. It is, therefore, advantageous to find fuel forms having the highest possible operational temperature capability. As was the case for commercial power reactors, the fissionable nuclide usually proposed for use in nuclear rocket fuels is ^{235}U. However, because it is quite desirable that nuclear rocket engines be small and relatively lightweight, the fuel enrichment in nuclear rockets will likely be higher than that used in commercial nuclear reactors since generally speaking, the higher the enrichment of ^{235}U, the smaller the critical size of the reactor. NERVA reactors used an enrichment is 93% ^{235}U which is high enough to be considered weapons grade. Since such high ^{235}U enrichments require a considerable amount of security to prevent diversion, most proposed nuclear reactor systems for nuclear rockets anticipate using enrichments of 20%. Enrichments of 20%, which is not considered to be bomb grade, still allows for the design of fairly compact reactors but without the need for the heightened security required for fully enriched (93%) reactors. Another nuclide often considered for use in nuclear rockets is ^{239}Pu. ^{239}Pu performs somewhat better from a nuclear standpoint than ^{235}U as a result of its higher fission cross section; however, political nuclear proliferation concerns coupled with its high toxicity probably preclude its use in the near term. Occasionally, ^{242m}Am is suggested as a potential nuclear rocket fuel since it has the highest known fission cross section of any nuclide (almost an order of magnitude higher than ^{235}U); however, because this nuclide is extremely rare and difficult to breed in quantity, it is doubtful that ^{242m}Am will ever be used in nuclear rocket engines except possibly in certain limited applications.

During rocket engine firing, the reactor operates at high power levels resulting in the production of a small, but nonnegligible, amount of fission products. In order to prevent the release of these fission products into the rocket exhaust, it is necessary to encase the fissionable isotopes of the fuel in some type of material so that they cannot escape the fuel element and get entrained in the propellant. In the NERVA program, the fuel elements tested consisted primarily of small fuel particles about 150 μm in diameter composed of a kernel of UO_2 or UC_2 surrounded by an inner layer of porous pyrolytic carbon followed by an outer layer of high-density nonporous pyrolytic carbon. These fuel particles (called *BISO* particles for Buffered ISOtopic) were bound together in a matrix using a graphite binder substrate. The porous graphite layer was designed to accommodate fission products produced in the kernel during operation to prevent excessive pressure from building within the fuel particle and possibly causing it to crack. The outer high-density pyrolytic graphite layer was designed to be a barrier that would prevent fission products residing in the buffer layer from migrating into the surrounding graphite substrate. The structure of these BISO type fuel particles is illustrated in Fig. 15.1.

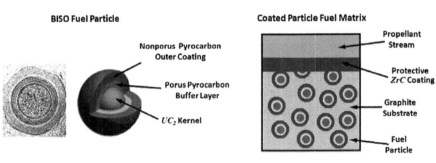

FIGURE 15.1

BISO fuel particles in a NERVA-type fuel element.

The resulting fuel form was then extruded into fuel elements containing 19 coolant holes as depicted in Fig. 1.3. The coolant holes themselves were coated with ZrC to provide an additional protective barrier to impede the release of the fission fragments into the exhaust stream. These efforts were not entirely successful, as testing revealed that the fuel particles tended to crack during engine firing thereupon releasing the fission products into the graphite substrate. Cracking also occurred in the ZrC coating on the fuel propellant channels thus allowing a path for the fission products in the graphite substrate to diffuse into the propellant stream and escape into the environment through the engine exhaust.

In the particle bed reactor (PBR), a more advanced fuel particle was fabricated to prevent the release of fission products into the propellant stream. These fuel particles were similar to the BISO particles in that they had a kernel of UC_2 surrounded by a layer of porous pyrolytic carbon followed by a layer of high-density nonporous pyrolytic carbon. In these fuel particles, however, there was another layer on the outside of the high-density pyrolytic carbon layer composed of ZrC. The ZrC layer acted as an additional high strength barrier to prevent the release of the fission products, plus it protected the inner graphite layers from attack by the hot hydrogen propellant stream which in this reactor configuration would be in contact with the fuel particles. These fuel particles are called *TRISO* particles (for TRIiSOtopic) since they consist of three protective fission product barrier layers. While these fuel particles generally performed quite well, the high thermal gradients present in the particle bed reactors did induce the UC_2 kernels to gradually migrate through their protective barrier coatings eventually causing particle failure. This kernel migration through the protective coatings was called the ameba effect and is illustrated in the photomicrographs shown in Fig. 15.2.

Toward the end of the NERVA program, the fuel particle matrix concept was abandoned in favor of what was called a composite fuel design. The composite fuel consisted of (U, Zr)C heterogeneously distributed throughout a graphite fuel element. These fuel elements were geometrically quite similar to the particle fuel elements and also had propellant holes coated with ZrC. It was felt that this fuel design would be better able to resist the significant hydrogen erosion which had plagued the fuel particle

Unirradiated TRISO Fuel Particle   **Irradiated TRISO Fuel Particle**

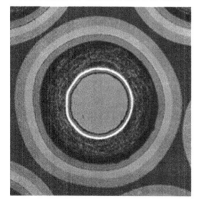 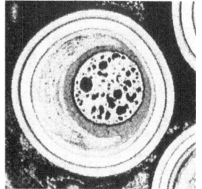

FIGURE 15.2

TRISO fuel particles illustrating the "Amoeba Effect."

design. These fuel elements did perform better than the fuel particle elements in nuclear furnace test, but only marginally [1].

At the end of the NERVA program, research was beginning on the use of solid-solution carbide fuel elements. These fuel elements contained no free carbon and were thus much less susceptible hydrogen erosion. Two fuel elements of this design were tested in the Nuclear Furnace. These fuel elements showed little erosion by the hot hydrogen, although they did suffer fairly extensive cracking [1]. Since the days of NERVA testing work has continued on carbide fuel elements. Researchers in the former Soviet Union experimented with tricarbide fuel forms such as (U, Zr, Nb)C which showed promise in achieving exhaust temperatures >3100 K. This Soviet work [2] on tricarbide fuels has continued in recent years at a low level in the United States and seems to be quite promising. In the photomicrographs [3] shown in Fig. 15.3, it can be seen that the pressing and sintering used in producing the samples resulted in samples which have near theoretical density with just a small amount of porosity, The white streaks are UC indicating that there is some heterogeneity in the material as processed. Note that the longer sintering time results in less porosity and greater homogeneity.

Besides the carbide base fuels just described, there is another class of materials which has been considered in the fabrication of nuclear fuels called cermets. Cermet (for ceramic/metallic) materials for nuclear rockets would typically be composed of $W-UO_2$ or $Mo-UO_2$ mixtures fabricated by cold pressing and sintering at temperatures between 2400 and 2600 K. The cermet materials are usually clad with refractory metal alloys composed of tungsten, tantalum, molybdenum, etc. before being fabricated into fuel elements. Often stabilizers such as $GdO_{1.5}$, $DyO_{1.5}$, $YO_{1.5}$, etc., are added to the cermet material to reduce the tendency of the uranium to dissociate and diffuse into the cladding material at high temperatures and thus weaken the fuel assembly. Much of the early work on cermets was performed as part of the GE 710 program [4] from 1962 to 1968. Other work on cermet nuclear materials for nuclear rockets was also performed by Argonne National Laboratory [5]. Illustrations of cermet materials considered suitable for use in nuclear rockets are presented in Fig. 15.4. It has been estimated that these materials would be sufficiently robust to allow the achievement of propellant exit temperatures in the range of 2800–2900 K. A good overview of the progress of the different cermet development programs can be found in a compilation gathered by the Idaho National Laboratory [6].

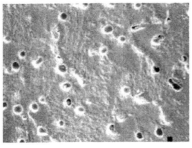
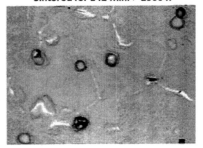

FIGURE 15.3

Microstructure of ($U_{0.1}$, $Zr_{0.45}$, $Nb_{0.45}$) tricarbide fuel.

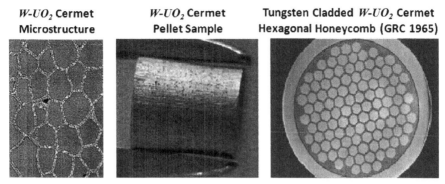

FIGURE 15.4

Physical characteristics of cermet fuel samples.

2. MODERATORS

Since the purpose of neutron moderators is to slow down neutrons into an energy regime where they may be easily captured into fissile materials having large thermal neutron fission cross sections, it is imperative that moderators have low absorption cross sections, yet relatively high scattering cross sections. In addition, it is helpful if the moderator requires only a minimal number of scattering interactions to thermalize the neutrons. Such a requirement implies that the moderating material also have a low mass number. These limitations severely restrict the number of materials available for use as moderators. The only substances which might possibly be used as moderators, therefore, is limited to a few compounds containing hydrogen, beryllium, or carbon.

In the NERVA program, most of the neutron moderation in the core was performed by carbon in the form of high-density impermeable graphite. Graphite is not quite as good a moderator as beryllium or some of the various hydrogen-containing compounds, but it is inexpensive, inert and has relatively good thermal and mechanical properties. It is also stable throughout a large usable temperature range extending to about 3200°C. Graphite does not melt, but sublimes at about 3650°C. The graphite crystals have hexagonal symmetry with large flat crystalline planes that are set relatively far apart. Since these planes are generally oriented is some preferred direction, the thermal, mechanical, and electrical properties of graphite tend to be anisotropic and can vary appreciably depending upon the direction over which the properties are measured. An illustration of the crystalline structure of graphite is presented in Fig. 15.5.

At low temperatures, high levels of neutron irradiation can result in lattice displacements in the crystalline structure that causes dimensional changes to occur in the graphite along with decreases in thermal conductivity. At elevated temperatures, however, annealing processes occur in the graphite that tend to mitigate these radiation-induced property changes. Typical properties of reactor grade graphite are shown in Table 15.1.

Another material of interest for use as a moderator in nuclear rockets is beryllium. Beryllium is a somewhat better moderator than graphite due primarily to its lower mass number. It does have several drawbacks, however, which tend to limit its use. The primary drawback of beryllium is that it is

FIGURE 15.5

Graphite crystalline structure showing unit cell configuration.

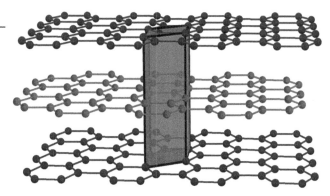

Table 15.1 Typical Properties of Reactor-Grade Graphite

Temperature (K)	Density (g/cm³)	Thermal Expansion Coefficient (1/K)	Specific Heat (J/g/K)	Thermal Conductivity (W/cm/K)	Ultimate Tensile Strength (MPa)	Compressive Strength (MPa)	Young's Modulus (MPa)
1000	1.605	0.0000040	1.80	0.57	17	58	14,100
1500	1.595	0.0000050	2.00	0.43	21	–	13,800
2000	1.580	0.0000055	2.10	0.35	26	–	13,100
2500	1.570	0.0000058	2.15	0.30	28	–	–
3000	1.555	0.0000060	2.18	–	–	–	–
3500	1.540	0.0000063	2.30	–	–	–	–

extremely toxic, especially in powder form. Inhalation of beryllium dust causes extreme irritation of the respiratory tract and exposure to the skin can cause rashes and other irritations. Death can even occur to individuals exposed to sufficient concentrations of beryllium. The material is also quite expensive.

Beryllium does have a relatively high melting point of 1280°C; however, it also has a fairly large grain size which results in the metal being somewhat brittle and difficult to machine without causing surface damage and subsurface cracking. Exposure of beryllium to fast neutrons results in the production of helium through (n, α) reactions and tritium through $(n, 2n)$ reactions. If the neutron fluence is high enough (eg, $> 10^{21}$ nvt), these gases can collect within the bulk metal and cause the material to swell. Typical properties of beryllium metal are shown in Table 15.2.

Because hydrogen is a gas except at cryogenic temperatures, practical neutron moderators using hydrogen must incorporate the element into compounds which are solid (or at least liquid) at elevated temperatures. For many terrestrial power reactors, the hydrogen compound of choice is water which besides its ability to be used as a neutron moderator, can also be used as a reactor coolant. It is doubtful

Table 15.2 Typical Properties of Beryllium Metal

Temperature (K)	Density (g/cm³)	Thermal Expansion Coefficient (1/K)	Specific Heat (J/g/K)	Thermal Conductivity (W/cm/K)	Ultimate Tensile Strength (MPa)	Compressive Strength (MPa)	Young's Modulus (MPa)
200	1.855	–	1.00	0.30	–	269	291,000
400	1.840	0.0000150	2.30	0.16	385	–	283,000
600	1.825	0.0000218	2.55	0.13	337	–	275,000
800	1.810	0.0000205	2.80	0.11	300	–	266,000
1000	1.788	0.0000205	3.10	0.09	–	–	255,000
1200	1.760	0.0000225	3.25	0.08	–	–	–

that liquid moderators, such as water, would be practical as a neutron moderator for nuclear rocket engines because of the inherent differences between nuclear rockets and power reactors; however, there are various solid hydrocarbon compounds such as organic polyphenyls or polyethylene which could be used for that purpose. These compounds have moderating characteristics which are generally similar to water and are fairly stable at moderately high temperatures. They are also relatively stable under fairly high levels of neutron irradiation; however, they do suffer more damage under similar levels of irradiation than graphite or beryllium.

A very interesting moderating material may be produced if the isotopes used in the hydrocarbon synthesis are deuterium and carbon 13. These nuclides have exceptionally low neutron absorption cross sections while retaining moderately high neutron scattering cross sections. For more conventional solid core nuclear reactors, the neutronic characteristics of these deuterated hydrocarbons have only a small effect on the core reactivity; however, if practical gaseous core reactors can be designed, deuterated carbon 13 hydrocarbon reflectors can have a quite large effect on the core reactivity. This core reactivity effect is significant in that the deuterated hydrocarbon reflectors allow criticality to be achieved in moderately sized gaseous core reactors at reasonable pressure levels, whereas if conventional hydrocarbon reflectors were used instead, the pressures required to achieve criticality would become unmanageably large.

Another hydrogen-bearing compound which has been used successfully in the past as a neutron moderator is zirconium hydride. This compound is quite stable under neutron irradiation; however, it decomposes at about 800°C so its application to nuclear rocket engines may be somewhat limited unless the fuel design is such that the zirconium hydride may be kept fairly cool during reactor operation.

3. CONTROL MATERIALS

The primary criteria for control materials for nuclear reactors is that they be strong neutron absorbers, be able to withstand fairly high levels of neutron induced heating, and be dimensionally stable under irradiation. Fortunately, there are a fairly large number of materials that meet these criteria very well. Probably, the most common material used for reactivity control is boron in the form of a compound, such as boron carbide (B_4C) or as a dispersion in stainless steel or copper. In NERVA reactors, boron copper alloy plates were attached to one side of rotating control drums located in the reflector region of the reactor. This configuration controlled the reactivity level of the reactor by in effect varying the neutron leakage fraction in the system through the positioning of the boron plates relative to the core. The main problem with the use of boron in reactor control systems is due to the fact that helium gas released through (n, α) reactions can result in swelling and cracking in the material if the neutron fluence levels are high enough. As mentioned previously, since the neutron fluence levels in nuclear rocket engines are expected to be fairly low, it does not seem likely that helium induced failures in the boron material will be a significant problem.

Hafnium has several isotopes which have quite large resonance capture cross sections in the epithermal resonance energy which make it quite effective in capturing neutrons in thermal reactors. Hafnium carbide has also an extremely high melting point, which makes it very attractive for use in nuclear rocket engines where the high neutron flux environment will likely subject the control material to high heating rates. Gadolinium has the highest neutron absorption cross section of any stable isotope (Xenon-135 has a larger thermal neutron absorption cross section, but it is a gas and is unstable), but it has not been used to date as part of an active reactivity control system; however, it has occasionally been used as a burnable poison. Other materials which have been used from time to time as neutron absorber materials include silver, indium, cadmium, and dysprosium among others. A plot of the absorption cross sections of several of these neutron absorber materials is presented in Fig. 15.6.

Table 15.3 presents some physical characteristics of several materials which have previously been used in nuclear reactors to provide reactivity control through neutron absorption. The list of materials in the table is by no means exhaustive but does present a sampling of materials, which could potentially be suitable for use in nuclear rocket systems.

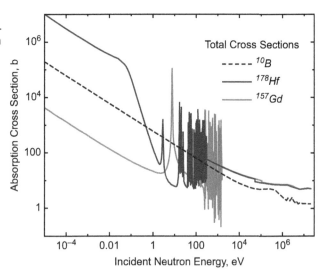

FIGURE 15.6

Absorption cross sections of several common neutron absorber materials.

Table 15.3 Properties of Several Neutron Absorbers

Material	Formula	Density (g/cm³)	Melting Point (K)	Thermal Conductivity (W/cm/K)
Boron carbide	B_4C	2.52	2445	35
Hafnium carbide	HfC	12.2	3890	293
Gadolinium	Gd	7.9	1312	10.6

4. STRUCTURAL MATERIALS

The harsh environment, which will be typical of nuclear rocket engines, presents unique challenges with regard to selecting the appropriate structural materials from which to construct the engine system. Structural materials for nuclear rocket engines must not only be easy to fabricate and of high strength and low weight, but they must also be compatible with the reactor nuclear radiation environment. Such requirements will obviously limit the number of materials which can be considered suitable for use in nuclear rockets. The location and function of the individual reactor components will also affect the selection of the most appropriate structural material. For instance, materials used to construct the reactor containment vessel are likely to be different from those materials used to support the core fuel elements.

Generally, the largest structural component in the nuclear rocket system and the one subjected to the highest mechanical stresses is the reactor containment vessel. Fortunately, since the reactor vessel is located outside of the core and reflector regions of the reactor, its influence on the nuclear operation of the reactor is minimal. The main criteria for the containment vessel from a nuclear standpoint becomes its resistance to activation from the relatively high neutron flux levels to which it is exposed. Materials usually considered for fabricating the containment vessel are stainless steel, aluminum, magnesium and superalloys such as Haynes 230. If stainless steel is to be used for the containment vessel, it is important that an alloy be chosen that contains very little cobalt since the cobalt has a high thermal neutron capture cross section and transmutes into cobalt-60 which is highly radioactive, emitting both gamma rays and beta particles. In Tables 15.4–15.6, the mechanical properties for several of the materials appropriate for use in nuclear rocket containment vessels are given.

Table 15.4 Properties of 316L Stainless Steel

Temperature (K)	Density (g/cm³)	Thermal Expansion Coefficient (1/K)	Specific Heat (J/g/K)	Thermal Conductivity (W/cm/K)	Ultimate Tensile Strength (MPa)	2% Offset Yield Strength (MPa)	Young's Modulus (MPa)
300	7.96	0.0000171	0.49	0.13	578	290	194,000
500	7.87	0.0000156	0.56	0.16	554	235	177,000
700	7.79	0.0000167	0.60	0.19	524	190	160,000
900	7.70	0.0000183	0.63	0.22	415	149	143,000
1100	7.60	0.0000198	0.67	0.24	177	114	126,000

Table 15.5 Properties of T6061 Aluminum

Temperature (K)	Density (g/cm³)	Thermal Expansion Coefficient (1/K)	Specific Heat (J/g/K)	Thermal Conductivity (W/cm/K)	Ultimate Tensile Strength (MPa)	2% Offset Yield Strength (MPa)	Young's Modulus (MPa)
300	2.70	0.0000225	1.02	1.55	311	276	72,100
400	2.68	0.0000242	1.11	1.70	263	240	69,800
500	2.66	0.0000261	1.16	1.79	93	69	66,000
600	2.64	0.0000274	1.21	1.84	31	18	58,900
700	2.62	0.0000296	1.25	1.82	–	–	47,700

Table 15.6 Properties of Haynes 230

Temperature (K)	Density (g/cm³)	Thermal Expansion Coefficient (1/K)	Specific Heat (J/g/K)	Thermal Conductivity (W/cm/K)	Ultimate Tensile Strength (MPa)	2% Offset Yield Strength (MPa)	Young's Modulus (MPa)
300	9.05	0.0000127	0.398	0.089	864	395	210,900
500	8.98	0.0000134	0.438	0.129	802	350	200,390
700	8.90	0.0000141	0.467	0.169	740	305	188,390
900	8.81	0.0000149	0.510	0.209	680	272	175,390
1100	8.72	0.0000158	0.599	0.249	502	271	162,120
1200	8.67	0.0000162	0.611	0.269	339	200	155,120
1300	8.61	0.0000166	0.619	0.289	201	117	148,120
1400	8.56	0.0000170	0.627	0.309	108	58	141,120

The determination as to which material is most appropriate for use in the pressure vessel or propellant tank depends upon many factors such as wall temperature, operating pressure, and radiation resistance; however, it is possible to gain some insight as to the most likely candidates by simply considering which material yields the lightest weight vessel for a given operating temperature and pressure. From a strength of materials analysis, it can be determined that for thin-walled pressure vessels the stress level may be represented as

$$\sigma = \frac{pR}{t} \quad \text{cylindrical shell} \tag{15.1}$$

$$\sigma = \frac{pR}{2t} \quad \text{spherical shell} \tag{15.2}$$

where p = pressure vessel operating pressure; T = temperature of the pressure vessel; t = pressure vessel thickness; R = radius of pressure vessel; and σ = pressure vessel stress level.

The stress levels in the pressure vessel walls must be kept below the yield stress limits in the material of which they are constructed. Normally, a factor of safety is also included when computing the stresses to account for uncertainties in manufacturing processes, material properties, etc. Assuming

4. STRUCTURAL MATERIALS

that the pressure vessel is cylindrical with hemispherical heads, the weight of the vessel may be approximated using Eqs. (15.1) and (15.2) such that:

$$M \approx \rho(T)[A_{cyl}t_{cyl} + A_{sph}t_{sph}] \approx \rho(T)\left[2\pi RLt_{cyl} + 4\pi R^2 t_{sph}\right]$$
$$\approx \rho(T)\left[2\pi RL\frac{pR}{\sigma_y(T)} + 4\pi R^2 \frac{pR}{2\sigma_y(T)}\right] \approx \frac{2\pi R^2 \rho(T)p}{\sigma_y(T)}(L+R) \quad (15.3)$$

where M = mass of the pressure vessel; L = length of the cylindrical portion of the pressure vessel; A_{cyl} = area of the cylindrical portion of the pressure vessel; A_{sph} = area of the spherical portion of the pressure vessel; and $\rho(T)$ = density of the pressure vessel material at temperature "T."

It is important to note that the calculated mass in Eq. (15.3) will generally be somewhat low, especially if the pressure vessel contains many penetrations for piping, drive mechanisms, etc. These penetrations induce additional stresses in the vessel which must be accounted for by adding material in these areas to strengthen the overall structure. In Fig. 15.7, the mass of a pressure vessel is calculated for various materials as a function of temperature assuming no penetrations.

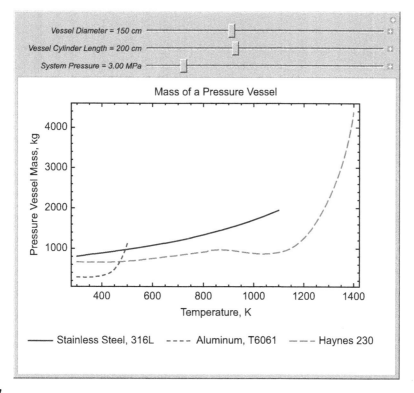

FIGURE 15.7

Pressure vessel characteristics for different structural material.

246 CHAPTER 15 MATERIALS FOR NUCLEAR THERMAL ROCKETS

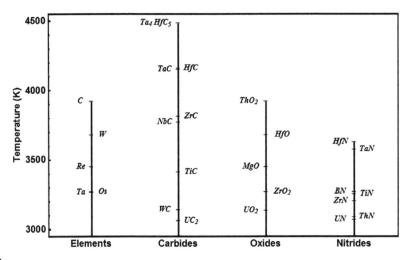

FIGURE 15.8

Melting points of various materials.

Note also in Fig. 15.7 that for temperatures less than about 490 K, aluminum pressure vessels lighter than either stainless steel or Haynes 230 due to its low temperature strength; however, aluminum loses strength rapidly above about 400 K.

For other locations in the space vehicle, environmental conditions will no doubt dictate material selections different from those just described. Because of the wide variety of structural and other requirements in a space vehicle employing a nuclear rocket engine and the wide variety of potential materials available to fulfill those requirements, it is impossible to generalize in a short amount of space the most appropriate materials to use in any particular situation. Nevertheless, to illustrate a few of the materials potentially available for use in the high temperature portions of nuclear rocket engines, Fig. 15.8 shows a number of materials with melting points above 3000 K. The final determination of the most appropriate material for use in a given situation will obviously depend on criteria other than melting point alone such as expected stress levels, chemical environment, and vibrational conditions.

REFERENCES

[1] L.L. Lyon, Performance of (U, Zr)C-graphite (Composite) and of (U, Zr)C (Carbide) Fuel Elements in the Nuclear Furnace 1 Test Reactor, Los Alamos Informal Report, LA-5398-MS, 1973.

[2] E. D'Yakov, M. Tischenku, Manufacturing and Testing of Solid Solution Ternary Uranium Carbides [(U, Nb, Zr)C and (U, Zr, Ta)C] Fuel in Hot Hydrogen, High Temperature Technology and Design Division Research Institute of the Scientific and Industrial Association LUTCH, Informal Report, July 1974.

[3] T. Knight, S. Anghaie, in: M.S. El-Genk (Ed.), Development and Characterization of Solid Solution Tri-carbides, Space Technology and Applications International Forum (STAIF), 2001.

[4] General Electric, 710 High-Temperature Gas Reactor Program Summary Report: Volume III-fuel Element Development, GEMP-600–V3, 1968.
[5] Argonne National Laboratory, Nuclear Rocket Program Terminal Report, ANL-7236, June 1966.
[6] D.E. Burkes, et al., Overview of current and past W-UO cermet fuel fabrication technology, in: INL/CON-07–12232 for Space Nuclear Conference 2007, Paper 2027, Boston, Massachusetts, June, 24–28 2007, 2007.

CHAPTER 16

NUCLEAR ROCKET ENGINE TESTING

Before any flight test of a nuclear engine can take place, considerable testing of the engine will have to be performed both at the component and at the engine level. During the days of the NERVA (nuclear engine for rocket vehicle application) program in the United States, open air testing of the engine was performed with little regard to the containment of fission products in the engine exhaust. Even though the contamination of the environment surrounding the NERVA test area was minimal, in today's regulatory environment, it is unlikely that such testing would be allowed again. As a result, the testing of nuclear rocket engines will probably require a variety of test facilities ranging from small nonnuclear facilities for component testing to very large facilities to test the entire engine assembly for long durations with total engine exhaust containment.

1. GENERAL CONSIDERATIONS

The testing of nuclear rocket engines is expected to be more complicated than the testing of conventional chemical rocket engines. This complexity arises from the fact that there will be significant amounts of radiation present during engine operation, and also as a result of the presence of biologically dangerous amounts of radioactive fission products in the engine assembly itself after testing has been completed. In addition, the current environmental regulations do not permit the testing of nuclear engines in the manner they were conducted in the 1960s and 1970s. The current National Emission Standards for Hazardous Air Pollutants (NESHAP 40 CFR61.90), states, "Emissions of radionuclides to the ambient air from Department of Energy facilities shall not exceed those amounts that would cause any member of the public to receive in any year an effective dose equivalent of 10 mrem/yr." It is extremely unlikely, therefore, that in today's regulatory environment tests like the "transient nuclear test" or TNT illustrated in Fig. 16.1 would be permitted. In this test, the engine was intentionally destroyed by forcing the reactor to go into a super prompt critical state.

Conventional chemical rocket engine testing requires that a number of different types of tests be followed in a logical step-by-step sequence [1]. These tests range from those which are fairly basic to those which are extremely complicated and include:

1. Inspection and testing of manufactured components to provide assurance that the components meet the required tolerances with regard to strength, fit, leak rates, and so on.
2. Component functional tests to assure proper operation of valves, drive assemblies, and so on.

250 CHAPTER 16 NUCLEAR ROCKET ENGINE TESTING

FIGURE 16.1

NERVA transient nuclear test. (Video available on companion website http://booksite.elsevier.com/9780128044742).

3. Static rocket system tests to establish engine operability underrated and off-design conditions.
4. Static vehicle tests wherein the engine installed in a stationary nonflying vehicle.
5. Flight testing on a vehicle which is heavily instrumented so as to demonstrate the operational characteristics of the vehicle under actual flight conditions.

In addition to the above test requirements, the testing of a nuclear engine will require extensive nonnuclear and nuclear testing of the constituent nuclear fuel assemblies in specially designed facilities designed for that purpose and zero-power criticals where the entire nuclear reactor system is operated at very low power levels to verify core power distributions and operation of the reactivity control system.

Besides the normal parameters which must be measured during testing such as temperatures, pressures, forces, flow rates, stresses, vibrations, etc., nuclear rocket engine testing will also require the monitoring of the radiation environment especially with regard to the possible release of fission products from the fuel. Typically, various types of transducers are used to take the required measurements which are then fed into a computer system which analyzes the data to determine the current operational state of the engine. If anomalous behavior is detected, the computer can command the engine to make appropriate control changes or shutdown the engine if required. Often much of the data which is gathered during the test is studied in detail long after the test is completed in posttest analyses. These analyses are extremely useful to engineers in guiding them with regard to making design or operational changes for future tests.

It is important to realize that the measurements taken during testing are not exact but include certain errors. These errors generally fall into several categories. These categories include:

1. Static errors which are usually errors of a fixed nature due to variations in the manufacturing process. These errors are often accounted for by applying appropriate correction factors to the measurements.
2. Drift errors which are measurement errors that occur over time due to variations in the surrounding conditions. Instrument calibration to a fixed standard on a periodic basis is required to eliminate these errors.
3. Dynamic response errors which are measurement errors resulting from such things as vibrations, electrical interference, etc. and can distort the true value of the quantity being measured. These types of errors can be reduced through proper electrical shielding of the instrumentation wiring and general design of the total system
4. Maximum frequency response errors which are measurement errors resulting in the inability of a measurement system (transducer, computer, and so on) to respond quickly enough to changes in the quantity being measured. These measurement errors may be eliminated by using a measurement system which has a frequency response above the natural frequency of the quantity being measured.
5. Linearity errors which are measurement errors resulting from situations where the ratio of the input quantity being measured to the value of the output signal from the instrument varies over the range of the instrument. Typically, instrumentation gives the most accurate readings in the central portion of their ranges (eg, from approximately 20–80% of full scale). Proper choice of instrumentation such that the device covers all parameter ranges of interest can mitigate these types of errors.
6. Hysteresis errors which are measurement errors resulting from the direction the quantity being measured is approached. This type of error is commonly associated with pressure systems wherein the instrumentation output at any particular pressure varies depending upon whether the pressure is approached from increasing pressure or decreasing pressure.
7. Sensitivity errors which are measurement errors resulting from the inability of an instrument to detect small changes in the quantity being measured.

2. FUEL ASSEMBLY TESTING

Probably, the most critical component in a nuclear rocket engine is the nuclear fuel itself. The fuel assemblies are routinely subjected to extremely harsh operating conditions since the performance constraints of nuclear engines are mainly the result of temperature limitations on the fuel coupled with the fuel's limited ability to withstand chemical attack by the hot hydrogen propellant. For the engine to operate at maximum efficiency, fuel forms are desired which can withstand the extremely hot, hostile environment characteristic of nuclear engines for at least several hours. The simulation of such an environment thus requires experimental facilities, which can simultaneously approximate the nuclear, power, flow, and temperature conditions, which nuclear fuel elements would encounter during operation. Such simulations are crucial in enabling the needed detailed studies of the fuel behavior under reactor-like conditions to be assessed.

Not all of the reactor environment simulation facilities need be nuclear. A great deal of information on fuel element behavior can be obtained from nonnuclear facilities which simulate only the thermal and fluid flow environment to which the fuel elements will be subjected. Such testing can be quite valuable since a great deal of information can be gleaned on fuel performance at very low cost when compared to nuclear testing. The results of this nonnuclear testing is also expected to be very representative of that which would be obtained from nuclear testing since the radiation dose and hence radiation damage to the fuel is expected to be quite modest. This lack of significant radiation damage is due to the fact that while the neutron flux levels in nuclear engines is expected to be quite high, the time duration over which the engine operates is expected to be very short. The total dose (proportional to power density times time) to the fuel is, therefore, fairly low. In fact, the dose to the fuel sustained in a nuclear rocket engine is over an order of magnitude lower than that which is normally accumulated by the fuel of a commercial nuclear reactor.

An example of a nonnuclear fuel element test facility is the nuclear thermal rocket element environmental simulator (NTREES) at the Marshall Space Flight Center [2]. The NTREES facility is designed to perform realistic nonnuclear testing of nuclear thermal rocket fuel elements and fuel materials. Although the NTREES facility cannot mimic the neutron and gamma environment of an operating nuclear rocket fuel element, it can simulate the thermal hydraulic environment to provide critical information on material performance and compatibility. NTREES is capable of testing fuel elements and fuel materials in flowing hydrogen at pressures up to 1000 psi at temperatures of around 3000 K and at near-prototypical power densities. NTREES is also capable of testing potential fuel elements with a variety of propellants besides hydrogen. The facility uses induction heating to simulate fission processes by enabling the fuel element to self-generate its own power through induced electrical eddy currents. Since no radiation is involved in the testing, the fuel elements can be safely tested to failure without any of the potentially catastrophic consequences which might be incurred during similar nuclear testing. Fig. 16.2 illustrates the NTREES facility and a prototypical fuel element under test. Facilities, such as NTREES, are useful in that many potential fuel element designs can be tested quite economically, with the best designs as determined from the testing then further developed for testing in nuclear facilities designed for that purpose.

 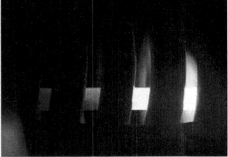

FIGURE 16.2

Nuclear thermal rocket element environmental simulator. (Video available on companion website http://booksite.elsevier.com/9780128044742).

2. FUEL ASSEMBLY TESTING

Once a particular fuel element has been selected for further development, it will be tested in a nuclear facility which has been prepared such that it can reproduce not only the thermal fluid environment which the fuel element will encounter, but the nuclear environment as well. Such testing will be quite expensive but will provide more assurance that the chosen fuel element will behave as expected when it is incorporated into a nuclear rocket engine.

During the last days of NERVA testing, such a facility was constructed at the Nuclear Rocket Development Station (NRDS) at Jackass Flats, Nevada. This facility called Nuclear Furnace 1 [3] was a water-cooled reactor with a closed-loop effluent cleanup system. It was designed to test candidate nuclear fuel elements at gas temperatures of 2444 K for at least 90 min and had an operating power level of 44 MW. The nuclear furnace is an example of a test facility constructed exclusively for the testing of nuclear rocket fuel elements. Depending upon the experimental objectives of the fuel element testing, it may also be possible to use existing nuclear facilities to perform some of the required nuclear tests. In particular, during the Air Force testing of the particle bed reactor fuel elements (PIPE 1 and PIPE 2) [4], the Annular Core Research Reactor (ACRR) [5] at Sandia National Labs was the facility chosen to carry out the tests. Fig. 16.3 illustrates the ACRR. If fuel elements can be tested in existing facilities, the cost of testing will no doubt be less than if dedicated fuel element test reactors are deemed necessary to carry out the required fuel qualification tests.

Generally, dedicated reactors that would be designed to test limited numbers of fuel elements for nuclear rocket engines would have cores that would consist of two separate regions. In one region, the core would be comprised of driver fuel elements whose purpose would be to supply most of the reactivity required for core criticality and to condition the neutron flux such that it would have the desired spatial distribution and energy spectrum. The driver fuel elements could also be cooled by some fluid other than that used for the propellant. For example, in the nuclear furnace, the driver fuel elements were water cooled. In all likelihood, the driver fuel elements

FIGURE 16.3

Annular Core Research Reactor at Sandia National Labs.

would make up the majority of the core with the remainder of the core consisting of the fuel elements being tested. These test fuel elements would be cooled using the propellant of choice (typically hydrogen) as the working fluid. Because of the complexity of these test reactors with their multiregion cores, it is probable that these facilities will be quite expensive to build and operate. Nevertheless, it is quite possible that dedicated fuel element test reactors will ultimately be required to provide the necessary assurance that the fuel elements designed for use in nuclear rocket engines driving crewed spaceships will operate in the safe and efficient manner for which they were designed.

3. ENGINE GROUND TESTING

The ground testing of nuclear rocket engines will be crucial for qualifying them for use in space, especially if these engines will be propelling spacecraft carrying human crews. Chemical rocket engines routinely undergo months and even years of testing before they are certified as being space worthy. It is, therefore, not unreasonable to expect that nuclear rocket engines will be required to complete a similarly rigorous testing regimen before they are qualified for use in space.

During NERVA testing, virtually all of the engine tests discharged the rocket exhaust directly to the atmosphere with no filtering whatsoever. As noted earlier, it is highly unlikely that this type of open air nuclear rocket engine testing will be permitted in the regulatory environment which currently exists. During the time when the NERVA engine tests were being performed, the facilities at the NRDS [6] consisted only of buildings for the assembly and disassemble of reactor and engine components, test stands, and operation and control areas. While similar test facilities will be required in the future if nuclear rocket engine tests resume, additional facilities will also be required whose purpose it will be to capture and scrub the rocket engine exhaust before it is released to the environment. It is interesting to note that in spite of the fact that all of the NERVA engine tests discharged the exhaust products directly to the atmosphere, even those tests which resulted in reactor failures did not incur any personnel injuries or over exposures to radiation. Several of the NERVA test facilities at the NRDS are illustrated in Fig. 16.4.

One possible method which has been proposed to carry out nuclear rocket testing without the release of unfiltered rocket engine exhaust into the atmosphere is to fire the engine into a borehole which was designed to accommodate nuclear weapons testing, but never used for that purpose. This technique called subsurface active filtering of exhaust (SAFE) [7] relies on the porosity of the alluvium soil (soil which is loose and unconsolidated) surrounding the borehole to absorb and filter the exhaust from the rocket engine. In this concept, the engine is located on top of the borehole and is surrounded by a steel and concrete containment structure. As the engine fires, the exhaust pressure in the hole increases until it reaches a point where the amount of exhaust gas driven into the porous rock equals the propellant mass flow rate from the rocket engine. In theory, the rocket engine could be operated for long periods of time over a wide range of power levels before the soil would become saturated by the rocket exhaust products. Whether the pressure levels in the borehole can be kept to reasonable levels during this type of testing is currently still somewhat uncertain. A diagram illustrating the SAFE concept is illustrated in Fig. 16.5.

Another interesting technique which could allow nuclear rocket testing to proceed without the release of unfiltered exhaust products into the atmosphere is one wherein the rocket engine exhaust is

3. ENGINE GROUND TESTING   255

FIGURE 16.4

Nuclear rocket development station at Jackass Flats, Nevada.

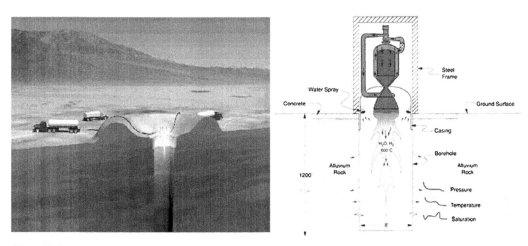

FIGURE 16.5

Nuclear rocket engine testing using subsurface active filtering of exhaust.

totally captured and confined in a relatively small volume. In this concept, the rocket engine exhaust is first directed into a water-cooled diffuser which creates a subsonic hydrogen flow that is then mixed with liquid oxygen and subsequently burned at high temperatures to yield water vapor and excess oxygen. A steam condenser cools the oxygen rich water vapor along with any radioactive contaminants that happen to be present in the exhaust. The contaminated water is then sent to storage tanks where the

256 CHAPTER 16 NUCLEAR ROCKET ENGINE TESTING

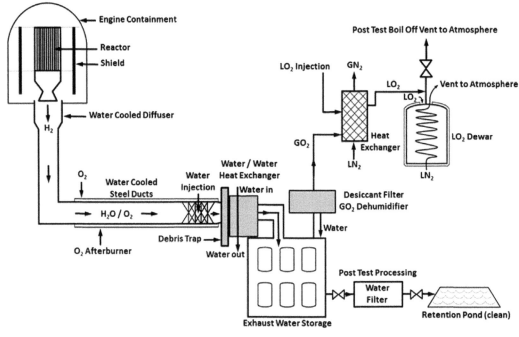

FIGURE 16.6

Nuclear rocket engine testing with total exhaust containment [8].

water is filtered and decontaminated before being discharged to the environment in outside water retention ponds. The contaminated mixture of water vapor and gaseous oxygen is directed into a desiccant filter which removes and condenses the water vapor, discharging the resulting liquid water to the water storage tanks. The contaminated oxygen is introduced into an oxygen condensing unit where the oxygen is liquefied and decontaminated. Finally, the decontaminated oxygen is released to the atmosphere to complete the process. In this manner, the entire rocket engine exhaust is captured and completely decontaminated. A schematic diagram of the nuclear rocket exhaust containment test facility is illustrated in Fig. 16.6.

REFERENCES

[1] G.P. Sutton, Rocket Propulsion Elements, An Introduction to the Engineering of Rockets, fifth ed., John Wiley & Sons, New York, 1986, ISBN 0-471-80027-9.
[2] W.J. Emrich, Nuclear thermal rocket element environmental simulator (NTREES), in: M.S. El-Genk (Ed.), Space Technology and Applications International Forum: 25th Symposium on Space Nuclear Power and Propulsion, Albuquerque, NM, 2008, pp. 541–548.
[3] W.L. Kirk, Nuclear Furnace-1 Test Report, Los Alamos Scientific Laboratory LA-5189-MS, March 1973.
[4] M.E. Vernon, PIPE Series Experiment Plan, Sandia National Laboratories, Albuquerque, New Mexico, 1988.

[5] U. S. Department of Energy, Operational Readiness Review of the Annular Core Research Reactor, Technical Area V, Sandia National Laboratories, Albuquerque, New Mexico, June 1998. Phase 1 Report.
[6] D.E. Bernhardt, R.B. Evans, R.F. Grossman, NRDS Nuclear Rocket Effluent Program 1959–1970, U. S. Environmental Protection Agency, Las Vegas, Nevada, June 1974. NERC-LV-539–6.
[7] S.D. Howe, et al., Ground testing a nuclear thermal rocket: design of a sub-scale demonstration experiment, in: 48th AIAA/ASME/SAE/ASEE Joint Propulsion Conference, Atlanta, GA, Paper AIAA 2012-3743, (July 30 to August 1, 2012) 2012.
[8] Coote, D., Power, K., Gerrish, H., and Doughty, G., "Review of nuclear thermal propulsion ground test options", AIAA Propulsion and Energy Forum, Orlando, FL, Paper AIAA 2015-3773, July 27–29, 2015.

CHAPTER 17

ADVANCED NUCLEAR ROCKET CONCEPTS

Due to the material limitations of nuclear thermal rockets having solid cores, the maximum practical specific impulse achievable by these engines is in the range of 900 s. To achieve significantly higher specific impulses, much higher propellant temperatures will be required, thus necessitating a radically different nuclear core design. One method which has been considered in the past to achieve these higher specific impulses is to use a nuclear pulse system. In this concept, a space vehicle is propelled forward by riding the series of blast waves which result from a succession of properly timed external nuclear explosions. This concept was actually tested in the 1960s under the project name Orion using chemical rather than nuclear explosives. Another method of achieving higher specific impulses is to employ a gaseous fissioning core so as to eliminate the problem of fuel melting at extremely high temperatures. As might be expected, a number of significant design challenges exist with this concept primarily with regard to designing a feasible means of transferring heat from the gaseous fissioning core to the gaseous propellant.

1. PULSED NUCLEAR ROCKET (ORION)

In the pulsed nuclear rocket concept, small nuclear bombs are ejected from the rear of a spacecraft and detonated after they have traveled a suitable distance away from the vehicle. The vehicle itself is designed such that a portion of the blast wave resulting from the nuclear detonation is intercepted by a specially designed "pusher plate" attached to the body of the spacecraft by giant shock absorbers. These shock absorbers act to moderate the spacecraft jerk (eg, time rate of change of acceleration) resulting from the impinging blast wave such that the acceleration of the main body of the vehicle is reduced and smoothed to levels which can be tolerated by the crew. Such a pulsed nuclear propulsion program was actually initiated in the United States during the late 1950s and early 1960s under the project name *Orion* [1]. The Orion program was led by Ted Taylor of General Atomic and physicist Freeman Dyson of Princeton who together with a small team of scientists and engineers built and flew several small-scale proof-of-principle models called *Putt-Putts or Hot Rods*. These proof-of-concept vehicles used chemical rather than nuclear explosives as the propulsive medium, and after several failures, one of the vehicles finally achieved stable flight and flew to an altitude of about 100 ft. A view of this test can be seen in Fig. 17.1.

In addition to the proof-of-concept vehicle testing undertaken as part of the Orion project, several designs for conceptual interplanetary vehicles were also completed. These conceptual vehicle designs covered a wide range of configurations suitable for a wide variety of mission types, even including vehicles suitable for interstellar missions. A drawing of one of these conceptual Orion spacecraft is presented in Fig. 17.2.

FIGURE 17.1

Orion flight testing using Putt-Putts. (Video available on companion website http://booksite.elsevier.com/9780128044742).

FIGURE 17.2

Conceptual Orion spacecraft.

The performance characteristics of the nuclear explosives employed in a pulsed nuclear rocket will be estimated through the use of the point kinetics equations derived earlier. In these derivations described later, it should be noted that many of the assumptions made are quite crude. Nuclear bomb analysis is quite a complex undertaking that requires detailed computer simulations that are calibrated to actual test data to achieve acceptable designs. Nevertheless, the analyses should show qualitatively correct trends and yield ballpark results. Typically, the nuclear core will be surrounded by a reflecting layer of dense material called a tamper which serves to keep the core together longer during the detonation process, thus increasing the yield of the explosion. The tamper plus the core in a nuclear explosive is called a pit.

1. PULSED NUCLEAR ROCKET (ORION)

In order to detonate a nuclear device, a pit is surrounded by a high explosive chemical charge which upon detonation creates a converging shock wave that compresses the pit. As the pit compresses the reactivity increases, eventually reaching a super prompt critical condition (first criticality). Assuming that a suitable neutron source has been provided to the device, the fission power generated will begin to increase rapidly. At some point, the power level and internal pressure within the device will become sufficiently high so as to terminate the compression phase of the detonation process. The point of maximum compression is also the point of maximum excess reactivity. As the power level continues to rise, the pressure and temperature inside the device also increase and the pit begins to expand. At the outside edge of the pit, the pressure abruptly drops to near zero resulting in a huge pressure gradient in the small layer of material at the outside edge of the pit. This pressure gradient is so extreme that it causes the outer surface layer of the pit to blow off and accelerate rapidly away from the main body of the device resulting in a rarefaction wave that propagates back into the pit. The propagation of this rarefaction wave into the pit limits the blow off rate of the outer layers of the pit to the speed of sound of the pit material. Eventually, the combination of the surface material blows off and the expansion of the pit itself causes the reactivity of the device to drop below super prompt critical (second criticality) resulting in the termination of the exponential growth in power and the end of the detonation process.

The main design characteristic to be determined in the derivation which follows will be that of the explosion efficiency of the device which is basically the percentage of fissionable material which actually fissions during the detonation. In this derivation, the main assumptions are:

1. The super-prompt critical condition of the device will terminate when the rarefaction wave reaches the critical radius of the pit.
2. The super-prompt critical reactivity value remains constant until the rarefaction wave reaches the critical radius of the pit whereupon the reactivity goes to zero.
3. The pit consists of a bare spherical core with no tamper.
4. The temperature in the pit is sufficiently high that it may be treated as a photon gas (eg, radiation pressure dominates kinetic gas pressure).
5. There is no energy lost to the surroundings during the detonation (eg, the device is adiabatic).

Using assumption 1, the time interval over which the device is super prompt critical may be determined from:

$$r_i - r_c = \int_0^{t_f} c(t)\, dt \tag{17.1}$$

where $c(t)$ = speed of sound at time "t"; r_i = radius of the pit core at maximum compression (maximum density); r_c = critical radius of the pit core at maximum density; and t_f = length of time required for the rarefaction wave to travel from "r_i" to "r_c."

As noted previously, the speed of sound in a fluid may be given by

$$c = \sqrt{\gamma RT} = \sqrt{\frac{4}{3}RT} \tag{17.2}$$

where γ = specific heat ratio of the vaporized pit ($=\frac{4}{3}$ using assumption 4); R = specific gas constant of the pit; and T = absolute temperature of the pit.

CHAPTER 17 ADVANCED NUCLEAR ROCKET CONCEPTS

Using assumption 5 along with the kinetic theory of gases and the ideal gas law to relate the internal energy of the vaporized pit to pressure, note that:

$$P = \rho RT \Rightarrow RT = \frac{P}{\rho_p} = \frac{U}{3\rho_p} = \frac{E_d}{3\rho_p V} = \frac{E_d}{3m} \quad (17.3)$$

where E_d = total internal energy of pit at the end of the detonation; U = specific internal energy of pit at the end of the detonation; m = mass of the pit; V = volume of pit; and ρ_p = density of pit.

Incorporating Eq. (17.3) into Eq. (17.2) then yields:

$$c = \frac{2}{3}\sqrt{\frac{E_d}{m}} \quad (17.4)$$

From the point kinetics equations derived earlier and using assumption 2, Eq. (11.33) gives an expression for the reactor power as a function of time for a given step increase in reactivity may be given by

$$P(t) \approx P_0 \left[\underbrace{\frac{\rho}{\rho - \beta} e^{\frac{\rho-\beta}{\Lambda}t}}_{\approx 1} - \underbrace{\frac{\beta}{\rho - \beta} e^{\frac{-\rho\lambda}{\rho-\beta}t}}_{\text{small}} \right] \quad (17.5)$$

If it is assumed that the reactivity insertion is quite large (eg, highly super prompt critical with $\rho \gg \beta$), then the second term (delayed neutron term) in Eq. (17.5) may be neglected and the expression may be reduced to

$$P(t) \approx P_0 e^{\frac{\rho}{\Lambda}t} \quad (17.6)$$

To determine the total amount of energy produced in the pit at a time "t" after the pit goes super prompt critical, Eq. (17.6) must be integrated over the time interval of interest such that:

$$E_d(t) = P_0 \int_0^t e^{\frac{\rho}{\Lambda}t'} dt' = \frac{P_0 \Lambda}{\rho} e^{\frac{\rho}{\Lambda}t} = \frac{E_1}{\rho} e^{\frac{\rho}{\Lambda}t} \quad (17.7)$$

where $E_1 = P_0 \Lambda$ energy released at $t = 0$ (initiating fission event). Here assumed to be the energy released by one fission.

The total amount of energy released during the detonation may thus be determined from Eq. (17.7) yielding:

$$E_d = \frac{E_1}{\rho} e^{\frac{\rho}{\Lambda}t_f} \quad (17.8)$$

Substituting Eqs. (17.4) and (17.7) into Eq. (17.1) then yields:

$$r_i - r_c = \int_0^{t_f} \frac{2}{3}\sqrt{\frac{E_1}{\rho m} e^{\frac{\rho}{\Lambda}t}} \, dt = \frac{2}{3}\sqrt{\frac{E_1}{\rho m}} \int_0^{t_f} e^{\frac{\rho}{2\Lambda}t} \, dt = \frac{4\Lambda}{3\rho}\sqrt{\frac{E_1}{\rho m}}\left(e^{\frac{\rho}{2\Lambda}t_f} - 1 \right) \quad (17.9)$$

1. PULSED NUCLEAR ROCKET (ORION)

Rearranging the terms in Eq. (17.9) then yields an expression for the exponential term such that:

$$e^{\frac{r_f}{\Lambda}} = \frac{9}{16}\left(\frac{\rho}{\Lambda}\right)^2 \frac{\rho m}{E_1}(r_i - r_c)^2 \tag{17.10}$$

The total amount of energy released during the detonation may now be rewritten by substituting the results obtained from Eq. (17.10) into Eq. (17.8) giving:

$$E_d = \frac{E_1}{\rho}\frac{9}{16}\left(\frac{\rho}{\Lambda}\right)^2\frac{\rho m}{E_1}(r_i - r_c)^2 = \frac{9}{16}\left(\frac{\rho}{\Lambda}\right)^2 m(r_i - r_c)^2 \tag{17.11}$$

If a change of variables is now made such that $r_i = r_c(1 + \Delta r)$, then Eq. (17.8) may be rewritten to yield:

$$E_d = \frac{9}{16}\left(\frac{\rho}{\Lambda}\right)^2 m r_c^2 \Delta r^2 \tag{17.12}$$

To calculate the efficiency of the detonation, it is first necessary to determine the total amount of energy which could potentially be released from the device assuming all of the fissionable material actually fissions. This maximum possible energy release can be represented by

$$E_{\max} = E_2 m \tag{17.13}$$

where E_2 = energy released per unit mass of fissionable material.

By dividing Eq. (17.12) by Eq. (17.13), the efficiency (ε) of the detonation may be determined such that:

$$\varepsilon = \frac{E_d}{E_{\max}} = \frac{1}{E_2 m}\frac{9}{16}\left(\frac{\rho}{\Lambda}\right)^2 m r_c^2 \Delta r^2 = \frac{9}{16 E_2 \Lambda^2}r_c^2 \Delta r^2 \rho^2 \tag{17.14}$$

At this point, it should be noted that the form of the detonation efficiency relation as represented by Eq. (17.14) is not terribly useful since the equation contains two unknowns which are dependent upon one another. These unknowns are the reactivity of the pit (ρ) and the amount of pit compression (Δr). To determine a relationship between these two quantities, note that:

$$\rho = 1 - \frac{1}{k_{\text{eff}}} = 1 - \frac{1}{\frac{v\Sigma_f}{DB^2 + \Sigma_a}} = 1 - \frac{\Sigma_a}{v\Sigma_f} - \frac{DB^2}{v\Sigma_f} \tag{17.15}$$

$$\rho_{\text{crit}} = 0 = 1 - \frac{1}{k_{\text{crit}}} = 1 - \frac{1}{\frac{v\Sigma_f}{DB_c^2 + \Sigma_a}} = 1 - \frac{\Sigma_a}{v\Sigma_f} - \frac{DB_c^2}{v\Sigma_f} \tag{17.16}$$

and

$$\rho_\infty = 1 - \frac{1}{k_\infty} = 1 - \frac{1}{\frac{v\Sigma_f}{\Sigma_a}} = 1 - \frac{\Sigma_a}{v\Sigma_f} \tag{17.17}$$

Rewriting Eqs. (17.15) and (17.16) in terms of Eq. (17.17) then yields:

$$\rho = \rho_\infty - \frac{DB^2}{v\Sigma_f} \Rightarrow \frac{D}{v\Sigma_f} = \frac{\rho_\infty - \rho}{B^2} \tag{17.18}$$

and

$$0 = \rho_\infty - \frac{DB_c^2}{v\Sigma_f} \Rightarrow \frac{D}{v\Sigma_f} = \frac{\rho_\infty}{B_c^2} \qquad (17.19)$$

Setting the far right terms in Eqs. (17.18) and (17.19) equal to one another and rearranging then yields an equation which relates the reactivity to only geometric and material parameters such that:

$$\frac{D}{v\Sigma_f} = \frac{\rho_\infty - \rho}{B^2} = \frac{\rho_\infty}{B_c^2} \Rightarrow \rho = \rho_\infty \left[1 - \left(\frac{B}{B_c}\right)^2\right] \qquad (17.20)$$

Assuming that the pit is spherical, the geometric buckling may be determined from Table 8.2 yielding:

$$B^2 = \left(\frac{\pi}{r}\right)^2 \qquad (17.21)$$

Incorporating Eq. (17.21) into Eq. (17.20) and using the change of variable for "r_i," described earlier, then yields for the pit reactivity:

$$\rho = \rho_\infty \left[1 - \left(\frac{r_c}{r_i}\right)^2\right] = \rho_\infty \left\{1 - \left[\frac{r_c}{r_c(1+\Delta r)}\right]^2\right\} = \rho_\infty \left[1 - \left(\frac{1}{1+\Delta r}\right)^2\right] \qquad (17.22)$$

Replacing the reactivity term in Eq. (17.14) then yields for the detonation efficiency an equation of the form:

$$\varepsilon = \frac{9}{16E_2\Lambda^2} r_c^2 \rho_{max}^2 \left[\Delta r - \frac{\Delta r}{(1+\Delta r)^2}\right]^2 \qquad (17.23)$$

Noting that for $0 < \Delta r < 1$, the term in brackets in Eq. (17.23) may be approximated by an expression such that:

$$\left[\Delta r - \frac{\Delta r}{(1+\Delta r)^2}\right]^2 \approx 0.6\Delta r^3 \qquad (17.24)$$

Incorporating Eq. (17.24) into Eq. (17.23) then yields a somewhat more useful expression for the pit detonation efficiency of the form:

$$\varepsilon = \frac{9}{16E_2\Lambda^2} r_c^2 \rho_{max}^2 (0.6\Delta r^3) = \frac{0.338}{E_2}\left(\frac{r_c\rho_\infty}{\Lambda}\right)^2 \Delta r^3 \qquad (17.25)$$

The expression for the detonation efficiency presented in Eq. (17.25) is a slightly modified form of the Serber equation [2] originally derived by Robert Serber in the spring of 1943 to estimate the yields of the atomic bombs dropped on Japan at the end of World War II. The modifications presented here to the original Serber equation are only to the constant term in the equation and reflect some suggestions made by Carey Sublett in her Nuclear Weapons Frequently Asked Questions internet archive plus some notational changes which were made to more clearly illustrate the effect of the point kinetics parameters have on the form of the Serber equation. Fig. 17.3 uses Eq. (17.25) to detail how the detonation efficiency varies as a function of explosive shock compression.

1. PULSED NUCLEAR ROCKET (ORION)

To minimize the degree to which the pit must be compressed to achieve a given detonation efficiency, the initial configuration of the pit will need to be such that the radius of the sphere of fissile material is just slightly smaller than the critical radius prior to shock compression. In other words, the pit must initially be in a state which is just barely subcritical. One of the most significant limitations in the efficiency relation given in Eq. (17.25) is that fissile burnup is not taken into account; therefore, the efficiency equations are valid only for relatively low burnups (eg, low efficiencies). As a result, these equations become increasingly inaccurate at high burnup levels and can, in fact, give efficiencies greater than 100%.

The energy required to compress the pit to the high densities required to initiate the desired divergent chain reaction is not insignificant. The graph illustrated in Fig. 17.4 shows the differences in amount of energy required for both isentropic and shock compression of uranium. Note that the energy required for shock compression of uranium is much higher than that required for gentle zero entropy isentropic compression of uranium. Unfortunately, isentropic compression of the pit is not a feasible means of instigating the desired explosive chain reaction since the nuclear reactions initiated by isentropic compression cannot be made to occur quickly enough to create any kind of significant detonation process. Shock compression on the other hand causes the pit to compress quite rapidly and, as a result, can initiate the desired rapidly divergent nuclear chain reactions necessary to cause highly energetic nuclear detonations. By its very nature, however, shock compression is highly irreversible and results in large dissipative heating effects in the pit. Shock compression, therefore, is much less efficient as compared to isentropic compression and results in significantly higher energy requirements to achieve similar degrees of pit compression.

If the type of chemical explosive used to initiate shock compression in the pit is known, then it is possible to calculate the amount of explosive charge necessary to compress the pit to the degree required to achieve a desired explosive yield. Note that even in well-designed implosion systems, due to inefficiencies in the compression process resulting from small asymmetries in the implosion, and so

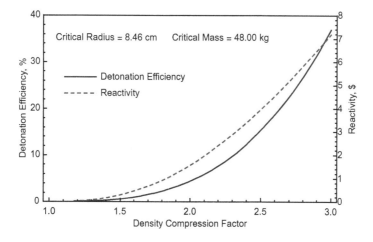

FIGURE 17.3

Pulse unit explosive yields for ^{235}U.

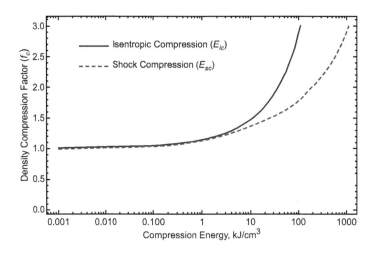

FIGURE 17.4

Specific energy required to compress uranium.

on, it is difficult to achieve implosion efficiencies much more than 30%. If the design of the implosion system is not well designed, the implosion efficiency could easily be much less than 30%. By knowing the mass of the pit, the type of chemical explosive being used to compress the pit, and the energy per unit mass required to compress the pit by a given amount, the mass of chemical explosive required to achieve a given degree of compression may be determined such that:

$$m_e = \frac{m_c E_{sc}(f_c)}{\eta \varepsilon_c} \tag{17.26}$$

where $E_{sc}(f_c)$ = energy per unit mass of chemical explosive required to shock compress a pit by a factor of "f_c"; ε_c = compression efficiency of the implosion process; m_e = mass of chemical explosive required to shock compress a pit by a factor of "f_c"; and η = conversion factor (energy per unit mass of chemical explosive).

If, for example, TNT ($\eta = 4184$ J/g) is used as the chemical explosive employed to compressively implode a pit composed of uranium, then from Eq. (17.26), the amount of explosive required to achieve a given degree of pit compression may be found from Fig. 17.5:

Since the impulsive nuclear detonations are initiated some distance away from the spacecraft, only a portion of the energy from the detonations actually intercepts the spacecraft. The fraction of energy which does intercept the spacecraft is primarily a function of the distance between the detonation point and the pusher plate. If shaped nuclear charges are used, the energy distribution in the detonation must also be taken into account; however, for the analyses to be presented here, it will be assumed that the detonations are spherically symmetric. To begin, consider the situation presented in Fig. 17.6. The detonation occurs a distance "L" behind the spacecraft resulting in a blast wave which impinges the spacecraft pusher plate and pushes the spacecraft as a whole forward. Note that only that portion of the blast wave impinging the pusher plate which is directed parallel to the flight velocity vector actually contributes to the forward motion of the spacecraft.

1. PULSED NUCLEAR ROCKET (ORION) 267

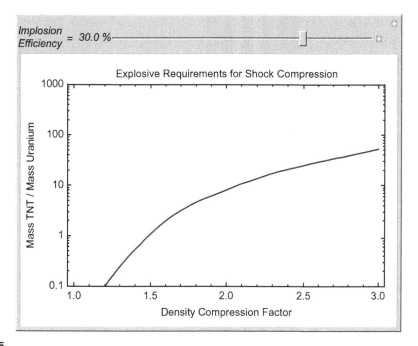

FIGURE 17.5
Explosive requirements to shock compress uranium.

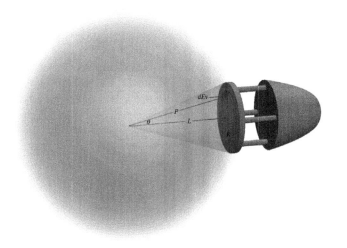

FIGURE 17.6
Nuclear pulse vehicle.

268 CHAPTER 17 ADVANCED NUCLEAR ROCKET CONCEPTS

The energy flux resulting from the nuclear detonation may be represented as

$$Q = \frac{E_d}{4\pi P^2} \qquad (17.27)$$

where Q = energy flux resulting from the detonation and P = distance from the point of detonation to a point on the vehicle pusher plate.

The energy normally directed on a differential ring of the pusher plate may then be determined to be:

$$dE_v = Q(2\pi r\, dr)\cos(\theta) = \frac{E_d}{4\pi P^2}(2\pi r\, dr)\frac{L}{P} = \frac{E_d L r}{2 P^3} dr \qquad (17.28)$$

where E_v = that portion of the blast energy imparted to the vehicle which results in an increase in the vehicle velocity.

From the Pythagorean theorem, it is observed that:

$$P = \sqrt{L^2 + r^2} \qquad (17.29)$$

Incorporating Eq. (17.29) into Eq. (17.28) and integrating over the entire pusher plate, the total normally directed energy imparted to the vehicle may be determined to be:

$$E_v = \int_0^R dE_v = \int_0^R \frac{E_d L r}{2(L^2 + r^2)^{3/2}} dr = \frac{E_d L}{2}\left(\frac{1}{L} - \frac{1}{\sqrt{L^2 + R^2}}\right) \qquad (17.30)$$

Eq. (17.30) may now be used to determine the velocity increment given to the spacecraft as a result of a single detonation. Assuming that the efficiency of the detonation in converting the blast energy into vehicle kinetic energy is "ε_{KE}" it is found that:

$$\varepsilon_{KE} E_v = \frac{1}{2} m_v v_v^2 = \varepsilon_{KE} \frac{E_d L}{2}\left(\frac{1}{L} - \frac{1}{\sqrt{L^2+R^2}}\right) \Rightarrow v_v = \sqrt{\varepsilon_{KE} \frac{E_d}{m_v}\left(1 - \frac{L}{\sqrt{L^2+R^2}}\right)} \qquad (17.31)$$

where m_v = mass of the vehicle and v_v = velocity increment given to the vehicle by a single detonation.

By knowing the velocity increment given to the vehicle as a result of a single detonation pulse, it is now possible to determine the effective specific impulse of the vehicle given the mass of a pulse unit. Using the conservation of momentum, therefore, the specific impulse may be determined from the effective velocity of the blast pulse using the following relationship:

$$m_v v_v = m_{pu} v_{pu} \Rightarrow v_{pu} = \frac{m_v v_v}{m_{pu}} = g_c I_{sp} \Rightarrow I_{sp} = \frac{m_v v_v}{g_c m_{pu}} \qquad (17.32)$$

where m_{pu} = mass of the pulse unit = mass of pit + mass of compression charge (eg, TNT + ... and v_{pu} = effective velocity of blast wave.

Assuming that the spacecraft velocity increment represented by Eq. (17.31) occurs instantaneously and that the spacecraft is accelerated by a series of such pulses occurring at appropriate regular time intervals, the velocity of the spacecraft as a function of time may be given by

$$V_v(t) = v_v \sum_{i=0}^{n} U(t - i\Delta t) = \sqrt{\varepsilon_{KE}\frac{E_d}{m_v}\left(1 - \frac{L}{\sqrt{L^2+R^2}}\right)} \sum_{i=0}^{n} U(t - i\Delta t) \qquad (17.33)$$

where $V_v(t)$ = velocity of the vehicle as a function of time; n = number of detonation pulses; Δt = time interval between detonation pulses; and $U(x)$ = unit step function.

By taking the time derivative of Eq. (17.33), the overall spacecraft acceleration as a function of time may also be determined such that:

$$a_v(t) = v_v \sum_{i=0}^{n} \delta(t - i\Delta t) = \sqrt{\varepsilon_{KE} \frac{E_d}{m_v} \left(1 - \frac{L}{\sqrt{L^2 + R^2}}\right)} \sum_{i=0}^{n} \delta(t - i\Delta t) \qquad (17.34)$$

where $a_v(t)$ = acceleration of the vehicle as a function of time and $\delta(x)$ = Dirac delta function.

The average vehicle acceleration may be determined simply by dividing the velocity increment resulting from a single detonation as expressed by Eq. (17.31) by the time interval between detonations. Thus the average vehicle acceleration becomes:

$$a_{ave} = \frac{v_v}{\Delta t} = \frac{1}{\Delta t} \sqrt{\varepsilon_{KE} \frac{E_d}{m_v} \left(1 - \frac{L}{\sqrt{L^2 + R^2}}\right)} \qquad (17.35)$$

As mentioned earlier, the design of the nuclear pulse vehicle is such that it incorporates large shock absorbers to moderate the impulsive shock on the main body of the spacecraft resulting from the nuclear detonations. For the shock absorbers to operate properly, nuclear detonations must be timed such that they are matched to the frequency response characteristics of the vehicle as a whole. In the analysis which follows, the vehicle will be treated as a simple two body spring—mass—damper system which is exposed to a series of impulsive loads to the pusher plate. The illustration presented in Fig. 17.7 highlights the details of dynamic vehicle configuration which is investigated.

Since the vehicle configuration is basically a two body problem, two equations of motion are required to represent the dynamic response of the vehicle. It is assumed that the mass of the shock absorbers themselves contribute little to the vehicle dynamics so their masses are ignored in the

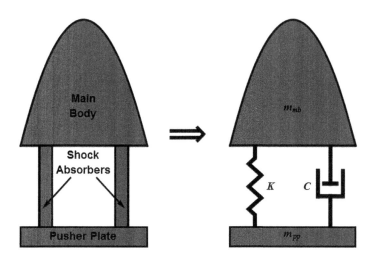

FIGURE 17.7

Simplified dynamic model of a nuclear pulse vehicle.

analyses. The equation of motion of the main body of the vehicle due to a single nuclear pulse may then be given by

$$m_v \frac{f}{f+1}\frac{d^2 z_{mb}}{dt^2} + C\left(\frac{dz_{mb}}{dt} - \frac{dz_{pp}}{dt}\right) + K(z_{mb} - z_{pp}) = 0 \qquad (17.36)$$

where $f = \frac{m_{mb}}{m_{pp}}$; m_{mb} = mass of the main body of the vehicle; m_{pp} = mass of the vehicle pusher plate; z_{mb} = position of the main body of the vehicle; z_{pp} = position of the vehicle pusher plate; C = damping coefficient of the vehicle shock absorber; and K = spring constant of the vehicle shock absorber.

If Eq. (17.36) is recast on a specific mass basis, the equation of motion of the main body of the vehicle may be represented as

$$\frac{f}{f+1}\frac{d^2 z_{mb}}{dt^2} + c\left(\frac{dz_{mb}}{dt} - \frac{dz_{pp}}{dt}\right) + k(z_{mb} - z_{pp}) = 0 \qquad (17.37)$$

where $c = \frac{C}{m_v}$ and $k = \frac{K}{m_v}$.

The equation of motion of the vehicle pusher plate due to a single nuclear pulse may be given by

$$m_v \frac{1}{f+1}\frac{d^2 z_{pp}^j}{dt^2} - C\left(\frac{dz_{mb}^j}{dt} - \frac{dz_{pp}^j}{dt}\right) - K\left(z_{mb}^j - z_{pp}^j\right) = \tau \delta(t - j\Delta t) \qquad (17.38)$$

where τ = impulsive force on the vehicle pusher plate resulting from the previous nuclear detonation; j = total number of prior detonations; and Δt = time interval between detonations.

If Eq. (17.37) is also recast on a specific mass basis, the equation of motion of the vehicle pusher plate may be represented as

$$\frac{1}{f+1}\frac{d^2 z_{pp}^j}{dt^2} - c\left(\frac{dz_{mb}^j}{dt} - \frac{dz_{pp}^j}{dt}\right) - k\left(z_{mb}^j - z_{pp}^j\right) = \frac{\tau}{m_v}\delta(t - j\Delta t) = \frac{v_v}{dt}\delta(t - j\Delta t) \qquad (17.39)$$

Solving the differential equations of motion for a single detonation as described by Eqs. (17.37) and (17.39) then yields:

$$z_{mb}^j(t) = \left\{t - j\Delta t + \frac{v_v}{\xi}e^{\frac{-c(f+1)^2(t-j\Delta t)}{2f}}\sin[\xi(j\Delta t - t)]\right\}U(t - j\Delta t) \qquad (17.40)$$

and

$$z_{pp}^j(t) = \left\{t - j\Delta t - \frac{fv_v}{\xi}e^{\frac{-c(f+1)^2(t-j\Delta t)}{2f}}\sin[\xi(j\Delta t - t)]\right\}U(t - j\Delta t) \qquad (17.41)$$

where $\xi = \frac{(f+1)\sqrt{4fk - c^2(f+1)^2}}{2f}$ = resonant frequency of vehicle.

Knowing the equations of motion from Eqs. (17.40) and (17.41), it is now possible to determine the time-dependent shock absorber length of travel during a single detonation pulse such that:

$$H^j(t) = z_{mb}^j(t) - z_{pp}^j(t) = \frac{v_v(f+1)}{\xi}e^{\frac{-c(f+1)^2(t-j\Delta t)}{2f}}\sin[\xi(j\Delta t - t)]U(t - j\Delta t) \qquad (17.42)$$

1. PULSED NUCLEAR ROCKET (ORION)

To determine the shock absorber length of travel resulting from a series of detonation pulses, the equations of motion for each individual pulse are simply superimposed on one another to yield:

$$H(t) = \sum_{i=0}^{n} H^i(t) = \sum_{i=0}^{n} \left[z_{mb}^i(t) - z_{pp}^i(t) \right] = \sum_{i=0}^{n} \frac{v_v(f+1)}{\xi} e^{\frac{-c(f+1)^2(t-i\Delta t)}{2f}} \sin[\xi(i\Delta t - t)] U(t - i\Delta t)$$

(17.43)

The acceleration on the main body of the vehicle where the crew would be quartered may be found by first differentiating Eq. (17.40) with respect to time to first determine the velocity of the main body of the vehicle resulting from a single detonation pulse such that:

$$v_{mb}^j(t) = v_v \left\{ 1 - e^{\frac{-c(f+1)^2(t-j\Delta t)}{2f}} \left(\cos[\xi(j\Delta t - t)] + \frac{c(f+1)^2}{2f\xi} + \frac{c(f+1)^2}{2f\xi} \sin[\xi(j\Delta t - t)] \right) \right\} U(t - j\Delta t)$$

(17.44)

By now differentiating the velocity expression found in Eq. (17.44), the acceleration experienced by the crew resulting from a single detonation pulse is found to be:

$$a_{mb}^j(t) = v_v e^{\frac{-c(f+1)^2(t-j\Delta t)}{2f}} \left\{ \frac{c(1+f)^2}{f} \cos[\xi(j\Delta t - t)] + \left(\frac{c^2(1+f)^4}{4f^2\xi} - \xi \right) \sin[\xi(j\Delta t - t)] \right\} U(t - j\Delta t)$$

(17.45)

The net acceleration experienced by the crew from a series of detonation is again, simply the superposition of the individual accelerations given by Eq. (17.45) such that:

$$a_{mb}(t) = \sum_{i=0}^{n} a_{mb}^i(t) = \sum_{i=0}^{n} v_v e^{\frac{-c(f+1)^2(t-i\Delta t)}{2f}} \left\{ \frac{c(1+f)^2}{f} \cos[\xi(i\Delta t - t)] + \left[\frac{\left(c^2(1+f)^4\right)}{4f^2\xi} - \xi \right] \sin[\xi(i\Delta t - t)] \right\} U(t - i\Delta t)$$

(17.46)

Fig. 17.8 illustrates the vehicle design characteristics and dynamic response as represented by Eqs. (17.43) and (17.46) resulting from a series of nuclear acceleration pulses. Note that contrary to what might be expected, the accelerations experienced by the crew during the time period over which the detonations are occurring can be made quite tolerable, if somewhat uncomfortable. Also note that the transient acceleration swings and shock absorber displacements that occur during start-up are somewhat larger than those which occur later on after the initial transient response terms have died away. As a result, it is probably desirable to adjust the amount of energy released from the initial detonations somewhat to minimize the effects of the start-up transients on the total vehicle dynamic response. The last few detonations at the end of the acceleration period will also probably have to be adjusted for the same reason. The vehicle velocity per pulse is determined by the force on the pusher plate which is a function of the detonation distance from the vehicle, the diameter of the pusher plate, and the energy released in the detonation pulse. These relationships were presented earlier in Eq. (17.31). The energy

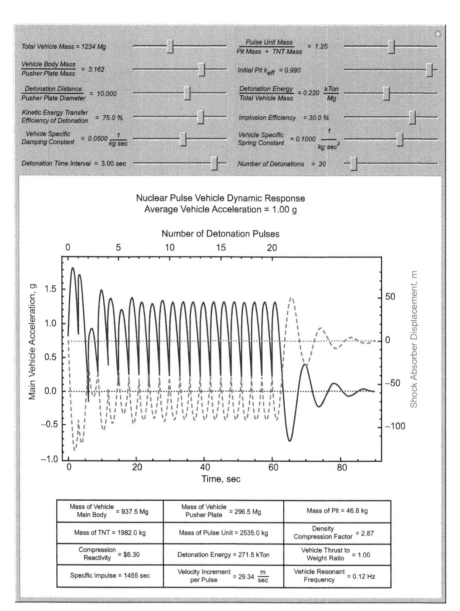

FIGURE 17.8

Dynamic response of a nuclear pulse vehicle.

released per detonation, in turn, is a function of the pit core diameter and the compression factor as expressed in Eq. (17.10). The compression factor required to achieve the desired yield determines the amount of compressive charge needed to achieve the desired detonation energy release. The mass of the compressive charge from Fig. 17.5 along with the mass of the pit then largely determines the mass of the pulse unit.

2. OPEN CYCLE GAS CORE ROCKET

The primary factor limiting performance in solid core nuclear rocket concepts results from the fact that the fuel temperature must be maintained at a low enough temperature that fuel structural integrity can be assured during engine operation. Such temperature limitations restrict the maximum specific impulse attainable by solid core nuclear engines to a value somewhere in the range of 900 s. In the open cycle gas core rocket concept, these temperature limitations on the fuel do not apply since the fuel is maintained in a gaseous state. Other problems manifest themselves in the gas core concept, however, which make the feasibility of this type of nuclear engine questionable. Chief among these feasibility questions is the issue of keeping the gaseous fissioning core from escaping through the nozzle at an unacceptably high rate. To be practical, the gas core rocket must maintain its gaseous core in a stable critical state, while minimizing the loss rate of fissionable material through the nozzle while simultaneously maximizing the heat transfer rate to the hydrogen propellant and allowing it only to escape through the nozzle. Such stringent requirements will be no doubt be difficult to achieve in practice.

An illustration of the gas core nuclear rocket concept which will be analyzed is presented in Fig. 17.9. In this configuration, a fissile material is injected into the core where it is subsequently vaporized due to the high temperatures present there. The hydrogen propellant is injected at an angle at the outside edge of the core through porous walls in the reflector. By injecting the hydrogen in this manner, the reflector may be kept reasonably cool while at the same time setting up a stabilizing fluid rotation in the gaseous fissioning core.

2.1 NEUTRONICS

As illustrated in Fig. 17.9, the configuration that is analyzed is a three-region spherical reactor having a gaseous core, a hydrogen propellant layer, and a solid reflector. The solid reflector in this concept also serves as the containment vessel.

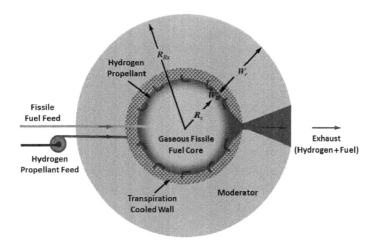

FIGURE 17.9

Gas core nuclear rocket.

One group nuclear diffusion theory in spherical geometry will be used to perform the neutronics analysis with the nozzle region and other inhomogeneities being neglected in the criticality calculations. The density of the uranium will be assumed to remain constant throughout the core region. The power distributions determined from the criticality calculations will subsequently be used in thermal fluid analyses to determine the engine performance characteristics.

Beginning the neutronics analysis with the gaseous uranium core, the one group nuclear diffusion equation in spherical geometry may be written as

$$0 = D_c \nabla^2 \phi + \frac{\nu \Sigma_f}{k_{eff}} \phi - \Sigma_a \phi = \frac{D_c}{r^2} \frac{d}{dr}\left(r^2 \frac{d\phi}{dr}\right) + \frac{\nu \Sigma_f^c}{k_{eff}} \phi - \Sigma_a^c \phi = \frac{1}{r^2}\frac{d}{dr}\left(r^2 \frac{d\phi}{dr}\right) + \alpha^2 \phi \quad (17.47)$$

where $\alpha^2 = \dfrac{\frac{\nu \Sigma_f^c}{k_{eff}} - \Sigma_a^c}{D_c} = \dfrac{\frac{\nu \rho \sigma_f^c}{k_{eff}} - \rho \sigma_a^c}{\frac{1}{3\rho \sigma_{tr}^c}}$ = buckling in the core.

Assuming that the temperatures in the core are high enough that the ideal gas law holds, the core buckling may consequently be represented in terms of the core temperature and pressure such that:

$$\alpha^2 = 3\rho^2 \left(\frac{\nu \sigma_f^c}{k_{eff}} - \sigma_a^c\right)\sigma_{tr}^c = \frac{3P^2\left(\frac{\nu \sigma_f^c}{k_{eff}} - \sigma_a^c\right)\sigma_{tr}^c}{R^2 T^2} \quad (17.48)$$

where P = core pressure; T = core temperature; R = gas constant for the uranium gas.

Solving the above neutron diffusion differential equation represented by Eq. (17.47) then yields:

$$\phi(r) = C_1 \frac{\sin(\alpha r)}{\alpha r} + C_2 \frac{\cos(\alpha r)}{\alpha r} \quad (17.49)$$

Noting that due to the symmetry of the problem, a boundary condition at the reactor center may be taken as

$$\frac{d\phi}{dr} = 0 \quad \text{at} \quad r = 0 \quad (17.50)$$

If the derivative of the neutron flux expressed in Eq. (17.49) is now taken with respect to the radial position, the result is an equation of the form:

$$\frac{d\phi}{dr} = \frac{(C_1 \alpha r - C_2)\cos(\alpha r)}{\alpha r^2} - \frac{(C_1 + C_2 \alpha r)\sin(\alpha r)}{\alpha r^2} \quad (17.51)$$

Now, by applying boundary condition of Eq. (17.50) to the derivative of the core neutron flux expressed in Eq. (17.51), it is found that:

$$0 = (C_1 \alpha 0 - C_2)\cos(\alpha 0) - (C_1 + C_2 \alpha 0)\sin(\alpha 0) \Rightarrow C_2 = 0 \quad (17.52)$$

The neutron flux in the core can now be found within an arbitrary constant by using the value for "C_2" from Eq. (17.52) in the general expression for the core neutron flux from Eq. (17.49) to yield an expression of the form:

$$\phi_c(r) = C_1 \frac{\sin(\alpha r)}{\alpha r} \quad (17.53)$$

In the seeded hydrogen propellant layer, the macroscopic cross sections are so small that they can be assumed to be zero; therefore, the governing equations become:

$$0 = D_H \nabla^2 \phi - \Sigma_a \phi = \frac{D_H}{r^2}\frac{d}{dr}\left(r^2\frac{d\phi}{dr}\right) - \Sigma_a^H \phi = \frac{1}{r^2}\frac{d}{dr}\left(r^2\frac{d\phi}{dr}\right) - \underbrace{\mu^2}_{0}\phi = \frac{d}{dr}\left(r^2\frac{d\phi}{dr}\right) \quad (17.54)$$

Integrating Eq. (17.54) then yields an expression for the flux in the hydrogen propellant layer of the form:

$$\phi_H(r) = \frac{C_3}{r} + C_4 \quad (17.55)$$

Since the neutron flux must be continuous at the interface between the core and the hydrogen propellant layer, the following relationship must hold:

$$\phi_c(R_c) = C_1\frac{\sin(\alpha R_c)}{\alpha R_c} = \phi_H(R_c) = \frac{C_3}{R_c} + C_4 \quad (17.56)$$

Similarly, the neutron current must be continuous at the interface between the core and the hydrogen propellant layer, therefore:

$$J_c(R_c) = D_c\frac{d\phi_c}{dr}\bigg|_{r=R_c} = C_1\frac{D_c\cos(\alpha R_c)}{R_c} - C_1\frac{D_c\sin(\alpha R_c)}{\alpha R_c^2} = J_H(R_c) = D_c\frac{d\phi_H}{dr}\bigg|_{r=R_c} = -C_3\frac{D_c}{R_c^2} \quad (17.57)$$

Solving Eqs. (17.56) and (17.57) simultaneously for the arbitrary constants "C_3" and "C_4" then yields:

$$C_3 = C_1\left[\frac{\sin(\alpha R_c)}{\alpha R_c} - \cos(\alpha R_c)\right]R_c \quad \text{and} \quad C_4 = C_1\cos(\alpha R_c) \quad (17.58)$$

Using the expressions for the arbitrary constants found in Eq. (17.58) in the expression for the neutron flux in the hydrogen propellant layer from Eq. (17.55), a new expression for the neutron flux in the hydrogen propellant layer involving only the arbitrary constant "C_1" may be determined such that:

$$\phi_H(r) = C_1\left[\frac{\sin(\alpha R_c)}{\alpha R_c} - \cos(\alpha R_c)\right]\frac{R_c}{r} + C_1\cos(\alpha R_c) \quad (17.59)$$

In the reflector region, the governing differential equation may be written as

$$0 = D_r\nabla^2\phi - \Sigma_a^r\phi = \frac{D_r}{r^2}\frac{d}{dr}\left(r^2\frac{d\phi}{dr}\right) - \Sigma_a^r\phi = \frac{1}{r^2}\frac{d}{dr}\left(r^2\frac{d\phi}{dr}\right) - \beta^2\phi. \quad (17.60)$$

where $\beta^2 = \frac{\Sigma_a^r}{D_r}$ = materials buckling in the reflector.

Solving the reflector neutron diffusion equation from Eq. (17.60) yields:

$$\phi_r(r) = C_5\frac{\sinh\left[\beta(R_{Rx}^* - r)\right]}{\beta r} + C_6\frac{\cosh\left[\beta(R_{Rx}^* - r)\right]}{\beta r} \quad (17.61)$$

where $R_{Rx}^* = R_c + W_H + W_r + 2D_r = R_c + W_H + W_r^* =$ extrapolated reflector radius.

Noting that at the reactor extrapolation distance the neutron flux goes to zero, Eq. (17.61) yields:

$$\phi_r(R_{Rx}^*) = 0 = C_5 \frac{\sinh(0)}{\beta R_{Rx}^*} + C_6 \frac{\cosh(0)}{\beta R_{Rx}^*} \Rightarrow C_6 = 0 \tag{17.62}$$

Applying the results from Eq. (17.62) to Eq. (17.61) for the neutron flux distribution in the reflector, a final expression for the neutron flux in the reflector may be determined such that:

$$\phi_r(r) = C_5 \frac{\sinh\left[\beta(R_{Rx}^* - r)\right]}{\beta r} \tag{17.63}$$

At the interface between the hydrogen propellant layer and reflector the neutron flux must again be continuous, therefore, from Eqs. (17.59) and (17.63), it is found that:

$$\phi_H(R_c + W_H) = \phi_r(R_c + W_H) \Rightarrow C_1 \left[\frac{\sin(\alpha R_c)}{\alpha R_c} - \cos(\alpha R_c)\right] \frac{R_c}{R_c + W_H} + C_1 \cos(\alpha R_c)$$

$$= C_5 \frac{\sinh(\beta W_r^*)}{\beta(R_c + W_H)} \tag{17.64}$$

Similarly, at the interface between the hydrogen propellant layer and reflector, the neutron current must also be continuous; therefore, from Eqs. (17.59) and (17.63), it can be shown that:

$$J_H(R_c + W_H) = J_r(R_c + W_H) \Rightarrow D_c \frac{d\phi_H}{dr}\bigg|_{r=R_c+W_H} = D_r \frac{d\phi_r}{dr}\bigg|_{r=R_c+W_H}$$

$$\Rightarrow \frac{C_1[\sin(\alpha R_c) - \alpha R_c \cos(\alpha R_c)]D_c}{\alpha(R_c + W_H)^2} \tag{17.65}$$

$$= C_5 \left[\frac{\cosh(\beta h_r^*)}{R_c + W_H} + \frac{\sinh(\beta W_r^*)}{\beta(R_c + W_H)^2}\right] D_r$$

If Eq. (17.64) is now divided into Eq. (17.65), it is possible to eliminate all of the arbitrary constants resulting in a criticality equation of the form:

$$D_c \frac{\tan(\alpha R_c) - \alpha R_c}{\tan(\alpha R_c) + \alpha W_H} = D_r \left[1 + \frac{\beta(R_c + W_H)}{\tanh(W_r^*)}\right] \tag{17.66}$$

Rearranging Eq. (17.64) to solve for "C_5" in terms of "C_1" and other geometric terms then yields:

$$C_5 = C_1 \frac{\beta[\alpha W_H \cos(\alpha R_c) + \sin(\alpha R_c)]}{\alpha \sinh(\beta W_r^*)} \tag{17.67}$$

By incorporating the results from Eq. (17.67) into Eq. (17.63), the reflector neutron flux may at last be found to be:

$$\phi_r(r) = C_1 \frac{\alpha W_H \cos(\alpha R_c) + \sin(\alpha R_c)}{\sinh(\beta W_r^*)} \frac{\sinh\left[\beta(R_{Rx}^* - r)\right]}{\alpha r} \tag{17.68}$$

2. OPEN CYCLE GAS CORE ROCKET

The arbitrary constant "C_1" may be determined from the core average power density to yield absolute values for the neutron flux and local power density. Assuming that the core average power density is known, the value for "C_1" may be determined by integrating Eq. (17.53) over the core volume such that:

$$\text{Total core power} = Q = q_{\text{ave}} \left(\frac{4}{3} \pi R_c^3 \right) = C_1 \int_0^{R_c} \frac{\sin(\alpha r)}{\alpha r} \left(4\pi r^2 \right) dr$$

$$= \frac{4\pi}{\alpha^3} [\sin(\alpha R_c) - \alpha R_c \cos(\alpha R_c)] \Rightarrow C_1 = \frac{q_{\text{ave}} \alpha^3 R_c^2}{3[\sin(\alpha R_c) - \alpha R_c \cos(\alpha R_c)]} = q_0 \quad (17.69)$$

Summarizing now the results from Eqs. (17.53), (17.59), and (17.68), the neutron flux in all regions of the gas core reactor may be given by

$$\phi(r) = \begin{cases} q_0 \dfrac{\sin(\alpha r)}{\alpha r}: & 0 \leq r \leq R_c \\[1em] q_0 \left\{ \left[\dfrac{\sin(\alpha R_c)}{\alpha R_c} - \cos(\alpha R_c) \right] \dfrac{R_c}{r} + \cos(\alpha R_c) \right\}: & R_c \leq r \leq R_c + W_H \\[1em] q_0 \dfrac{\alpha W_H \cos(\alpha R_c) + \sin(\alpha R_c)}{\sinh(\beta W_r^*)} \dfrac{\sinh[\beta(R_{Rx}^* - r)]}{\alpha r}: & R_c + W_H \leq r \leq R_c + W_H + W \end{cases} \quad (17.70)$$

2.2 CORE TEMPERATURE DISTRIBUTION

At the extremely high temperatures which the core will be operating, thermal radiation from the hot fissioning uranium gas will be the dominant mode of heat transfer. Because the gas is not completely opaque, however, the radiation emitted at any point in the core will gradually be attenuated from its point of origin until it eventually leaves the core. The degree of attenuation experienced by radiation will be determined by the gaseous uranium's *opacity* which is dependent upon the frequency of the radiation being emitted and the uranium gas density. The radiation will be attenuated according to *Beer's law* which states that:

$$\frac{dI_v(x)}{dx} = -\kappa_{v,U} \rho I_v(x) \Rightarrow I_v(x) = I_v(0) e^{-\kappa_{v,U} \rho x} = I_v(0) e^{-\kappa_{v,U} \frac{P}{RT} x} \quad (17.71)$$

where $I_v(x)$ = frequency-dependent radiation intensity as a function of distance from its point of origin; $I_v(0)$ = frequency-dependent radiation intensity at its point of origin; $\kappa_{v,U}$ = uranium gas opacity as a function of frequency "\[Upsilon]"; and x = distance the radiation has traveled from its point of origin.

It is important to note that the thermal radiation emitted by the fissioning gas does not occur at a single frequency, but rather is emitted over a spectrum of frequencies defined by *Plancks's law*, which describes the characteristics of electromagnetic radiation emitted by a *black body*. Planck's law can be described by an equation of the form:

$$B_\upsilon(T) = \frac{2h\upsilon^3}{c^2} \frac{1}{e^{\frac{h\upsilon}{kT}} - 1} \qquad (17.72)$$

where $B_\upsilon(T)$ = Planck's black body radiation energy density distribution function; h = Planck's constant; k = Boltzmann constant; and c = speed of light.

The light spectrum resulting from Planck's law is illustrated in Fig. 17.10. Note that at the temperatures at which gas core rocket engines operate, a large fraction of the power generated in the core is emitted in the ultraviolet (UV) region of the spectrum. It is also interesting to note that at the surface temperature of the sun (about 10,000 K), the peak of the spectrum in the visible region of the spectrum.

Using Planck's law from Eq. (17.72) to weight the gas opacity, it is possible to obtain a spectrum averaged opacity as a function of temperature which can be used in calculations to determine the attenuation characteristics of the thermal radiation emitted by the fissioning gas core. Because the uranium gas is generally fairly opaque, the weighting scheme appropriate for use in this particular case uses a diffusion approximation to the radiative transport equation. The resulting weighted average opacity is called the Rosseland opacity. Rosseland opacity is valid for optically thick gases where the mean free path of the thermal radiation is less than some characteristic dimension of the system (eg, in

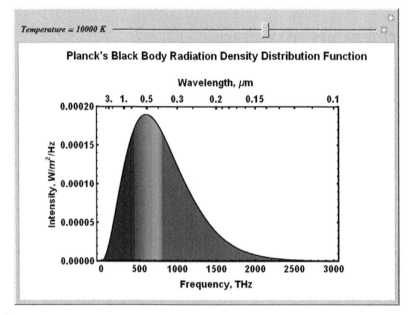

FIGURE 17.10

Planck's black body radiation spectrum.

this case, the core radius) and the radiation field is fairly isotropic. The weighting expression used to calculate the Rosseland opacity may be described by

$$\frac{1}{\kappa_U} = \frac{\int_0^\infty \frac{1}{\kappa_{v,U}} \frac{\partial B_v(T)}{\partial T} dv}{\int_0^\infty \frac{\partial B_v(T)}{\partial T} dv} \tag{17.73}$$

where κ_U = Rosseland opacity for the uranium gas.

The Rosseland opacity of the uranium gas has been calculated at various temperatures and pressures using measured optical data [3] with the results being presented in Fig. 17.11.

To simplify the analysis of the core temperature distribution, convection processes within the gaseous fissioning uranium sphere will be neglected and all heat transfer through the gas will be assumed to occur through conduction only. Using the Rosseland averaged opacity, a pseudothermal conductivity may be defined [4] which is valid for optically thick gases. In this approximation, the thermal conductivity of the uranium gas may be represented by

$$k_u = \frac{16\sigma T^3}{3\kappa_U} \tag{17.74}$$

where σ = Stefan–Boltzmann constant and k_u = thermal conductivity of the uranium gas.

Incorporating the core power distribution determined from Eq. (17.53) and the Rosseland thermal conductivity relationship from Eq. (17.74) into Poisson's heat transfer from Eq. (9.10), it is possible to derive a differential equation which when solved yields the temperature distribution in the gaseous uranium core such that:

$$\nabla^2 T + \frac{q}{k_u} = \frac{d^2 T}{dr^2} + \frac{2}{r}\frac{dT}{dr} + q_0 \frac{3\kappa_U}{16\sigma T^3} \frac{\sin(\alpha r)}{\alpha r} = 0 \tag{17.75}$$

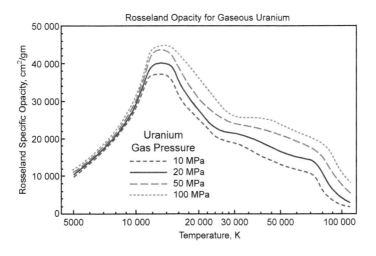

FIGURE 17.11

Uranium Rosseland opacity.

where q = local core power density.

Because of the nonlinearity of Eq. (17.75), it is not possible to solve this expression analytically, so a numerical solution is required for its evaluation.

2.3 WALL TEMPERATURE CALCULATION

By using the heat flux at the outside edge of the core coupled with a knowledge of the radiation absorption characteristics of the propellant layer, the wall temperature at the inside edge of the reflector may be calculated. Since the hydrogen propellant layer cannot always be assumed to be optically opaque as was the case for the gaseous fissioning core, a different approach must be taken to properly evaluate the modes of heat transfer to the propellant and in calculating the wall temperature at the propellant/reflector interface.

As the radiant energy from the core strikes the hydrogen propellant layer, the energy is gradually absorbed as it travels through the propellant, in the process heating it to high temperatures. At these high temperatures, the propellant is sufficiently hot so as to also radiate away a significant amount of the heat which it absorbs.

The first step in the analysis is to determine the amount of heat transferred by radiation to the propellant. Referring to Fig. 17.12, for a differential volume in the propellant layer, the radiation heat balance equation may be written as

$$\frac{d[4\pi r^2 I(r)]}{dr} = \left[-\underbrace{\kappa_H(T)I(r)}_{\text{Absorption}} + \underbrace{\kappa_H(T)J(r)}_{\text{Emission}} \right] (4\pi r^2) \tag{17.76}$$

where $J(r) = \sigma T^4$ = Black body thermal radiation emitted from the hydrogen propellant and $\kappa_H(T)$ = Planck opacity of the hydrogen propellant.

In Eq. (17.76), the *Planck* opacity is used rather than the Rosseland opacity because while the radiation field in the propellant layer is radiative equilibrium, it is neither isotropic nor is it generally optically thick as was the case for the gaseous uranium core. The weighting expression used to calculate the Planck opacity directly weights the frequency-dependent opacity of the hydrogen gas by the Planck distribution function such that:

$$\kappa_H(T) = \frac{\int_0^\infty \kappa_{v,H} B_v(T)\, dv}{\int_0^\infty B_v(T)\, dv} \tag{17.77}$$

Mathematically, Eq. (17.76) is the spherical geometry representation of what is known as *Schwarzschild's equation*. A form of this equation is often used by researchers to determine solar energy absorption in the atmosphere. In this particular case, however, it will be used to calculate radiative heat transfer in the propellant region of the gas core reactor. Rearranging Eq. (17.76) into a more conventional form yields:

$$\frac{dI(r)}{dr} + \left[\frac{2}{r} + \kappa_H(T)\right] I(r) = \kappa_H(T)\sigma T^4 \tag{17.78}$$

2. OPEN CYCLE GAS CORE ROCKET

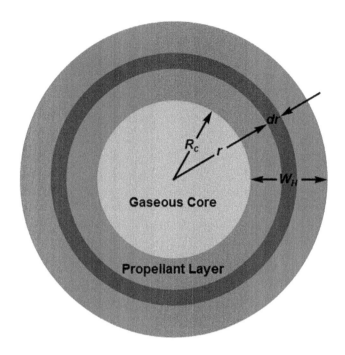

FIGURE 17.12

Heat transfer from the gaseous core to the propellant layer.

Using the method of integrating factors, it is possible to obtain an analytical solution to Eq. (17.78) of the form:

$$I(r) = \frac{e^{-\int_{R_c}^{r} \kappa_H(T)dr'}}{r^2} \int_{R_c}^{r} \kappa_H(T)\sigma T^4 e^{\int_{r'}^{R_c+W_H} \kappa_H(T)dr''} r'^2 \, dr' + C\frac{e^{-\int_{R_c}^{r} \kappa_H(T)dr'}}{r^2} \quad (17.79)$$

To determine the arbitrary constant "C", the heat flux at the core propellant interface will be used as a boundary condition such that:

$$I(R_c) = q_{int} = \frac{e^{-\int_{R_c}^{R_c} \kappa_H(T) \, dr'}}{R_c^2} \int_{R_c}^{R_c} \kappa_H(T)\sigma T^4 e^{\int_{R_c}^{R_c} \kappa_H(T)dr''} r'^2 \, dr' + C\frac{e^{-\int_{R_c}^{R_c} \kappa_H(T) \, dr'}}{R_c^2}$$

$$= C\frac{1}{R_c^2} \Rightarrow C = q_{int}R_c^2 \quad (17.80)$$

where q_{int} = heat flux at the core/propellant boundary.

Substituting the expression for the arbitrary constant "C" from Eq. (17.80) into the relationship for the radiation intensity in the hydrogen propellant layer from Eq. (17.79) at the propellant/reflector boundary and integrating over the hydrogen propellant layer then yields for the wall heat flux an expression of the form:

$$I(R_c + W_H) = q_s$$

$$= q_{int} R_c^2 \frac{e^{-\int_{R_c}^{R_c+W_H} \kappa_H(T)\, dr'}}{(R_c + W_H)^2} + \frac{e^{-\int_{R_c}^{R_c+W_H} \kappa_H(T)\, dr}}{(R_c + W_H)^2} \int_{R_c}^{R_c+W_H} \kappa_H(T) \sigma T^4 e^{\int_{r'}^{R_c+W_H} \kappa_H(T)\, dr''} r'^2\, dr' \quad (17.81)$$

where q_s = heat flux at the surface of the reflector.

Because the hydrogen temperature as a function of position in the propellant layer is unknown, Eq. (17.81) cannot be integrated in its present form. As a consequence, in order to solve the problem, an assumption will be made that the temperature of the propellant varies linearly between the temperature at the interface between the fissioning gaseous core boundary and the propellant layer on one side and the wall surface temperature on the other side. The average temperature of the propellant (propellant temperature at the center of the propellant layer) will be defined to be some desired propellant outlet temperature, therefore:

$$T = 2T_{ave} - T_s - \frac{2}{W_H}(T_{ave} - T_s)(r - R_c) \quad (17.82)$$

where T_s = temperature at the surface of the reflector and T_{ave} = average outlet temperature of the propellant.

Substituting Eq. (17.82) into Eq. (17.81) at the core boundary then yields:

$$I(R_c + W_H) = q_s = q_{int} R_c^2 \frac{e^{-\int_{R_c}^{R_c+W_H} \kappa_H(T)\, dr'}}{(R_c + W_H)^2} + \frac{e^{-\int_{R_c}^{R_c+W_H} \kappa_H(T)\, dr}}{(R_c + W_H)^2}$$

$$\times \int_{R_c}^{R_c+W_H} \kappa_H(T)\sigma \left[2T_{ave} - T_s - \frac{2}{W_H}(T_{ave} - T_s)(r - R_c) \right]^4 \quad (17.83)$$

$$\times e^{\int_{r'}^{R_c+W_H} \kappa_H(T)\, dr''} r'^2\, dr'$$

Note that since Eq. (17.83) contains three unknowns, the heat flux at the surface of the reflector (q_s), the heat flux at the core/propellant boundary (q_{int}), and the temperature at the surface of the reflector (T_s), two more equations will be needed to solve for all three variables. One additional equation may be found by performing a heat balance at the reflector wall such that:

$$0 = \underbrace{q_s}_{\text{Radiant Heat Transfer to Wall}} + \underbrace{h_c(T_{ave} - T_s)}_{\text{Transpiration Heat Transfer from Wall to Hydrogen Propellant}} - \underbrace{\frac{\dot{m}}{A_s} c_p (T_s - T_i)}_{\text{Hear Absorbed by Hydrogen Propellant}} \quad (17.84)$$

where h_c = transpiration heat transfer coefficient; A_s = reflector wall surface area; $\dot{m}$ = total propellant mass flow rate through the reflector wall; and T_i = inlet temperature of the propellant.

2. OPEN CYCLE GAS CORE ROCKET

If the total core power is known along with the average outlet temperature of the propellant, it is found that:

$$Q = qV_{core} = \dot{m}c_p(T_{ave} - T_i) \Rightarrow \dot{m}c_p = \frac{qV_{core}}{T_{ave} - T_i} \quad (17.85)$$

where V_{core} = core volume.

Substituting Eq. (17.85) into Eq. (17.84) and rearranging terms then finally yields:

$$q_s = q\frac{V_{core}}{A_s}\frac{T_s - T_i}{T_{ave} - T_i} - h_c(T_{ave} - T_s) = q\frac{\frac{4}{3}\pi R_c^3}{4\pi(R_c + W_H)^2}\frac{T_s - T_i}{T_{ave} - T_i} - h_c(T_{ave} - T_s)$$

$$= q\frac{R_c^3}{3(R_c + W_H)^2}\frac{T_s - T_i}{T_{ave} - T_i} - h_c(T_{ave} - T_s) \quad (17.86)$$

For the last required equation, note that since the rate at which heat is generated within the core must equal to the rate at which energy is transferred across the core boundary (divergence theorem), it is found that:

$$Q = qV_{core} = q_{int}A_{int} \Rightarrow q_{int} = q\frac{V_{core}}{A_{int}} = q\frac{\frac{4}{3}\pi R_c^3}{4\pi R_c^2} = \frac{qR_c}{3} \quad (17.87)$$

where A_{int} = core surface area.

Substituting Eqs. (17.86) and (17.87) into Eq. (17.83) then finally yields an equation in which the only unknown term is the surface temperature of the reflector yielding:

$$q\frac{R_c^3}{3(R_c + W_H)^2}\frac{T_s - T_i}{T_{ave} - T_i} - h_c(T_{ave} - T_s) = \frac{qR_c^3}{3(R_c + W_H)^2}e^{-\int_{R_c}^{R_c+W_H}\kappa_H(T)\,dr} + \frac{e^{-\int_{R_c}^{R_c+W_H}\kappa_H(T)\,dr}}{(R_c + W_H)^2}$$

$$\times \int_{R_c}^{R_c+W_H}\kappa_H(T)\sigma\left[2T_{ave} - T_s - \frac{2}{W_H}(T_{ave} - T_s)(r - R_c)\right]^4$$

$$\times e^{\int_{r'}^{R_c+W_H}\kappa_H(T)\,dr''}r'^2\,dr' \quad (17.88)$$

Assuming that a temperature averaged value is used for the opacity of the hydrogen propellant, Eq. (17.88) can be integrated analytically to give a closed form symbolic solution for the surface temperature of the reflector. This equation is quite complicated and lengthy and will not be presented here since doing so would give little insight into the physical behavior of the energy absorption into the hydrogen propellant. The equation will be used later, however, in numerical calculations to determine the qualitative performance characteristics of a conceptual gas core nuclear rocket.

It should also be noted that the opacity of hydrogen is fairly small at temperatures below about 10,000°C [5]. As a result, it is normal to assume that a seed material of some kind will be added to the hydrogen propellant gas to increase its opacity. This seed would typically consist of a tungsten aerosol composed of small tungsten particles in the range of 0.02–0.5 μm in diameter. Experiments show that

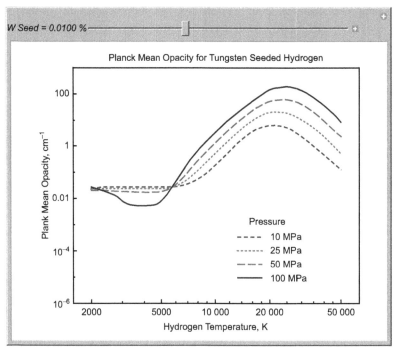

FIGURE 17.13

Planck mean opacity for tungsten seeded hydrogen.

less than 1% tungsten seed can have a profound effect on the opacity of the hydrogen propellant [6]. Fig. 17.13 illustrates how the hydrogen opacity varies as tungsten seed material is introduced into the hydrogen propellant. Note that the experimental seeded opacity values at the lower temperatures illustrated in the plot are quite scattered and are probably only qualitatively correct.

2.4 URANIUM LOSS RARE CALCULATIONS

Thus far all the analyses of the gas core nuclear rocket have assumed that there is no mixing between the gaseous uranium core and the hydrogen propellant layer. In practice, however, this will not be the case. Turbulent mixing between the two regions will inevitably occur resulting in fluid oscillations at the interface. Under certain circumstances, these oscillations may prove to be unstable. These instabilities will generally be of two types, those being the Kelvin—Helmholtz instability and the acoustic instability. Kelvin—Helmholtz instabilities occur when two fluids of different densities move across one each other with different velocities. In the gas core nuclear rocket, the Kelvin—Helmholtz instability can occur as a result of the faster moving hydrogen propellant layer moving across the slower moving or stationary gaseous uranium core. An illustration of this instability is given in Fig. 17.14.

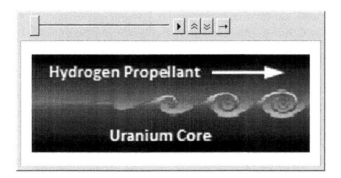

FIGURE 17.14

Kelvin–Helmholtz instability.

Under the influence of a gravitational force and assuming that velocity of the uranium core is zero, the instability condition for the gas core nuclear rocket may be given by [7]

$$V_\infty^2 > \frac{g}{\xi_v} \frac{\rho_U^2 - \rho_H^2}{\rho_U^2 \rho_H^2} \Rightarrow \xi_v^{\min} = \frac{g}{V_\infty^2} \frac{\rho_U^2 - \rho_H^2}{\rho_U^2 \rho_H^2} \qquad (17.89)$$

where g = gravitational acceleration; ξ_v = wave number.

The wave growth rate of the Kelvin–Helmholtz instability may be represented by an equation of the form:

$$\zeta_{KH} = V_\infty \xi_v^{\min} \sqrt{\frac{\rho_H}{\rho_U}} \qquad (17.90)$$

where ζ_{KH} = wave growth rate for Kelvin–Helmholtz instability.

Using the wave number from Eq. (17.89) and wave growth rate from Eq. (17.90), the diffusion coefficient for uranium transport into the hydrogen propellant flow stream may be represented by

$$D_{KH} = \frac{\zeta_{KH}}{\left(\xi_v^{\min}\right)^2} \qquad (17.91)$$

where D_{KH} = diffusion coefficient.

The transport of uranium from the core region into the hydrogen propellant layer may now be evaluated using the diffusion coefficient from Eq. (17.91) in the diffusion equation which is given by

$$F = D_{KH} \nabla \rho_U \qquad (17.92)$$

where F = uranium mass flux from the core region into the hydrogen propellant layer.

Assuming that the uranium density in the core is constant and recalling that the core is spherically symmetric, Eq. (17.92) becomes:

$$F = \frac{D_{KH}}{r^2} \frac{d}{dr}\left(r^2 \rho_U\right) = \frac{2 D_{KH} \rho_U}{r} \qquad (17.93)$$

To determine the total rate at which the uranium is transported out of the core and into the propellant layer due to the Kelvin–Helmholtz instability, the uranium mass flux from Eq. (17.93) is multiplied by the surface area of the core such that:

$$L_{KH} = 4\pi r_c^2 F = 4\pi r_c^2 \frac{2D_{KH}\rho_U}{r_c} = 8\pi r_c D_{KH}\rho_U \tag{17.94}$$

where L_{KH} = total uranium mass flow out of the core due to the Kelvin–Helmholtz instability.

Acoustic instabilities in a gas core nuclear rocket can occur due to temperature and density fluctuations in the uranium core region. In this instability, a standing sound wave will be found to be present in the core region of the engine. In those regions of the wave where the uranium density is high, increased fissions will occur, adding energy to the wave. Similarly, in the lower-density regions of the wave, the reduced uranium density will lead to a decrease in power. The net effect of the variations in the fission power between the regions is to increase the pressure gradient within the wave leading to a transfer of fission power into the wave. Some of this increase in power will be transported out of the core due to radiation. In addition, the diffusion of the radiation within the core will tend to smear out some of the temperature fluctuations between the regions. As the wavelength of the standing sound wave becomes shorter, the temperature fluctuations within the core will be diminished due to increases in the smearing effect. The net effect of all these competing processes is that there will be a critical wavelength of the sound wave below which the wave is stable. If the dimensions of the core are less than the critical wavelength, then the core will be stable against acoustic instabilities. If, on the other hand, the core is larger than the critical wavelength, then the ensuing acoustic instabilities will result in additional uranium loss from the core. A dispersion equation relating the wave number to the frequency of the acoustic wave has been solved [8] yielding a relationship of the form:

$$\zeta_A = \frac{2q_{ave} - \frac{\xi_v^2 k_U}{R}(V_s^2 - 2RT_0)}{6\rho_0 V_s^2} \tag{17.95}$$

where ζ_A = wave growth rate for acoustic instability; R = gas constant for gaseous uranium; ξ_v = acoustic wave number; V_s = speed of sound in the uranium core; ρ_0 = equilibrium average uranium density in the core; and T_0 = equilibrium average core temperature.

The speed of sound used in Eq. (17.95) may be determined from a form of Eq. (2.22) which has been modified to account for ionization effects in the uranium such that:

$$V_s = \sqrt{\gamma Z R T_0} \tag{17.96}$$

where Z = charge state of the uranium and γ = specific heat ratio for gaseous uranium.

Incorporating Eq. (17.96) into Eq. (17.95) then yields for the growth rate of acoustic instabilities an expression of the form:

$$\zeta_A = \frac{2q_{ave} - \xi_v^2 k_U(\gamma Z - 2)T_0}{6\rho_0 \gamma Z R T_0} \tag{17.97}$$

Noting that positive values for the growth rate indicate an unstable condition in which the acoustic waves increase in amplitude with time and negative values for the growth rate indicate that the acoustic waves decrease in amplitude or are damped with time, it can be inferred that a growth rate of zero represents the limit for which stability can be maintained against acoustic waves. Assuming that the growth rate for acoustic instabilities is equal to zero, Eq. (17.97) can be solved to yield an expression for the critical wave number such that:

$$\xi_v^{crit} = \sqrt{\frac{2q_{ave}}{k_U(\gamma Z - 2)T_0}} \qquad (17.98)$$

Substituting Eq. (17.98) into Eq. (17.97) and rearranging terms, then finally yields for the growth rate of acoustic instabilities an equation of the form:

$$\zeta_A = \frac{k_U(\gamma Z - 2)}{6\rho_0 \gamma Z R}\left[\left(\xi_v^{crit}\right)^2 - \xi_v^2\right] \qquad (17.99)$$

Using the critical wave number determined from Eq. (17.98), a critical wavelength may be evaluated such that:

$$\lambda_A = \frac{2\pi}{\xi_v^{crit}} \qquad (17.100)$$

If the critical wavelength exceeds the dimensions of the gaseous uranium core, it will be impossible for standing acoustic waves to exist and the system will be stable against acoustic instabilities. On the other hand, if the dimensions of the core exceed the critical wavelength, standing acoustic waves are possible within the system and the core will likely experience acoustic instabilities. To determine the growth rate for these acoustic instabilities, a wave wavelength equal to the core radius is used to determine a wave number such that:

$$\xi_v = \frac{2\pi}{r_c} \qquad (17.101)$$

This wave number from Eq. (17.101) is then used in Eq. (17.99) to determine the growth rate for acoustic instabilities in the core such that:

$$\zeta_A = \frac{k_U(\gamma Z - 2)}{6\rho_0 \gamma Z R}\left[\left(\xi_v^{crit}\right)^2 - \left(\frac{2\pi}{r_c}\right)^2\right] \qquad (17.102)$$

The loss rate of uranium from the core due to acoustic instabilities may now be estimated in a manner similar to that used to calculate the loss rate due to Kelvin–Helmholtz instabilities. First, a diffusion coefficient is calculated from Eq. (17.91) using the wave number from Eq. (17.101) and the growth rate from Eq. (17.102) yielding:

$$D_A = \frac{\zeta_A}{\xi_v^2} \qquad (17.103)$$

Finally, the total rate at which the uranium is transported out of the core and into the propellant layer due to acoustic instabilities may now be determined by inserting the diffusion coefficient from Eq. (17.103) in the uranium loss rate expression from Eq. (17.94) such that:

$$L_A = 8\pi r_c D_A \rho_U \qquad (17.104)$$

where L_A = total uranium mass flow out of the core due to acoustic instabilities.

288 CHAPTER 17 ADVANCED NUCLEAR ROCKET CONCEPTS

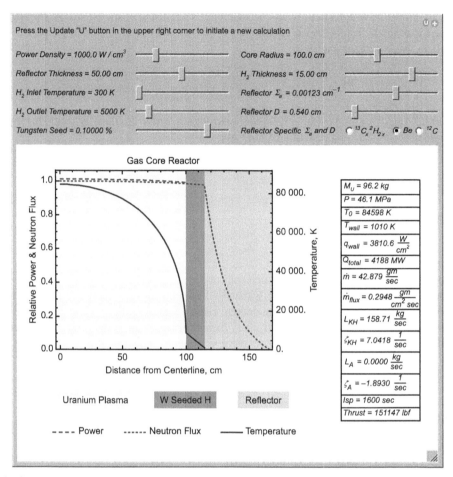

FIGURE 17.15

Gas core nuclear rocket performance characteristics.

Fig. 17.15 qualitatively illustrates the performance characteristics of a gas core nuclear rocket as represented by Eqs. (17.70), (17.75), (17.88), (17.90), (17.94), (17.102), and (17.103). Note the flatness of the power distribution presented in the figure. This flatness is due to the extremely long mean free path for neutrons in the uranium gas. For many cases, neutrons can often travel completely across the core without undergoing a single interaction.

Also note that while the gas core nuclear rocket generally appears to be stable against acoustic instabilities, uranium losses from the Kelvin–Helmholtz instability can be quite high. For the given default case, the entire core is lost about every half second. Such high uranium loss rates make the viability of the type of rocket quite questionable. Not only will the required uranium replacement inventory be quite high, but the rocket performance itself will be severely compromised due to the high effective molecular weight of the exhaust stream and its consequent impact on the specific impulse of the engine. Obviously, some kind of alternative flow configuration will be required which reduces the severity of these instabilities and minimizes the migration of uranium into the hydrogen flow stream by maintaining an effective and stable equilibrium separation between the two gas layers.

3. NUCLEAR LIGHT BULB

While the open cycle gas core rocket described in the previous section can theoretically give impressive performance, it became apparent that various flow instabilities characteristic of the concept would result in unacceptably high uranium loss rates. To address this deficiency, another nuclear engine concept called the nuclear light bulb was conceived. In the nuclear light bulb, the gaseous uranium is confined in closed transparent containers which allow radiant energy from the core to be transmitted through the container where the energy is absorbed by a seeded hydrogen propellant which flows on the outside of the container. This concept has the obvious advantage of containing 100% of the nuclear fuel; however, it also introduces a whole new set of design challenges. Chief among these design challenges is the problem of maintaining the structural integrity of the transparent core containment vessel in the presence of an extremely harsh temperature environment while simultaneously allowing the transmission of vast amounts of radiant energy through its walls. These design challenges were addressed in a program at United Technologies [9] in the 1960s when the company had an active program underway to develop a nuclear light bulb rocket engine. A diagram of the engine concept developed by the company is illustrated in Fig. 17.16.

In this design, the extremely hot fissioning uranium plasma is prevented from touching the transparent containment vessel by a vortex flow of seeded neon gas which acts as a buffer between the containment walls and the uranium plasma. It might be supposed that the uranium plasma would be denser than the neon gas due to its higher atomic weight and thus due to centrifugal forces be disposed to reside on inside edge of the containment vessel where it would melt the containment walls. This, however, is not the case. Because the uranium plasma is at a considerably higher temperature than the

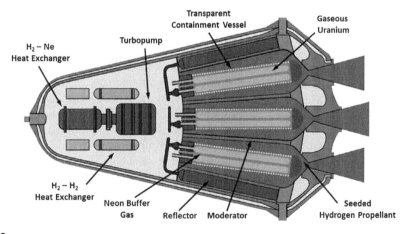

FIGURE 17.16

"Nuclear light bulb" rocket engine.

neon buffer gas, application of the ideal gas law indicates that at equal pressures, the neon gas will actually have the higher density and it, rather than the uranium plasma, will reside at the inside edge of the containment vessel where it properly serves its purpose as a buffer gas. The neon gas (along with some entrained uranium) is continually extracted from the edge of the reactor core where it is separated from the uranium and cooled in a heat exchanger before being reinjected back into the containment vessel. The rejected heat from the neon is used to partially preheat the main hydrogen propellant stream. The separated uranium is also reinjected back into the containment vessel, thus preventing any uranium loss from occurring in the system.

The radiation (primarily UV light) emitted from the high-temperature fissioning uranium plasma, passes through the containment vessel's transparent walls and is absorbed in seeded hydrogen propellant which flows along the outside of the containment vessel. This hot hydrogen propellant is subsequently exhausted through nozzles to produce thrust. The transparent walls of the containment vessel are of particular concern in the nuclear light bulb design. Anyone who has accidentally come in contact with a bare-lighted bulb is painfully aware that not all the light from the filament passes through the light bulb, but that some of the light is absorbed by the glass itself. The same is true with the nuclear light bulb. Consequently, the material comprising the containment vessel walls must not only be highly transparent, but must also be actively cooled to prevent overheating and eventual vessel failure. Materials which have been considered for use in the containment vessel include fused silica which has a UV cutoff (eg, *UV* wavelength below which the material becomes opaque) of about 0.18 μm and single-crystal beryllium oxide (BeO) which has a UV cutoff of 0.12 μm. Any radiation with wavelengths below the UV cutoff which encounters the vessel wall will, therefore, be absorbed by the containment vessel and must be removed.

One method which has been considered to reduce the radiation below the UV cutoff to manageable levels is to seed the uranium plasma and neon buffer with a NO/O_2 mixture. The seeding has the effect of making the fuel opaque to light radiation having very small wavelengths. Radiation which is absorbed in the containment walls is removed by a dedicated cooling system using hydrogen as the working fluid. The cooling system rejects its heat to the main hydrogen propellant stream where it completes preheating the propellant before it enters the turbopump and heating chamber. A simplified flow diagram of the nuclear light bulb is illustrated in Fig. 17.17.

In addition to a number of theoretical studies performed by United Technologies on the nuclear light bulb concept, considerable experimental work was also performed. In these experiments, a number of program objectives were achieved. In the nuclear area, zero power critical experiments were performed with UF_2 to verify core pressure values required to achieve criticality and to study the neutronic characteristics of gaseous fissioning plasmas. Studies were also performed on inductively heated plasmas designed to simulate a gaseous fissioning core including the testing of the internally cooled transparent silica pressure vessel walls in a high radiant heat flux environment. The experimental facility used to perform the tests and the hot confined plasma is shown in Fig. 17.18.

Overall, the experimental program was divided into four parts with each part having very specific objectives. The first part of the experimental tests involved examining the behavioral characteristics of the inductively heated argon simulating the fissioning core. The second part of the tests involved buffer gas injection into the experimental cavity to examine the characteristics of the vortex flow as it moved tangentially to the pressure vessel walls. In the third part of the experimental program, the characteristics of the radiant energy transport through the vessel walls was investigated, and finally, the fourth

3. NUCLEAR LIGHT BULB 291

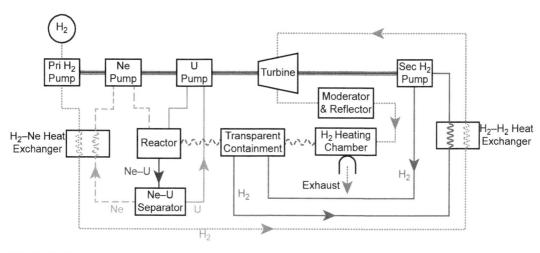

FIGURE 17.17

"Nuclear light bulb" simplified flow diagram.

FIGURE 17.18

Testing in United Technologies nuclear light bulb experimental facility.

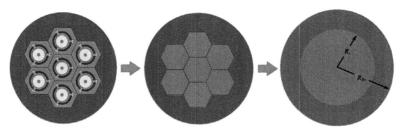

FIGURE 17.19

Nuclear light bulb reactor with smeared core.

part of the program involved pressure testing the filament wound containment vessel. The tests were remarkably successful and did much to validate the nuclear light bulb concept.

To analytically determine the performance characteristics of the nuclear light bulb engine, a number of simplifying assumptions will have to be made. First, since the reactor is geometrically complex, assumptions will have to be to reduce this complexity so as to allow for the calculation of the uranium gas density required to achieve criticality. A technique commonly employed in cases where the core is geometrically complex is to smear out the fine details of the individual fuel elements so as to reduce the number of regions which must be analyzed. This procedure is generally appropriate when the mean free path of neutrons in the core is large compared with the dimensions of the fuel element. Since gas core reactors generally have quite long mean free paths for neutrons, smearing out the individual fuel elements in the nuclear calculations should not introduce large errors in the uranium gas density determined in the criticality calculations provided appropriate smeared average nuclear cross sections are provided. For the nuclear light bulb, the core will be reduced to a two region problem containing just core and reflector regions. It will also be assumed that the core is cylindrical in shape and that neutron leakage out the ends of the reactor can be neglected. The reactor cross-section model to be analyzed is illustrated in Fig. 17.19.

3.1 NEUTRONICS

One group nuclear diffusion theory in cylindrical geometry incorporating the homogenized core and the reflector regions will be used to perform the neutronics analysis. The neutron density determined from the criticality calculations will subsequently be used in thermal fluid analyses to determine the engine performance characteristics.

Beginning the neutronics analysis with the homogenized uranium core, the one group nuclear diffusion equation in cylindrical geometry may be written as

$$0 = D_c \nabla^2 \phi + \frac{\nu \Sigma_f}{k_{\text{eff}}} \phi - \Sigma_a \phi = \frac{D_c}{r^2} \frac{d}{dr}\left(r^2 \frac{d\phi}{dr}\right) + \frac{\nu \Sigma_f^c}{k_{\text{eff}}} \phi - \Sigma_a^c \phi = \frac{1}{r}\frac{d}{dr}\left(r\frac{d\phi}{dr}\right) + \alpha^2 \phi \quad (17.105)$$

where $\alpha^2 = \dfrac{\nu \Sigma_f^c / k_{\text{eff}} - \Sigma_a^c}{D_c} = \dfrac{\frac{\nu \rho \sigma_f^c}{k_{\text{eff}}} - \rho \sigma_a^c}{\frac{1}{3\rho \sigma_{\text{tr}}^c}} =$ buckling in the core.

3. NUCLEAR LIGHT BULB

Assuming that the temperatures in the core are high enough that the ideal gas law holds, the core buckling may consequently be represented in terms of the core temperature and pressure such that:

$$\alpha^2 = 3\rho^2 \left(\frac{\nu\sigma_f^c}{k_{eff}} - \sigma_a^c\right)\sigma_{tr}^c = \frac{3P^2 \left(\frac{\nu\sigma_f^c}{k_{eff}} - \sigma_a^c\right)\sigma_{tr}^c}{R^2 T^2} \qquad (17.106)$$

where P = core pressure; T = core temperature; and R = gas constant for the uranium gas.

Solving the above neutron diffusion differential equation represented by Eq. (17.105) then yields a solution involving Bessel functions such that:

$$\phi(r) = C_1 J_0(\alpha r) + C_2 Y_0(\alpha r) \qquad (17.107)$$

Noting that due to the symmetry of the problem, a boundary condition at the reactor center may be taken to be:

$$\frac{d\phi}{dr} = 0 \quad \text{at} \quad r = 0 \qquad (17.108)$$

If the derivative of the neutron flux expressed in Eq. (17.107) is now taken with respect to the radial position, the result is an equation of the form:

$$\frac{d\phi}{dr} = -C_1 \alpha J_1(\alpha r) - C_2 \alpha Y_1(\alpha r) \qquad (17.109)$$

Now, by applying boundary condition of Eq. (17.108) to the derivative of the core neutron flux expressed in Eq. (17.109), it is found that:

$$0 = -C_1 \alpha J_1(0) - C_2 \alpha Y_1(0) \Rightarrow C_2 = 0 \qquad (17.110)$$

The neutron flux in the core can now be found within an arbitrary constant by using the value for "C_2" from Eq. (17.110) in the general expression for the core neutron flux from Eq. (17.107) to yield an expression of the form:

$$\phi_c(r) = C_1 J_0(\alpha r) \qquad (17.111)$$

In the reflector region, the governing differential equation may be written as:

$$0 = D_r \nabla^2 \phi - \Sigma_a^r \phi = \frac{D_r}{r}\frac{d}{dr}\left(r\frac{d\phi}{dr}\right) - \Sigma_a^r \phi = \frac{1}{r}\frac{d}{dr}\left(r\frac{d\phi}{dr}\right) - \beta^2 \phi \qquad (17.112)$$

where $\beta^2 = \frac{\Sigma_a^r}{D_r}$ = materials buckling in the reflector.

Solving the reflector neutron diffusion equation from Eq. (17.112) then yields:

$$\phi_r(r) = C_3 I_0(\beta r) + C_4 K_0(\beta r) \qquad (17.113)$$

Noting that at the reactor extrapolation distance the neutron flux goes to zero, Eq. (17.113) yields:

$$\phi_r(R_{Rx}^*) = 0 = C_3 I_0(\beta R_{Rx}^*) + C_4 K_0(\beta R_{Rx}^*) \Rightarrow C_4 = -C_3 \frac{I_0(\beta R_{Rx}^*)}{K_0(\beta R_{Rx}^*)} \qquad (17.114)$$

Applying the results from Eq. (17.114) to Eq. (17.113) for the neutron flux distribution in the reflector, a final expression for the neutron flux in the reflector may be determined such that:

$$\phi_r(r) = C_3 \left[I_0(\beta r) - \frac{I_0(\beta R_{Rx}^*)}{K_0(\beta R_{Rx}^*)} K_0(\beta r) \right] \quad (17.115)$$

Recalling that at the interface between the core and reflector regions the neutron flux must be continuous, it is found from Eqs. (17.111) and (17.115) that:

$$\phi_c(R_c) = \phi_r(R_c) \Rightarrow C_1 J_0(\alpha R_c) = C_3 \left[I_0(\beta R_c) - \frac{I_0(\beta R_{Rx}^*)}{K_0(\beta R_{Rx}^*)} K_0(\beta R_c) \right] \quad (17.116)$$

Similarly, at the interface between the core and reflector regions, the neutron current must also be continuous; therefore, from Eqs. (17.111) and (17.115), it can be shown that:

$$J_c(R_c) = J_r(R_c) \Rightarrow D_c \frac{d\phi_c}{dr}\bigg|_{r=R_c} = D_r \frac{d\phi_r}{dr}\bigg|_{r=R_c} \Rightarrow -C_1 \alpha D_c J_1(\alpha R_c)$$

$$= -C_3 \beta D_r \left[I_1(\beta R_c) + \frac{I_0(\beta R_{Rx}^*)}{K_0(\beta R_{Rx}^*)} K_1(\beta R_c) \right] \quad (17.117)$$

If Eq. (17.116) is now divided into Eq. (17.117), it is possible to eliminate all of the arbitrary constants resulting in a criticality equation of the form:

$$\alpha D_c \frac{J_1(\alpha R_c)}{J_0(\alpha R_c)} = \beta D_r \frac{K_0(\beta R_{Rx}^*) I_1(\beta R_c) + I_0(\beta R_{Rx}^*) K_1(\beta R_c)}{K_0(\beta R_{Rx}^*) I_0(\beta R_c) - I_0(\beta R_{Rx}^*) K_0(\beta R_c)} \quad (17.118)$$

3.2 FUEL CAVITY TEMPERATURE DISTRIBUTION

To determine the core temperature, a single-fuel element will be analyzed as illustrated in Fig. 17.20. Since the power distribution in gaseous core systems tend to be quite flat overall, little error will be introduced into the analysis if spatial variations in the power distribution within a single-fuel element are neglected. As in the analysis of the core temperature distribution in the open cycle gas core rocket, convection processes within the region of the fuel element where fissions are occurring will again be neglected and all heat transfer through the gas will be assumed to occur through conduction only.

Using the Rosseland averaged thermal conductivity from Eq. (17.74) and assuming a flat core power distribution, Poisson's heat transfer in cylindrical geometry from Eq. (9.10), then becomes:

$$\nabla^2 T + \frac{q}{k_u} = \frac{d^2 T}{dr^2} + \frac{1}{r} \frac{dT}{dr} + q_0 \frac{3\kappa_U}{16\sigma T^3} = 0 \Rightarrow T = T_U(r) \quad (17.119)$$

where q_0 = core average power density and $T_U(r)$ = radial temperature distribution in the uranium plasma.

To solve Eq. (17.119), two boundary conditions are required. One boundary condition may be determined by noting that due to symmetry, the temperature derivative at the fuel element centerline must be equal to zero. The other boundary condition may be determined by the requirement that at the

3. NUCLEAR LIGHT BULB

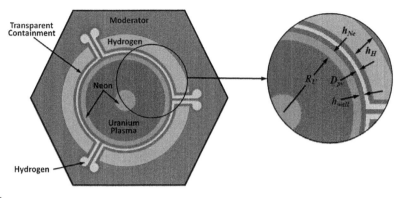

FIGURE 17.20

Nuclear light bulb fuel element cross section.

interface between the uranium plasma and the neon buffer gas, the temperature must be such that the heat radiated away at the interface (assuming a black body radiator) equals the power produced by the fissioning uranium plasma, thus:

$$\frac{q_0 V_U}{A_U} = q_0 \frac{\pi R_U^2 L_{cav}}{2\pi R_U L_{cav}} = \sigma [T_U(R_U)]^4 \Rightarrow T_U(R_U) = \left(\frac{q_0 R_U}{2\sigma}\right)^{\frac{1}{4}} \quad (17.120)$$

where L_{cav} = length of the fuel element cavity; V_U = volume of uranium plasma column; and A_U = surface area of uranium plasma column.

Again as was the case in the open cycle gas core rocket, the heat conduction relationship expressed by Eq. (17.119) is nonlinear such that it is not possible to solve this expression analytically. Numerical solutions are, therefore, required for its evaluation.

3.3 HEAT ABSORPTION IN THE NEON BUFFER LAYER

Again as was the case in the open-cycle gas core rocket, the heat conduction relationship expressed by Eq. (17.119) is nonlinear such that it is not possible to solve this expression analytically. Numerical solutions are, therefore, required for its evaluation.

From analyses performed by United Technologies [10], the gas temperature in the seeded neon buffer layer was seen to drop approximately exponentially by a factor which was dependent on the partial pressure of the NO/O_2 seed mixture. Such an assumption requires that there be no reradiation of energy by the neon gas. For a given temperature at the containment vessel wall and a given temperature at the interface between the uranium plasma and the neon buffer layer, the temperature distribution in the neon buffer layer, therefore, becomes:

$$T_{Ne}(r) = \frac{T_{wall} - T_{int} e^{-\xi h_{Ne}}}{1 - e^{-\xi h_{Ne}}} - \frac{T_{wall} - T_{int}}{1 - e^{-\xi h_{Ne}}} e^{-\xi(r - R_U)} \quad (17.121)$$

where $T_{Ne}(r)$ = radial temperature distribution in the seeded neon buffer layer; T_{wall} = temperature at the containment vessel wall; T_{int} = temperature at the interface between the uranium plasma and the neon buffer layer; and ξ = neon temperature distribution decay factor.

The neon temperature distribution decay factor used in Eq. (17.121) is a function of the partial pressure of the NO/O$_2$ seed mixture. The decay factor's functional relationship to the partial pressure of the NO/O$_2$ seed mixture is based on information derived from the United Technologies report and is presented in Fig. 17.21.

The power distribution in the neon buffer layer is also a function of the partial pressure of the NO/O$_2$ seed mixture plus, it is a function of the temperature of the uranium plasma at the interface between the uranium plasma and the seeded neon buffer layer. The NO/O$_2$ seed mixture in the neon generally only attenuates the light in the UV region of the spectrum at wavelengths above approximately 0.13 µm. Longer wavelengths of light travel through the neon buffer with little attenuation. The 0.13 µm wavelength is not a hard cutoff in the seeded neon but simply marks that region of the spectrum where light attenuation begins to become significant. The light attenuation characteristics through the NO/O$_2$ seed mixture have also been extracted from the United Technologies report and are presented in Fig. 17.22. Obviously, if other seed materials are used in the neon, the light absorption cutoff wavelengths will vary.

The light which does penetrate the seeded neon buffer layer will strike the containment vessel walls where some of the light will be transmitted through the walls to be absorbed in the seeded hydrogen propellant external to the containment vessel, and some of the light will be absorbed by the containment vessel itself. If properly designed, the seeded neon gas will absorb those wavelengths of light which would otherwise be absorbed by the containment vessel. Thus the seeded neon gas acts as a shield, protecting the containment vessel from absorbing significant amounts of the light from that portion of the spectrum which could cause overheating. Single-crystal BeO as a containment vessel wall material has been found to match fairly well with the NO/O$_2$ neon seed since it has a cutoff wavelength of 0.125 µm. In addition, the thermal conductivity of BeO is reasonably high resulting in

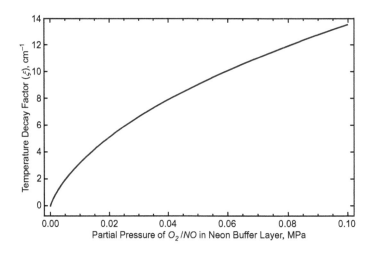

FIGURE 17.21

Neon temperature decay factor as a function of NO/O$_2$ partial pressure.

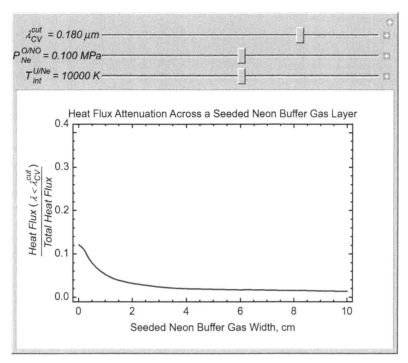

FIGURE 17.22

Heat flux attenuation characteristics in the seeded neon buffer layer.

fairly manageable temperature gradients across the containment vessel wall. Fused silica (SiO_2) has also been considered as a containment vessel wall material; however, since its cutoff wavelength for light absorption is 0.18 μm, it will absorb significantly more light energy than the BeO. This higher rate of light absorption is due to the fact that the seeded neon allows much of the light energy between 0.125 and 0.18 μm to pass through the buffer layer unattenuated. The SiO_2 also suffers from the fact that it has a relatively low thermal conductivity resulting in large temperature gradients across the containment vessel walls.

The amount of heat absorbed in the neon buffer layer is determined by a heat balance between the amount of heat entering the neon buffer layer at the interface between the uranium plasma and the seeded neon buffer layer and the amount of heat exiting the buffer layer at the containment vessel wall. Making use of the heat flux correlation from Fig. 17.22 then yields an equation of the form:

$$Q_{Ne} = 2\pi R_U L_{cav} q_{Ne}^{\lambda_{ONO}}(R_U) - 2\pi(R_U + h_{Ne})L_{cav} q_{Ne}^{\lambda_{ONO}}(R_U + h_{Ne}) = \dot{m}_{Ne} c_p^{Ne}(T_{Ne}^{ave} - T_{Ne}^{CV}) \quad (17.122)$$

where Q_{Ne} = heat absorbed in the seeded neon buffer layer; $q_{Ne}^{\lambda_{ONO}}(r)$ = heat flux at position "r" in the seeded neon buffer layer subject to being absorbed in the buffer gas (from Fig. 17.22); $\dot{m}_{Ne}$ = mass flow rate of neon; c_p^{Ne} = specific heat capacity of neon; T_{Ne}^{ave} = average temperature of the neon as it leaves the fuel element; and T_{Ne}^{CV} = temperature of the neon at the containment vessel wall (assumed to be the containment vessel wall temperature where it is injected into the fuel element).

The required mass flow rate of neon buffer gas needed to remove the heat absorbed in the gas by radiation can now be determined from Eq. (17.122) such that:

$$\dot{m}_{Ne} = \frac{Q_{Ne}}{c_p^{Ne}\left(T_{Ne}^{ave} - T_{Ne}^{CV}\right)} = \frac{2\pi L_{cav}[R_U q_{Ne}(R_U) - (R_U + h_{Ne})q_{Ne}(R_U + h_{Ne})]}{c_p^{Ne}\left(T_{Ne}^{ave} - T_{Ne}^{CV}\right)} \quad (17.123)$$

The average temperature of the neon used in Eq. (17.122) is determined by volume averaging the neon buffer layer temperature distribution calculated using Eq. (17.121) yielding:

$$T_{Ne}^{ave} = \frac{2\pi \int_{R_U}^{R_U+h_{Ne}} r T_{Ne}(r)\, dr}{\pi\left[(R_U + h_{Ne})^2 - R_U^2\right]} = \frac{T_{Ne}^{CV} - T_{int} e^{-\xi h_{Ne}}}{1 - e^{-\xi h_{Ne}}} + 2\frac{T_{int} - T_{CV}^{OD}}{1 - e^{-\xi h_{Ne}}}$$

$$\times \frac{1 + \xi R_U - [1 + \xi(h_{Ne} + R_U)]e^{-\xi h_{Ne}}}{\xi^2[h_{Ne}(h_{Ne} + 2R_U)]} \quad (17.124)$$

The heat absorbed by the neon buffer gas is dumped into the H_2/Ne heat exchanger where the heat is transferred to the main hydrogen propellant stream before being further heated in the H_2/H_2 heat exchanger. From a heat balance on the H_2/Ne heat exchanger, it is found that:

$$Q_{Ne} = \dot{m}_H^{prop} c_p^{H_2}\left(T_{H_2}^{NeHX} - T_{H_2}^{tank}\right) \Rightarrow T_{H_2}^{NeHX} = \frac{Q_{Ne}}{\dot{m}_H^{prop} c_p^{H_2}} + T_{H_2}^{tank} \quad (17.125)$$

where $\dot{m}_H^{prop}$ = mass flow rate of hydrogen propellant; $c_p^{H_2}$ = specific heat capacity of hydrogen propellant; $T_{H_2}^{tank}$ = temperature of the hydrogen propellant entering engine from its storage tank; and $T_{H_2}^{NeHX}$ = temperature of the hydrogen propellant after leaving the H_2/Ne heat exchanger.

3.4 HEAT ABSORPTION IN THE CONTAINMENT VESSEL

The light radiation which is above the containment vessel cutoff frequency and which is not absorbed in the seeded neon buffer layer will be absorbed in the containment vessel and must be removed to prevent overheating. In the United Technologies design, the containment vessel consisted of numerous circumferential tubes connected to a common header system and extending about a third of the way around the circumference of the containment vessel. Because of the short distance, the hydrogen coolant had to flow through the containment vessel before exiting to the H_2-H_2 heat exchanger; the temperature rise in the hydrogen coolant was kept small which minimized the temperature variations that the containment vessel had to withstand. The temperature drop across the containment vessel wall may be determined by applying Poisson's equation over one tube section of the containment vessel wall such that:

$$Q_{tube} = q_{CV}^{\lambda_{CV}} A_{tube}^{OD} = k_{CV} A_{tube} \frac{dT}{dr} \Rightarrow dT = \frac{q_{CV}^{\lambda_{CV}} A_{tube}^{OD}}{k_{CV} A_{tube}} dr \quad (17.126)$$

where Q_{tube} = heat absorbed in one tube section of the containment vessel wall; A_{tube}^{OD} = outside area of one containment vessel tube section exposed to light from the fissioning core; A_{tube} = area of one containment vessel tube section at a specific position within the tube wall; $q_{CV}^{\lambda_{CV}}$ = heat flux at the containment vessel wall subject to being absorbed by the containment vessel (from Fig. 17.22); and k_{CV} = thermal conductivity of the containment vessel.

Assuming that the length of a wall tube segment extends a third of the way around the containment vessel and that a third of the circumference of the tube comprising wall segment is exposed to the radiant heat emanating from the fissioning core, the areas of the wall tube segment in Eq. (17.126) may be calculated. If the resulting equation is then integrated over the thickness of the containment vessel wall, the temperature difference across the wall may be determined yielding:

$$\int_{T_{CV}^{ID}}^{T_{CV}^{OD}} dT = T_{CV}^{OD} - T_{CV}^{ID} = \Delta T_{wall} = \int_{\frac{D_c}{2}}^{\frac{D_c}{2}+h_{wall}} \frac{q_{CV}^{\lambda_{CV}}}{k_{CV}} \frac{\left[\frac{1}{3}\pi(Ru+h_{Ne})\right]\left[\frac{2}{3}\pi\left(\frac{D_c}{2}+h_{wall}\right)\right]}{\left[\frac{1}{3}\pi(Ru+h_{Ne})\right]\left[\frac{2}{3}\pi r\right]} dr$$

$$= \frac{q_{CV}^{\lambda_{CV}}}{k_{CV}}\left(\frac{D_c}{2}+h_{wall}\right)\int_{\frac{D_c}{2}}^{\frac{D_c}{2}+h_{wall}} \frac{dr}{r} = \frac{q_{CV}^{\lambda_{CV}}}{2k_{CV}}(D_c+2h_{wall})\ln\left(\frac{D_c+2h_{wall}}{D_c}\right) \quad (17.127)$$

where T_{CV}^{ID} = temperature on the inside wall of the containment vessel; T_{CV}^{OD} = temperature on the outside wall of the containment vessel; and ΔT_{wall} = temperature difference across the containment vessel wall.

The heat which is absorbed in the tube sections comprising the containment vessel walls is removed by hydrogen coolant which flows within the tubes. The rate of this hydrogen flow must be such that the maximum wall temperature seen by containment vessel is below the maximum allowable temperature of the material from which the containment vessel is fabricated. If the temperature rise in the hydrogen coolant is specified, the required flow hydrogen flow rate in individual containment vessel coolant tubes can be calculated from:

$$Q_{tube} = \dot{m}_{H_2}^{tube} c_p^{H_2} \Delta T_{H_2}^{tube} \Rightarrow \dot{m}_{H_2}^{tube} = \frac{Q_{tube}}{c_p^{H_2} \Delta T_{H_2}^{tube}} \quad (17.128)$$

where $\dot{m}_{H_2}^{tube}$ = hydrogen flow rate within a single coolant tube of the containment vessel and $\Delta T_{H_2}^{tube}$ = temperature rise of the hydrogen as it flows through the containment vessel coolant tubes

The maximum permissible outlet hydrogen temperature consistent with the containment vessel wall material temperature limitations can be calculated by noting that the heat which is absorbed in the containment vessel walls is transferred to the hydrogen flowing in the containment vessel coolant tubes such that:

$$Q_{tube} = \dot{m}_{H_2}^{tube} c_p^{H_2} \Delta T_{H_2}^{tube} = h_c^{CV} A_{wall}^{ID} \left(T_{CV}^{ID} - T_{H_2}^{out}\right) = h_c^{CV} A_{wall}^{ID} \left(T_{CV}^{max} - \Delta T_{wall} - T_{H_2}^{out}\right)$$
$$\Rightarrow T_{H_2}^{out} = T_{CV}^{max} - \Delta T_{wall} - \frac{Q_{tube}}{h_c^{CV} A_{wall}^{ID}} \quad (17.129)$$

where $T_{H_2}^{out}$ = temperature of the hydrogen as it exits the coolant tube of the containment vessel; T_{CV}^{max} = maximum allowable containment vessel wall temperature; h_c^{CV} = heat transfer coefficient of the hydrogen in the containment vessel coolant tubes; and A_{wall}^{ID} = total area for heat transfer area within a single containment vessel coolant tube.

The required inlet hydrogen coolant temperature to the containment vessel may then be calculated from Eq. (17.129) and the specified temperature rise in the hydrogen to yield:

$$T_{H_2}^{in} = T_{H_2}^{out} - \Delta T_{H_2}^{tube} \quad (17.130)$$

where $T_{H_2}^{in}$ = temperature of the hydrogen as it enters the coolant tube of the containment vessel.

The heat absorbed by the containment vessel hydrogen coolant is dumped into the H_2/H_2 heat exchanger where the heat is transferred to the main hydrogen propellant stream. The propellant streams are further heated before finally being injected into the main propellant heating chamber. From a heat balance on the H_2/H_2 heat exchanger, it is found that:

$$N_{tube} Q_{tube} = \dot{m}_H^{prop} c_p^{H_2} \left(T_{H_2}^{HHX} - T_{H_2}^{NeHX} \right) \Rightarrow T_{H_2}^{HHX} = \frac{N_{tube} Q_{tube}}{\dot{m}_H^{prop} c_p^{H_2}} + T_{H_2}^{NeHX} \tag{17.131}$$

where $T_{H_2}^{HHX}$ = temperature of the hydrogen propellant after leaving the H_2/H_2 heat exchanger; $T_{H_2}^{NeHX}$ = temperature of the hydrogen propellant after leaving the H_2/Ne heat exchanger; and N_{tube} = total number of coolant tubes in the reactor containment vessels.

3.5 HEAT ABSORPTION IN THE HYDROGEN PROPELLANT

The radiant energy not absorbed by the neon buffer gas or in the containment vessel wall is transmitted to the seeded hydrogen propellant external to the reaction cavity where it is absorbed and thereby heated prior to being expelled through the engine nozzle to produce thrust. This energy is given by

$$Q_{CV_0} = q_0 V_{tot} - N_{FE} Q_{Ne} - N_{tube} Q_{tube} \tag{17.132}$$

where Q_{CV_0} = total heat transmitted through the containment vessels and into the propellant cavity; V_{tot} = total nuclear reaction volume in the engine; and N_{FE} = number of fuel elements in the engine.

By knowing the heat flux at the outside wall of the containment vessel along with a knowledge of the radiation absorption characteristics of the seeded hydrogen propellant, the wall temperature at the inside edge of the fuel element moderator region may be calculated. Since the hydrogen propellant layer is generally not optically opaque, similar to the situation present in the open cycle gas core configuration, Schwarzschild's equation is again used to calculate the wall temperature at the propellant/moderator interface.

As the radiant energy from the core cavity strikes the hydrogen propellant layer, the energy is gradually absorbed as it travels through the propellant, in the process heating it to high temperatures. At these high temperatures, the propellant is sufficiently hot so as to also radiate away a significant amount of the heat which it absorbs.

The first step in the analysis is to determine the amount of heat transferred by radiation to the propellant. Referring to Fig. 17.20, for a differential volume in the propellant layer, the radiation heat balance equation may be written as

$$\frac{d[rI(r)]}{dr} = \left[-\underbrace{\kappa_H(T)I(r)}_{\text{Absorption}} + \underbrace{\kappa_H(T)J(r)}_{\text{Emission}} \right] r \tag{17.133}$$

where $I(r)$ = Black body thermal radiation intensity in the hydrogen propellant; $J(r) = \sigma T^4$ = Black body thermal radiation emitted from the hydrogen propellant; and $\kappa_H(T)$ = Planck opacity of the hydrogen propellant.

Rearranging Eq. (17.133) into a more conventional form yields:

$$\frac{dI(r)}{dr} + \left[\frac{1}{r} + \kappa_H(T) \right] I(r) = \kappa_H(T) \sigma T^4 \tag{17.134}$$

Using the method of integrating factors, it is possible to obtain an analytical solution to Eq. (17.134) of the form:

$$I(r) = \frac{e^{-\int_{R_{cav}}^{r} \kappa_H(T)dr'}}{r} \int_{R_{cav}}^{r} \kappa_H(T)\sigma T^4 e^{\int_{r'}^{R_{cav}+h_H} \kappa_H(T) dr''} r' \, dr' + C\frac{e^{-\int_{R_{cav}}^{r} \kappa_H(T) dr'}}{r} \quad (17.135)$$

To determine the arbitrary constant "C," the heat flux at the outside wall of the containment vessel will be used as a boundary condition such that:

$$I(R_{CVo}) = \frac{Q_{CVo}}{A_{CVo}} = q_{CVo} = \frac{e^{-\int_{R_c}^{R_c} \kappa_H(T) dr'}}{R_{CVo}} \int_{R_c}^{R_c} \kappa_H(T)\sigma T^4 e^{\int_{R_c}^{R_c} \kappa_H(T) dr''} r'^2 \, dr'$$

$$+ C\frac{e^{-\int_{R_c}^{R_c} \kappa_H(T)dr'}}{R_{CVo}} = C\frac{1}{R_{CVo}} \Rightarrow C = q_{CVo} R_{CVo} \quad (17.136)$$

where q_{CVo} = heat flux at the outside surface of the containment vessel; A_{CVo} = total outside surface area of the containment vessels; and R_{CVo} = containment vessels outer wall radius.

Substituting the expression for the arbitrary constant "C" from Eq. (17.136) into the relationship for the radiation intensity in the hydrogen propellant layer from Eq. (17.135) and integrating over the hydrogen propellant layer will yield an expression for the wall heat flux at the moderator surface of the form:

$$I(R_{CVo} + h_H) = q_{mod} = q_{CVo}\frac{R_{CVo}}{R_{CVo} + h_H} e^{-\int_{R_{CVo}}^{R_{CVo}+h_H} \kappa_H(T)dr} + \frac{e^{-\int_{R_{CVo}}^{R_{CVo}+h_H} \kappa_H(T)dr}}{R_{CVo} + h_H}$$

$$\times \int_{R_{CVo}}^{R_{CVo}+h_H} \kappa_H(T)\sigma T^4 e^{\int_{r}^{R_{CVo}+h_H} \kappa_H(T)dr'} r \, dr \quad (17.137)$$

where q_{mod} = heat flux at the surface of the moderator.

In order to analytically solve Eq. (17.137) the assumption will be made that the temperature of the propellant is constant throughout the propellant layer. This assumption will allow Eq. (17.137) to be integrated fairly simply and is probably fairly reasonable given that there should be a considerable amount of turbulent mixing within the propellant stream resulting in the temperature distribution being fairly uniform. Carrying out these integrations then yields:

$$q_{mod} = \frac{e^{-h_H \kappa_H}\left[\left(q_{CVo}R_{CVo} + h_H \sigma T_{prop}^4\right)\kappa_H - (1 - R_{CVo}\kappa_H)(1 - e^{-h_H \kappa_H})\sigma T_{prop}^4\right]}{\kappa_H(h_H + R_{CVo})} \quad (17.138)$$

In order to determine the temperature at the surface of the moderator (T_{mod}), one more equation is needed. This additional equation may be found by performing a heat balance at the moderator wall such that:

$$0 = \underbrace{q_{mod}}_{\text{Radiant Heat Flux at Moderator Wall}} + \underbrace{h_c^{prop}(T_{prop} - T_{mod})}_{\text{Convective Heat Transfer to Moderator Wall}} - \underbrace{\frac{\dot{m}_H^{prop}}{A_{mod}}c_p^{H_2}\left(T_{mod} - T_{H_2}^{HHX}\right)}_{\text{Transpiration Heat Transfer from Moderators Wall to Hydrogen Propellant}} \quad (17.139)$$

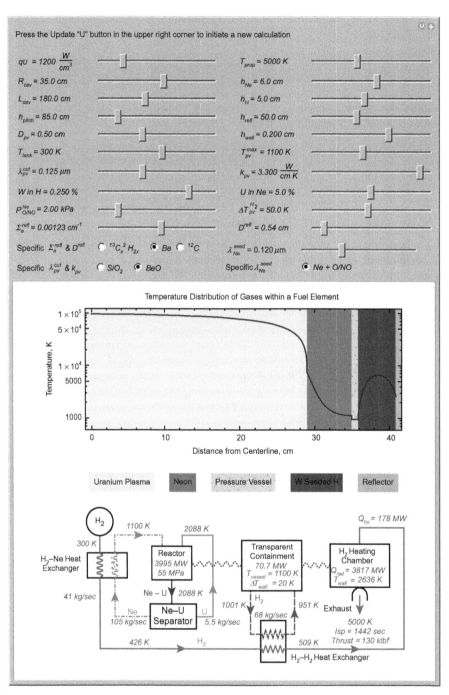

FIGURE 17.23

Nuclear light bulb performance characteristics.

Where h_c^{prop} = moderator wall transpiration heat transfer coefficient; A_{mod} = total moderator wall surface area.

If the total core power is known along with the average outlet temperature of the propellant, it is found that:

$$Q_{\text{CVo}} = q_{\text{CVo}} A_{\text{CVo}} = \dot{m}_H^{\text{prop}} c_p^{H_2} \left(T_{\text{prop}} - T_{H_2}^{\text{HHX}} \right)$$

$$\Rightarrow \dot{m}_H^{\text{prop}} c_p^{H_2} = \frac{q_{\text{CVo}} A_{\text{CVo}}}{T_{\text{prop}} - T_{H_2}^{\text{HHX}}} \quad (17.140)$$

Substituting Eq. (17.140) into Eq. (17.139) and rearranging terms then finally yields:

$$q_{\text{mod}} = q_{\text{CVo}} \left(\frac{T_{\text{mod}} - T_{H_2}^{\text{HHX}}}{T_{\text{prop}} - T_{H_2}^{\text{HHX}}} \right) \frac{A_{\text{CVo}}}{A_{\text{mod}}} - h_c^{\text{prop}} (T_{\text{prop}} - T_{\text{mod}}) \quad (17.141)$$

Equating Eqs. (17.138) and (17.141) and rearranging terms then yields an equation for the surface temperature of the moderator such that:

$$T_{\text{mod}} = \frac{q_{\text{CVo}} A_{\text{CVo}} T_{H_2}^{\text{HHX}} \kappa_H (h_H + R_{\text{CVo}}) - A_{\text{mod}} \left(T_{H_2}^{\text{HHX}} - T_{\text{prop}} \right) \left[h_c^{\text{prop}} T_{\text{prop}} \kappa_H (h_H + R_{\text{CVo}}) \right]}{\kappa_H (h_H + R_{\text{CVo}}) \left[q_{\text{CVo}} A_{\text{CVo}} + h_c^{\text{prop}} A_{\text{mod}} \left(T_{\text{prop}} - T_{H_2}^{\text{HHX}} \right) \right]} +$$

$$\frac{A_{\text{mod}} e^{-h_H \kappa_H} \left(T_{H_2}^{\text{HHX}} - T_{\text{prop}} \right) \left\{ \kappa_H q_{\text{CVo}} R_{\text{CVo}} + \sigma T_{\text{prop}}^4 \left[h_H \kappa_H - (1 - e^{-h_H \kappa_H})(1 - \kappa_H R_{\text{CVo}}) \right] \right\}}{\kappa_H (h_H + R_{\text{CVo}}) \left[A_{\text{CVo}} q_{\text{CVo}} + A_{\text{mod}} h_c^{\text{prop}} \left(T_{\text{prop}} - T_{H_2}^{\text{HHX}} \right) \right]}$$

$$(17.142)$$

As noted in the section on the open-cycle gas-core nuclear rocket, the opacity of hydrogen is fairly small at temperatures below about 10,000°C and that normally a seed material such as tungsten is added to the hydrogen propellant gas to increase its opacity. Fig. 17.13 illustrates how the hydrogen opacity varies as tungsten seed material is introduced into the hydrogen propellant.

Eq. (17.142) can now be used to determine the qualitative performance characteristics of a conceptual nuclear light bulb nuclear rocket. These calculations are presented in Fig. 17.23. Note that these calculations have neglected a number of important effects such as pumping power losses and various heat losses due to neutron and gamma radiation. Nevertheless, the parameters given below should trend properly and give a rough idea as to what level of performance could be expected from the engine.

REFERENCES

[1] General Atomic Division of General Dynamics, Nuclear Pulse Space Vehicle Study, 1964. GA-5009, vol. I thru IV, NASA/MSFC Contract NAS 8−11053.
[2] R. Serber, The Los Alamos Primer: The First Lectures on How to Build an Atomic Bomb, 1943. Report LA-1.
[3] D.E. Parks, G. Lane, J.C. Stewart, S. Peyton, Optical Constants of Uranium Plasma, 1968. NASA CR-72348 and Gulf General Atomic GA-8244.

[4] S. Rosseland, Theoretical Astrophysics: Atomic Theory and the Analysis of Stellar Atmospheres and Envelope, Clarendon Press, Oxford, 1936.
[5] R.W. Patch, Interim Absorption Coefficients and Opacities for Hydrogen Plasma at High Pressure, NASA Lewis Research Center, Cleveland, OH, October 1969. NASA TM X-1902.
[6] J.R. Williams, et al., Opacity of tungsten-seeded hydrogen to 2500 K and 115 atmospheres, in: 2nd Symposium on Uranium Plasmas: Research and Applications, Atlanta, GA, November 15–17, 1971.
[7] S. Chandrasekhar, Hydrodynamic and Hydromagnetic Stability, Dover Publications, New York, NY, 1961, ISBN 048664071X.
[8] T. Kammash, D.L. Galbraith, Fuel confinement and stability in the gas core nuclear propulsion concept, in: 28th Joint Propulsion Conference, Nashville, TN, Paper AIAA 92–3818, July 6–8, 1992.
[9] G.H. Mclafferty, Investigation of Gaseous Nuclear Rocket Technology \[Dash] Summary Technical Report, United Aircraft Research Laboratories, November 1969. Report H-910093–46, prepared under Contract NASw-847.
[10] R.J. Rogers, T.S. Latham, H.E. Bauer, Analytical Studies of Nuclear Light Bulb Engine Radiant Heat Transfer and Performance Characteristics, United Aircraft Research Laboratories, September 1971. Report H-910900–10, prepared under Contract SNPC-70.

Problems

CHAPTER 2

1. A mission to Mars has been determined to require a total velocity increment of 14.2 km/s. A vehicle incorporating a nuclear thermal rocket engine is to be used for the mission. Assume that the propellant is hydrogen and that its temperature upon leaving the reactor is 2845 K. Also assume that the rocket nozzle has an area ratio of 5 in its converging section and an area ratio of 300 in its diverging section. From this information, determine the engine-specific impulse and the vehicle mass fraction.

CHAPTER 3

1. An NTR engine system is to be designed around a turbopump assembly which has a rated hydrogen flow rate of 15 kg/s. It is estimated that because of shielding, cooling requirements, and strength of materials considerations, the weight of the engine system will grow with increasing hydrogen outlet temperatures according to the following equation:

$$\text{mass(kg)} = 2000 + \frac{500}{2.5 - \left(\frac{T_4}{2700}\right)^3} \text{ for } H_2 \text{ chamber exit temperatures } T_4 < 3600 \text{ K}.$$

Assume that a hot bleed cycle will be used for the engine cycle and that the bypass bleed flow can be neglected in your calculations. Also assume that the inlet hydrogen temperature to the reactor is:

$$T_3 = 300 \text{ K and } \gamma_{H_2} = 1.4.$$

For the stated conditions, determine the following parameters such that they yield an engine system which maximizes the engine thrust to weight ratio using only the information given:

a. Reactor exit (ie, chamber) temperature
b. Reactor power level
c. Specific impulse assuming a nozzle exit pressure of zero
d. Engine mass
e. Thrust level assuming a nozzle exit pressure of zero

$$\text{Note that } R_u = 8.314 \frac{J}{\text{mole K}}$$

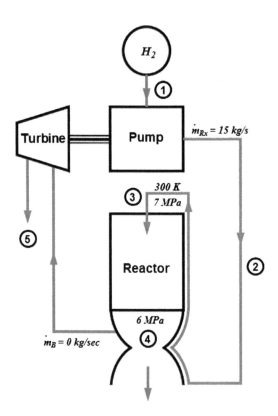

2. A Mars "hopper" is to be used to "hop" from place to place on the Martian surface using the atmosphere (primarily carbon dioxide) as the propellant in an NTR-based propulsion system using a cold bleed cycle. Assume the "hopper" is on the Martian surface and that carbon dioxide has been pumped into the propellant tanks. Based upon the given state conditions in the adjacent figure, determine:
   a. The reactor inlet thermodynamic conditions
   b. The minimum bleed power required
   c. The mass flow rate through the turbopump turbine
   d. The turbopump pumping power requirement
   e. The mass flow rate through the reactor
   f. The engine-specific impulse
   g. The Mach number at nozzle exit
   h. The nozzle exit thermodynamic conditions
   i. The engine thrust level insert

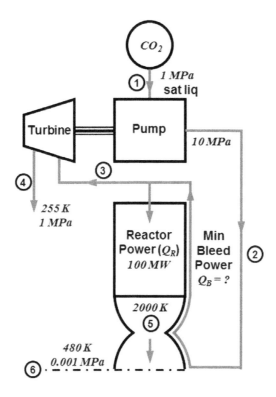

$$\text{Given:} \ \gamma = 1.184 \quad \frac{A_{Rx}}{A^*} = 8 \quad \frac{A_{exit}}{A^*} = 475$$

In addition, sketch the thermodynamic cycle on the carbon dioxide T–S diagram (see Appendix) and be sure to show all work and justify any assumptions you make. Do *not* assume turbine is isentropic.

3. A nuclear power system is being designed to energize an electric propulsion unit. Assume that an ideal Stirling cycle is to be used in the design. Determine the temperature (T_L) at which heat should be radiated to space in order to minimize the mass required to generate a given amount of electricity (eg, maximum system α), the mass of the reactor, the mass of the radiator, and the system α in kW$_e$/kg.

$$\text{Note:} \ \alpha = \frac{Q_e}{\text{Mass of reactor system} + \text{Mass of radiator}}$$

where the thermal to electric conversion efficiency of an ideal Stirling cycle is given by

$$\eta = \left(1 - \frac{T_L}{T_H}\right)\eta_{me}$$

and heat is rejected to space according to the Stefan–Boltzmann law: $Q_{rad} = \varepsilon \sigma A T_L^4$.

Parameter	Symbol	Value	Units
Electrical power output	Q_e	100	kW_e
Mechanical to electrical efficiency	η_{me}	0.9	
Fin emissivity	ε	0.85	
Temperature after heat addition	T_H	2000	K
Areal density of radiator	ρ_{rad}	1.5	gm/cm^2
Reactor thermal power density	Q	1	kW/kg
Stefan–Boltzmann constant	σ	5.67×10^{12}	W/cm^2 K^4

CHAPTER 4

1. Determine the travel time to Mars if a spacecraft is launched from earth with a velocity vector that is in the same direction as the earth's velocity vector around the sun and a solar orbit eccentricity of 0.6. Assume that the distance for earth to the sun is 1.497×10^8 km, and the distance from Mars to the sun is 2.25×10^8 km. Also assume that the solar gravitational constant is 1.327×10^{11} km^3/s^2.

2. Determine the engine specific impulse required for a one-way trip to Mars if the spacecraft described in Problem 1 is launched from a 200-km high earth orbit and captures into a 100-km high Mars orbit. Assume that the spacecraft has a vehicle mass fraction of 0.15 and that the radius of earth is 6378 km and the radius of Mars is 3393 km. Also assume that the earth's gravitational constant is 398,600 m^3/s^2 and that the Mars gravitational constant is 42,830 m^3/s^2.

CHAPTER 5

1. A nuclear rocket engine has a cylindrical reactor ($D = L = 2$ m) composed of fully enriched UC ($100\%^{235}$U) with hydrogen propellant flow holes (hole volume fraction $= 0.3$). The engine fires for a total of 100 min during a Mars mission and operates at 4000 MW:
   a. What average neutron flux levels would you expect?
   b. What fractional fuel burnup would be expected during the mission?

   Note: $\sigma_f = 577$ b $\rho_{UC} = 13{,}500$ kg/m^3.

2. The threshold for fission in ^{238}U is often taken to be 1.4 MeV since the fission cross section is small below this energy. Using this threshold, what fraction of fission neutrons are capable of producing fission in ^{238}U?

3. The prompt fission neutron spectrum is often represented by the function:

$$\chi(E) = Ce^{-aE} \sinh\sqrt{bE}$$

where C, a, and b are constants.
   a. If the function $\chi(E)$ is normalized to 1.0, that is, if:

$$\int_0^\infty \chi(E)dE = 1$$

Show that the constant "C" is given by

  b. Find a transcendental function for the most probable energy.
  c. Show that the average energy of a prompt fission neutron is given by

$$\overline{E} = \frac{3}{2a} + \frac{b}{4a^2}$$

4. The absorption mean free path for 2200 m/s (0.025 eV) neutrons in a material having a 1/V absorption cross section is 1.0 cm. The corresponding reaction rate is $10^{12} \frac{abs}{s\,cm^3}$. The material has an atomic mass of 10 amu and a density of 2.0 g/cm³. Find:
  a. The 2200 m/s neutron flux.
  b. The material's microscopic absorption cross section (σ_a) for 10 eV neutrons in barns.

CHAPTER 6

1. Suppose that during an inelastic scattering collision the target nucleus absorbs an amount of energy Q. Show that:

$$\frac{E'}{E} = \frac{A^2 + 1 + 2A\cos(\phi)}{(1+A)^2}$$

becomes

$$\frac{E'}{E} = \frac{A^2\tau^2 + 1 + 2A\tau\cos(\phi)}{(1+A)^2}$$

where

$$\tau = \sqrt{1 - \frac{Q}{E}\frac{A}{1+A}}$$

2. Show that the expression for the average loss in the logarithm of the energy during a collision:

$$\xi = 1 + \frac{\alpha}{1-\alpha}Ln(\alpha)$$

may be approximated by

$$\xi = \frac{2}{A + \frac{2}{3}} \quad \text{for large } A, \text{ where}: \quad \alpha = \left(\frac{A-1}{A+1}\right)^2.$$

Hint: Let $\lambda = \frac{1}{\xi}$ and $B = \frac{1}{A}$, then do a series expansion.

CHAPTER 7

1. Calculate the macroscopic transport cross section (Σ_{tr}) and the diffusion coefficient (D) for beryllium given that the nuclide has an atomic weight of 9, a density of 1.85 g/cm³, a microscopic absorption cross section (σ_a) of 0.0095 b, and a microscopic scattering cross section (σ_s) of 7 b.

CHAPTER 8

1. For a bare cylinder of a given constant volume "V" containing fissionable material, determine the height to diameter ratio (H/D) which will yield a maximum value for k_{eff}.

2. A nuclear rocket engine is configured as shown in the figure below. Cold hydrogen enters the engine through the reflector region, is heated to high temperatures in the core region, and finally exits the core where it is expelled as exhaust. The reactor power level is 100 MW.

Assumptions:
   a. The sides of the core and reflector are bare.
   b. The length of the core and the diameter of the core are equal.
   c. The length of the reflector can be considered effectively infinite.
   d. Include extrapolation distances on *both* the core and reflector when calculating their lengths.
   e. The neutron flux and power density are constant in the radial direction.

Parameter	Reflector	Core
D	0.088 cm	1.45 cm
Σ_a	0.0033 cm^{-1}	0.0044 cm^{-1}
$\nu\Sigma_f$	—	0.0080 cm^{-1}

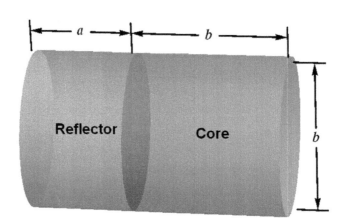

   f. Derive the one group critical equation for this reactor system.
   g. Plot k_{eff} as "b" goes from 0 to 500 cm.
   h. At what length "b" is the reactor exactly critical? Indicate on the above plot.
   i. Derive the value for k_∞ (eg, as $b \to \infty$) for the reactor? Indicate on the above plot.
   j. At what length "a" is the reflector considered infinite.
   k. Plot the neutron flux in the reactor as a function of axial position.
   l. Plot the power density (in W/cm^3) in the reactor as a function of axial position.
   m. What is the axial power peaking factor $\left(\frac{P_{max}}{P_{ave}}\right)$?

3. A thin-walled spherical container (the effects of which may be neglected in your calculations) contains a gaseous uranium compound being examined for possible use in a gas core nuclear rocket engine. Using a one energy group model and assuming the gaseous uranium compound behaves as an ideal gas:
   a. Derive an expression relating the gas pressure, gas temperature, and container radius to k_{eff}.
   b. Derive an expression for the one group flux as a function of "r."

Hint : Note that $\nabla^2 \phi = \frac{1}{r^2} \frac{d}{dr}\left(r^2 \frac{d\phi}{dr}\right)$ and use the solution : $\phi(r) = \frac{u(r)}{r}$.

4. A fissionable gas, assumed to be ideal, is contained in a cylinder that is pressurized by a piston which exerts a constant 200 kPa pressure on the gas. The absorption cross section of the gas varies as $1/V$ while the fission cross section and the diffusion coefficient remain constant. At STP (300 K and 101 kPa), the height of the gas in the cylinder is 43 cm. At what height and at what temperature is maximum criticality (eg, k_{eff}) attained? What is the maximum k_{eff}?

Parameter	Value
ν	2.5
$\sigma_a^{0.025\ eV''}$	600 b
σ_f	300 b
D	0.9 cm
k (Boltzmann constant)	$8.17 \times 10^{-5''}$ eV/K
R (gas constant)	$13,811 \dfrac{\text{b cm kPa}}{\text{atom K}}$

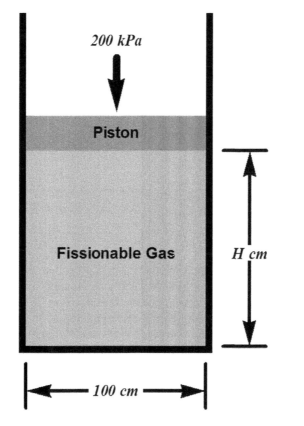

Note: From kinetic theory of gases ... $E = kT$.
Neglect extrapolation lengths in calculations.

5. A two-region symmetrical slab reactor is constructed as shown in the following figure. Derive expressions for the flux in each region of the reactor and k_{eff}. Sketch a plot of the expected flux shape. Use a one group model. Clearly state all assumptions. Assume the reactor is infinite in the "Y" and "Z" directions.

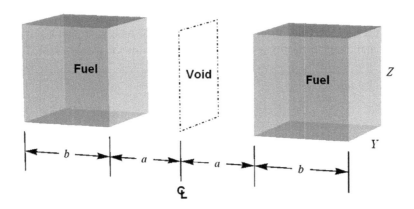

6. A nuclear reactor is being designed for a nuclear thermal propulsion system. Determine the core length and reflector length combination which will yield the minimum nuclear reactor mass. Assume that the reactor criticality equation is given by Eq. (8.41) and that the reactor dimensions (other than along the length) are effectively infinite.

Parameter	Core	Reflector
$\nu\Sigma_f(\text{cm}^{-1})$	0.00499	0
$\Sigma_a(\text{cm}^{-1})$	0.0042	0.0092
$D(\text{cm})$	1.23	0.78
$\rho(\text{gm/cm}^3)$	15	8

CHAPTER 9

1. In a NERVA-type nuclear rocket engine, the fuel element propellant channel walls often must be clad with some type of material to protect the underlying uranium fuel material from being corrosively attacked by the hot hydrogen propellant. It is, therefore, important that this protective cladding maintain its structural integrity during operation. Should there be a mismatch in the thermal expansion coefficients between cladding and the underlying fuel material, the differential thermal expansion which would occur during operation could cause the cladding to fail thus exposing the fuel material to the hot hydrogen propellant, potentially resulting in the failure of the fuel element. A likely place for this cladding failure to occur is where the channel wall surface

temperature gradient is maximum. Given below are typical operating conditions for a NERVA-type fuel element:

Parameter	Value
L	140 cm
r_i	0.125 cm
r_o	0.150 cm
c_p	14.5 W s/g K
k	0.0035 W/cm K
μ	0.00021 g/cm s
$\dot{m}$	0.2 g/s
T_{in}	300 K
P_{ave}	2.5 kW/cm^3
α	0.011 cm^{-1}

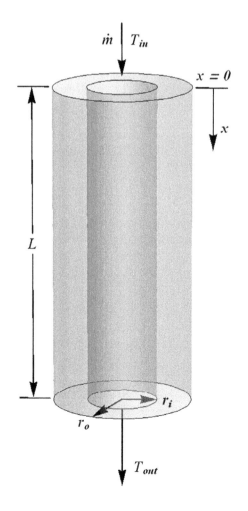

Assume the axial power profile varies according to the following equation (eg, from Eq. (9.32)):

$$P(z) = \frac{\alpha P_{ave} L}{2 \sin\left(\frac{\alpha L}{2}\right)} \cos\left[\alpha\left(\frac{L}{2} - z\right)\right]$$

    a. Derive an expression describing the location of the maximum channel wall temperature gradient.
    b. Plot the axial power distribution and the axial channel wall temperature distribution.
    c. Locate on the above plot, the point where the maximum channel wall temperature gradient occurs.
    d. Comment.

2. Derive a core power distribution "$q(x)$" which will yield a constant fuel temperature along the length of the core. This distribution is optimal with regard to minimizing the size of a nuclear rocket core since all of the fuel would be able to operate at its maximum allowable temperature. In the derivation, assume that the fuel temperature is constant everywhere (eg, the fuel temperature does not vary axially or radially) and neglect any calculations related to radial power variations. Do you think it is possible to practically achieve this power distribution within the core? Briefly discuss.

$$\text{Assume}: \quad q(x) = U(T_{fuel} - T_{propellant})$$

where U is a general conductance term and $q(x)$ is the heating rate per unit length.

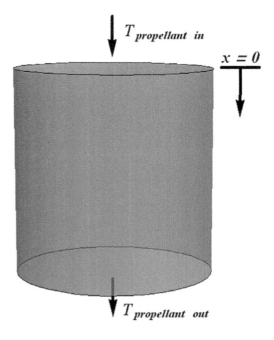

3. A nuclear rocket engine core was accidently misloaded such that fuel with an extremely high-fission cross section was inserted at a particular axial location yielding the power distribution illustrated below. Based on your best engineering judgment, sketch in the blank plot below the expected axial propellant and fuel temperature distributions resulting from the misloaded core. Explain your reasoning.

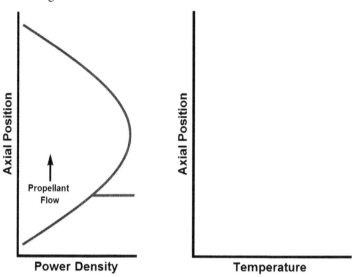

CHAPTER 10

1. A hydrogen turbopump for a nuclear rocket engine employing an expander cycle is to be designed. The inlet pressure to the pump from the propellant feed system will be 0.5 MPa and the liquid hydrogen is to be delivered to the engine system at a flow rate of 10 kg/s. The pump impeller and turbine rotor should have the same diameter and each should have the same rotational speed.

   For the stated conditions, determine the following turbopump parameters:
   a. Turbopump turbine inlet hydrogen temperature
   b. Reactor inlet hydrogen temperature
   c. Pumping power
   d. Turbopump turbine efficiency
   e. Turbopump pump efficiency
   f. Main shaft rotational speed
   g. Diameter of the pump impeller and turbine rotor

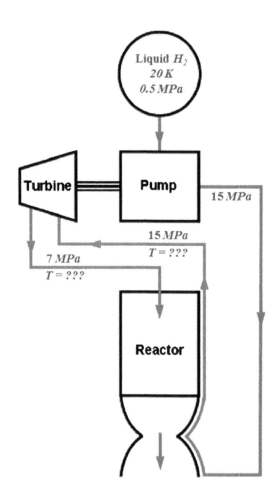

CHAPTER 11

1. A nuclear rocket has been firing for 2 h in route to Mars and is now to be shut down for a long coast period. Assume that the engine had been operating at 1500 MW and a reactivity insertion of −$1.00 was used to shutdown the engine. How much total decay energy must be dissipated during the vehicle coast phase?

$$\text{Assume}: \quad \beta = 0.0075, \quad \lambda = 0.08 \text{ s}^{-1}, \quad \Lambda = 0.00006 \text{ s}.$$

2. For super prompt critical transients, it is usually satisfactory to assume that λC in the point kinetics equations remains constant during the transient. With this assumption, show that if the

reactor is initially critical and at steady state with $n(0) = n_0$ and a reactivity insertion occurs such that $\rho = \beta + \alpha t$, then the reactor response is approximately:

$$\frac{n(t)}{n_0} = e^{\frac{\alpha}{2\Lambda}t^2}\left[1 + \beta\sqrt{\frac{\pi}{2\alpha\Lambda}}\mathrm{Erf}\left(t\sqrt{\frac{\alpha}{2\Lambda}}\right)\right]$$

3. A nuclear rocket has been firing for 1 h in route to Mars and is now to be shut down for a long coast period. Assume that the engine had been operating at a power density of 1500 W/cm³ and a reactivity insertion of −$1.00 is used to shutdown the engine. Also assume that 2 W/cm³ of auxiliary reactor cooling (ie, independent of the cooling resulting from the flow of propellant through the engine) is available to take care of engine decay heat after shutdown:

where: $\beta = 0.0075$, $\lambda = 0.08\ \mathrm{s}^{-1}$, $\Lambda = 0.00006\ \mathrm{s}$, $c_p^{\mathrm{fuel}} = 0.15\ \mathrm{J/g\ K}$,

$\rho_{\mathrm{fuel}} = 13.4\mathrm{g/cm}^3$

a. How long will some level of propellant flow need to be maintained in order that the reactor temperatures remain equal to or below their steady-state values?
b. If propellant flow is terminated at the time of engine shutdown, how much does the reactor temperature rise over the steady-state reactor temperature?

4. The point kinetics equations can be simplified considerably by employing what is called the "prompt jump approximation." This approximation assumes that the initial prompt jump which takes place immediately after a reactivity insertion, occurs instantaneously and thereafter only the stable reactor period controls the rate of change of the neutron population. The approximation assumes that the rate of change of the neutron population is small when compared to the other terms in the point kinetics equation, such that:

$$\frac{dn}{dt} = 0 = \frac{\rho - \beta}{\Lambda}n + \lambda C$$

Using the prompt jump approximation in the point kinetics equations, determine an expression which yields the time-dependent behavior of the neutron population.

5. The reactor in a nuclear rocket must reach full power in 30 s.

where: $\beta = 0.0075$, $\lambda = 0.08\ \mathrm{s}^{-1}$, $\Lambda = 0.00006\ \mathrm{s}$

a. What step increase in reactivity (in $) is needed to induce the required period?
b. If the power must increase by a factor of 10^5 during this time, estimate the stable period required.

CHAPTER 12

1. Assuming a constant k_∞ and B^2 along with the thermal neutron absorption cross section having a $1/V$ dependence, show that the nuclear temperature coefficient of reactivity in a thermal reactor is:

$$\frac{\partial \rho}{\partial T} = -\frac{DB^2}{2k_\infty \Sigma_a^0 \sqrt{T_0}}\frac{1}{\sqrt{T}} \quad \text{where:}\ \Sigma_a^0\ \text{is the 2200 m/s absorption cross section}$$

CHAPTER 13

1. For deep space robotic missions, thermoelectric generators are sometimes used to power the spacecraft systems. The power to run these generators often comes from general purpose heat source (GPHS) units which produce heat from the radioactive decay of ^{238}Pu, The ^{238}Pu used in these units is produced by the neutron bombardment of ^{237}Np in special production reactors designed for this purpose. ^{237}Np does not occur naturally, but is available in quantity from reprocessed spent nuclear reactor fuel. The ^{238}Pu production cycle consists of:

$$n + {}^{237}\text{Np} \rightarrow {}^{238}\text{Np} \xrightarrow[(2.1 \text{ day})]{\beta^-} {}^{238}\text{Pu} \xrightarrow[(89 \text{ years})]{\alpha} {}^{234}\text{U}$$

Assuming that only ^{237}Np is initially present, determine an expression which yields the time-dependent concentration of ^{238}Pu. Also determine an expression which yields the time required to reach a maximum concentration.

2. Assume that a reactor has been operating at steady state for a long period of time and the reactor "scrams." Derive a relationship that will yield the time at which the maximum xenon concentration occurs.

CHAPTER 14

1. To save shielding weight on a Mars mission, it is desired to take advantage of the shielding effects provided by the liquid hydrogen fuel tanks with regard to protecting the crew from the gamma radiation emanating from the propulsion system's nuclear reactor. The vehicle configuration is illustrated in the following figure:

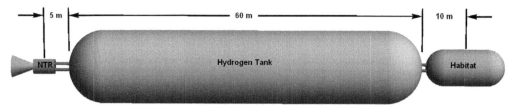

During engine operation, the liquid hydrogen (saturated at 20 K) in the tank is depleted such that the hydrogen level in the tank decreases at a rate of 2.5 cm/s. The amount of power produced by the NTR associated with the 6 MeV gamma rays is 2 kW. Assuming that the reactor can be treated as a point radiation source, determine:

    a. The whole-body dose (in REM) due to 6 MeV prompt fission gamma rays received by an astronaut located at the center of the habitat after the engine firing is complete.
    b. A plot of the 6 MeV prompt fission gamma ray dose (in REM) received by the astronaut as a function of time.
    c. The dose (in REM) due to 6 MeV prompt fission gamma rays which the astronaut would have received assuming that the hydrogen provided no shielding.
    Assume that the astronaut has a mass of 75 kg and an effective cross-sectional area of

8000 cm². Also assume the hydrogen in the fuel tank is fully depleted at the end of engine thrusting and that there is no radiation shielding other than the hydrogen in the tank. If any other assumptions are required to solve the problem, they should be clearly stated and justified. Comment on the effectiveness of using the hydrogen tank as a means of reducing the radiation dose to the crew.

CHAPTER 15

1. Derive an expression that for a given pressure vessel internal volume yields the L/D ratio which minimizes the pressure vessel mass. Assume that "L" is the length of the cylindrical portion of the pressure vessel and "D" is the diameter of the cylindrical portion of the pressure vessel.

DESIGN PROBLEM

A Mars vehicle is to be designed using a nuclear thermal rocket engine incorporating an expander cycle as its main propulsion system. The vehicle will be configured as illustrated below:

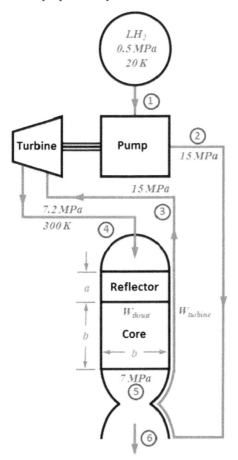
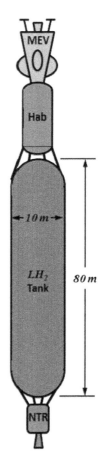

DETERMINE
- Critical core dimensions
- Reflector dimensions
- Total reactor power ($W_{total} = W_{thrust} + W_{turbine}$)
- Hydrogen mass flow rate
- Core outlet temperature (T_5)
- Nozzle outlet temperature and pressure (T_6 and P_6)
- Engine thrust level
- Engine-specific impulse
- Mass of NTR engine including core, reflector, vessel, and ancillary equipment
- Engine thrust to weight ratio
- Turbopump rotational speed
- Turbopump impeller/rotor diameter
- Turbopump pump power requirements
- Turbopump pump efficiency
- Turbopump turbine inlet temperature (T_3)
- Turbopump turbine efficiency
- Hydrogen tank mass
- Hydrogen mass
- Total vehicle mass
- Vehicle mass fraction $\left(\dfrac{m_{total}^{unfueled}}{m_{total}^{fueled}} \right)$
- Total time of engine thrusting
- One-way travel time assuming a round trip mission type

ASSUMPTIONS
- Reflector is effectively infinite.
- Neglect pressure drop through reflector.
- The reactor pressure vessel is made of stainless steel with hemispherical end caps.
- The propellant tank is made of aluminum and has hemispherical end caps.
- Turbopump pump and turbopump turbine have the same rotational speed.
- Turbopump turbine rotor diameter equals the turbopump pump impeller diameter.
- Calculate Mars mission details from Table 4.2.
- Mass of ancillary NTR components (eg, turbopump and nozzle) may be determined from:

$$m_{anc}(\text{kg}) = 1.5 \times W_{total}(\text{MW})$$

- The fuel and propellant temperatures in a cell may be related to one another through an equation of the form:

$$Q(z) = U[T_f(z) - T_p(z)]S$$

where $T_f(z)$ = axial fuel temperature distribution; $T_p(z)$ = axial propellant temperature distribution; $Q(z)$ = axial power density distribution; U = general conductance factor; and S = surface to volume ratio of the fuel.

Parameter	Symbol	Value	Units
Reflector diffusion coefficient	D_r	0.427	cm
Reflector absorption cross section	Σ_a^r	0.005	cm^{-1}
Core diffusion coefficient	D_c	1.274	cm
Core absorption cross section	Σ_a^c	0.010	cm^{-1}
Core fission cross section	Σ_f^c	0.0083	cm^{-1}
Neutrons per fission	ν	2.5	
Reference multiplication factor	k_{eff}	1.1	
Average reflector density	ρ_r	5.0	g/cm^3
Average core density	ρ_c	15.0	g/cm^3
Radius of propellant holes in fuel	r_i	0.125	cm
Equivalent radius of fuel cell	r_o	0.250	cm
Maximum fuel temperature	T_{max}	3100	K
General conductance factor	U	6.26	W/(cm^2 K)
Friction factor	f	0.01	
Specific heat ratio for hydrogen	γ	1.4	
Hydrogen specific heat	c_p	16.4	(W s)/(g K)
Universal gas constant	R_u	8.3145	(W s)/(K mol)
Gravitational acceleration	g	9.8067	m/s
Nozzle converging area ratio	A_c/A^*	5	
Nozzle diverging area ratio	A_d/A^*	300	
Reactor vessel temperature	T_v	700	K
Propellant tank wall temperature	T_t	300	K
Habitat mass (Hab)	m_{Hab}	35,000	kg
Mars excursion vehicle mass (MEV)	m_{MEV}	45,000	kg
Vehicle structural Mass	m_{VS}	15,000	kg
Earth parking orbit altitude	H_{Earth}	400	km
Mars parking orbit altitude	H_{Mars}	200	km

Appendix

Selected values for various physical constants.

Parameter	Symbol	Value	Units
Boltzmann constant	k	1.3807×10^{-23}	$\dfrac{J}{K}$
Electron mass	m_e	9.1094×10^{-31}	kg
Proton mass	m_p	1.6726×10^{-27}	kg
Neutron mass	m_n	1.6749×10^{-27}	kg
Pion mass	m_{pion}	135–140	$\dfrac{MeV}{c^2}$
Gravitational constant	G	6.6726×10^{-11}	$\dfrac{m^3}{s^2 kg}$
Plank's constant	h	6.6261×10^{-34}	J s
Speed of light in vacuum	c	2.9979×10^8	$\dfrac{m}{s}$
Planck constant × speed of light	hc	1.24	eV μm
Stefan–Boltzmann constant	σ	5.6705×10^{-8}	$\dfrac{W}{m^2 K^4}$
Temperature equivalence of 1 eV	T_{eV}	11,604	K
Avogadro's number	N	6.0221×10^{23}	
Universal gas constant	R_u	8.3145	$\dfrac{J}{K mole}$
Atomic mass unit (amu)	m_u	1.6605×10^{-27}	kg
Energy equivalent of 1 amu	E_u	1.4916×10^{-10}	J
Gravitational acceleration	g	9.8067	$\dfrac{m}{s}$
Standard temperature	T_0	273.15	K
Standard pressure	P_0	1.0133×10^5	Pa
Energy released per fission	E_f	$\sim 3.2 \times 10^{-11}$	J
Molar volume at STP	V_0	2.2414×10^{-2}	m^3
Astronomical Unit (AU)	AU	1.50×10^{11}	m
Solar intensity at 1 AU	f	1360	$\dfrac{W}{m^2}$

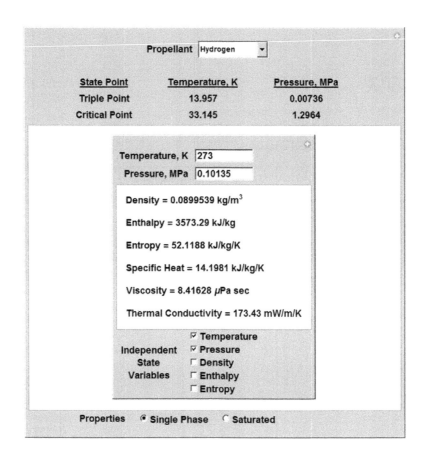
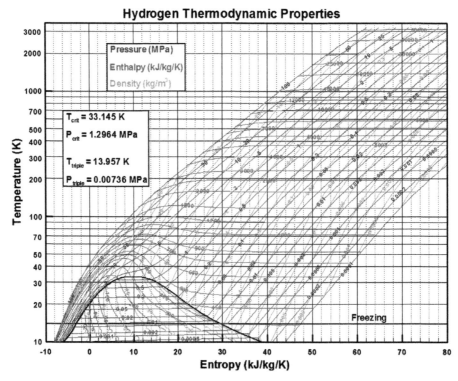

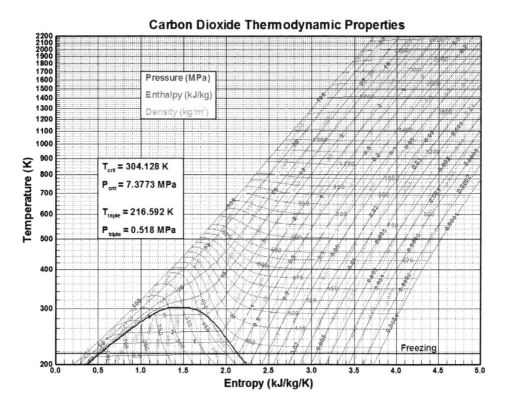

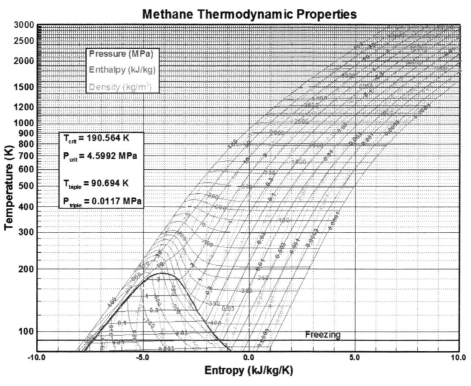

Index

'*Note*: Page numbers followed by "f" indicate figures, "t" indicate tables and "b" indicate boxes'.

A

A.A. Bochvar All-Union Research Institute of Inorganic Materials (ARIIM), 8
Absorption loss rate, 91
Acute radiation exposure, biological effects of, 231–232, 231t
Adults, individual dose limits regulations on, 232–233, 233t
^{242m}Am, 235–236
Annular Core Research Reactor (ACRR), 253, 253f
Apogee, 46
As Low As Reasonably Achievable (ALARA) principle, 233
Asymptotic rise, 173
Atomic Energy Commission (AEC), 3
Attenuation coefficient, 221, 222f
Axial flow geometry, nuclear reactor temperature and pressure distributions in, 128–138, 129f, 132f, 134f, 137f
Axial flow pumps, 151
Axial flow turbines, 157, 157f
    performance characteristics, 160f
    stress factors, 158f

B

Baikal-1, 8
Baryons, 56
Beer's law, 277
Beryllium, 66, 239–240
    properties of, 241t
Binding energy, 59–61, 60f
Biot number, 122, 177–179
BISO (Buffered ISOtopic) particles, 236, 236f
Bode plot, 191–195, 192f, 192b, 195f
Boron, 66
Boron carbide (B_4C), 241–242
Boundary layer, 123
Brayton cycle, 27–28, 27f
Breeder reactors, 203–204
Breit–Wigner formula, 71
Buckling
    geometric, 103–105, 105t
    materials, 101
Buildup factor, 222
Burnable poison, 215

C

Capture gamma, 217, 225–227
Carbide, 238, 238f
Cavitation, 152, 153f–154f
Centrifugal flow pumps, 151
Cermets, 238, 239f
Chemical rocket engine testing, 249–250
Chernobyl reactor disaster, 189–190
Chronic radiation exposure, biological effects of, 232, 232f
Cold bleed cycle, 23–24, 23f
Compressible flow relationships, 17f
Conduction, 117–122, 118f–119f, 121f
Conservation of momentum, 11, 11f
Containment vessel, heat absorption in, 298–300
Control materials, 241–242, 242f, 243t
Convection, 122–128, 123f, 126f
    processes, in nuclear reactor fuel elements, 122–128, 123f, 126f
Core, 106–110
Coulomb potential, 58–59, 58f
Criticality equation, 103–104

D

Darcy–Weisbach friction factor, 124–125
Decay heat removal considerations, 173–177, 174f, 176f
Delayed neutrons, 165–166, 168t
Depletion
    chain
        of samarium 149, 211f
        of xenon 135, 208f. *See also* Xenon 135 (^{135}Xe)
    effects, reactor operation with, 214f
Diffuser, 151
Diffusion coefficient, 95
Diffusion theory approximation, 94–95
Discrete ordinates technique, 92
Dittus–Boelter correlation, 124
Doppler broadening effect, 185, 189–190
    cross sections, 69–72
Dyson, Freeman, 259

327

E

Earth/Mars
  minimum transit times, 52f
  mission characteristics, 47f
Eigenvalue, 97, 103
Energy, 1–2
Engine ground testing, 254–256, 255f–256f
Ergun correlation, 199
Evaluated Nuclear Data File (ENDF), 67
Excited quantum nuclear states, 59, 59f
Expander cycle, 24–26, 24f
Extrapolation length, 103

F

Fast reactors, 98
Fast region, 64–67, 65f–66f
Fertile nuclides, 203–204
First law of thermodynamics, 13, 15
Fissile nuclides, 64, 203
Fission product buildup and transmutation, 203–207
Fission production rate, 90
Fission source range, neutron-scattering interactions in, 86
Flight time equations, 43–48, 44f, 47f, 48b, 52f, 53t
Flow
  subsonic, 15
  supersonic, 15
Fourier equation, 119–120
Fourier transform approach, 92
Francis-type pump, 151
Fuel. *See also* Propellant
  assembly testing, 251–254, 252f–253f
  burnup effects, on reactor operation, 213–215, 214f
  enrichment, 204
Functional tests, 249

G

Gadolinium, 242
Galactic cosmic radiation (GCR), 228–229
Gas constant, 30t
Gas core nuclear rocket, 273–289, 273f
  core temperature distribution, 277–280, 278f–279f
  neutronics, 273–277
  uranium loss rare calculations, 284–289, 285f, 288f
  wall temperature calculation, 280–284, 281f, 284f
Gell-Mann, Murray, 55
Geometric buckling, 103–105, 105t
Gibbs equation, 14
Gnielinski correlation, 124–125
Graphite, 66, 239
  crystalline structure, 240f
  reactor-grade, properties of, 240t
Grooved ring fuel elements (GRFEs), 139

H

Hadrons, 55
Hafnium, 242
Haynes 230, 243
  properties of, 244t
Head, 150
Health physics, radiation protection and, 228–233, 230t–231t, 232f, 233t
Heat absorption
  in containment vessel, 298–300
  in hydrogen propellant, 300–303, 302f
  in neon buffer layer, 295–298, 296f–297f
Heat conduction, in nuclear reactor fuel elements, 117–122, 118f–119f, 121f
Hohmann trajectory, 46
Hot bleed cycle, 21–22, 22f
Hydrogen propellant, heat absorption in, 300–303, 302f
Hyperbolic trajectories, 36–37

I

Ideal gas law, 14
Impeller, 151
Impulse turbines, 157, 158f
Inducer, 152–153
Inhour equation, 167–168, 168f
Intercooling, 28
International Commission on Radiological Protection (ICRP), 230
Interplanetary mission analysis, 31–53
  basic equations, 32–37, 35f
  flight time equations, 43–48, 44f, 47f, 48b, 52f, 53t
  patched conic equations, 37–43, 38f, 41f
Isentropic compression, 265

J

Jump. *See* Prompt jump

K

k-effective (k_{eff}), 103–104
Kelvin–Helmholtz instability, 284–289, 285f
Kepler orbits, 31
Kepler's equation, 46
Kiwi, 3
Kosberg nuclear engines, 8
Kurchatov' Atomic Energy Institute (AEI), 8

L

Lawson criterion, 77
Legendre differential equation, 92
Legendre polynomials, 92–94

Lethargy, 83
Linear dose-rate effectiveness factor model (linear DREF), 232
Linear quadratic model, 232
Linear threshold model, 232
Loss rate
    absorption, 91
    scattering, 91
Lumped parameter, 177—178

M

Mach—area relationship, 16—17, 19b
Mach number, 15—16, 149—150, 154—156
Mach—pressure relationship, 17
Marshall Space Flight Center, 252
Materials, for nuclear thermal rockets, 235—247
    control materials, 241—242, 242f, 243t
    fuels, 235—238, 236f—239f
    moderators, 239—241, 240f, 240t—241t
    structural materials, 243—246, 243t—244t, 245f—246f
Materials buckling, 101
Matter with neutron beams, interaction of, 72—74, 73f
Maxwellian distribution, 86—87, 87f—88f
Mesons, 56
Mixed flow pump, 151
Moderators, 239—241, 240f, 240t—241t
Monte Carlo method, 92
Moody chart, 125, 126f
Multigroup diffusion theory, 97—100, 98f, 99t
Multigroup neutron diffusion equations, 97—116
    multigroup diffusion theory. *See* Multigroup diffusion theory
    one group, one region neutron diffusion equation, 100—105, 101f, 105t
    one group, two region neutron diffusion equation, 106—110, 106f, 108f—109f
        core, 106—110
        reflector, 107—110
    two group, two region neutron diffusion equation, 110—116, 115f
Multilayer shield
    configuration, 219f
    radiation attenuation in, 227, 228f

N

NASA, 3, 232—233
National Council on Radiation Protection and Measurements (NCRP), 232—233
National Emission Standards for Hazardous Air Pollutants (NESHAP), 249
National Nuclear Data Center, 67, 225
Neon buffer layer, heat absorption in, 295—298, 296f—297f
Net-positive suction head (NPSH), 152—153

Neutron(s), 55
    attenuation, 219—220
    balance equation, 89—91
        absorption loss rate, 91
        fission production rate, 90
        leakage, 89—90, 90f
        scattering loss rate, 91
        scattering production rate, 90—91
        steady-state, 91
    beams with matter, interaction of, 72—74, 73f
    current, 89
    energy group structure, 99t
    fluence, 240—242
    flux, 67—69, 68f
    flux dose equivalents, 231, 231t
    flux energy distribution, 81—88
        energy distribution spectrum, 88
        in fission source range, 86
        neutron-scattering interactions, classical derivation of, 81—84, 82f—83f, 84t
        in slowing down range, 84—86, 85f
        in thermal energy range, 86—87, 87f—88f
    flux extrapolation, at reactor boundary, 102f
    leakage, 89—90, 90f
    mean free paths, 74t
Neutronics
    nuclear light bulb, 292—294
    open cycle gas core rocket, 273—277
Neutron-scattering interactions, classical derivation of, 81—84, 82f—83f, 84t
Newton's law of gravitation, 32
Newton's second law of motion, 12
Newton's third law of motion, 11
Nozzle
    characteristics, 12f
    isentropic calculations, 18t
    thermodynamics, 15—20, 17f—18f, 18t
Nuclear bomb analysis, 260
Nuclear cross sections, 63—67
    $1/V$ region, 64
    resonance region, 64
    unresolved resonance region or fast region, 64—67, 65f—66f
Nuclear electric thermodynamic cycles, 27—30
    Brayton cycle, 27—28, 27f
    Stirling cycle, 28—30, 29f
Nuclear Engine for Rocket Vehicle Applications (NERVA), 2—6, 8, 21, 23, 62—63, 117, 119—120, 119f, 121f, 122, 128—129, 138—139, 152—153, 192b, 195—197, 204, 235—239, 241—242, 249, 253—254
    applications, 4f
    core and fuel segment cluster, 6f

Nuclear Engine for Rocket Vehicle Applications (NERVA) (*Continued*)
　milestones of, 3
　test firing, 5f
　transient nuclear test, 250f
Nuclear fission, 59−63, 60f−61f, 218f, 249
Nuclear fusion, 74−80, 75f−76f, 78f−79f
Nuclear light bulb, 289−303
　containment vessel, heat absorption in, 298−300
　engine, 289f
　experimental facility, 291f
　flow diagram, 291f
　fuel cavity temperature distribution, 294−295, 295f
　hydrogen propellant, heat absorption in, 300−303, 302f
　neon buffer layer, heat absorption in, 295−298, 296f−297f
　neutronics, 292−294
　reactor with smeared core, 292f
Nuclear reactor kinetics, 165−181
　decay heat removal considerations, 173−177, 174f, 176f
　nuclear reactor transient thermal response, 177−181, 180f
　point kinetics equations
　　derivation of, 165−168, 168f, 168t
　　solution of, 169−173, 172f
Nuclear Regulatory Commission (NRC), 232−233
Nuclear Rocket Development Station (NRDS, Jackass Flats, Nevada), 253−254, 255f
Nuclear rocket engine testing, 249−257
　engine ground testing, 254−256, 255f−256f
　errors in, 251
　fuel assembly testing, 251−254, 252f−253f
　general considerations for, 249−251
Nuclear rocket stability, 183−201
　point kinetics equations, derivation of, 183−185
　reactor stability model, 185−195, 186f, 192f, 192b, 195f
　thermal fluid instabilities, 195−201, 196t, 198f, 199t, 200f
Nuclear structure, 55−59, 57f
Nuclear thermal rocket element environmental simulator (NTREES), 252, 252f
Nuclear thermal rockets (NTR), 8, 62−64, 66, 66f, 219, 227−229
　materials for, 235−247
　solid-core, 1−2
Nuclear thermal rocket thermodynamic cycles, 21−26
　cold bleed cycle, 23−24, 23f
　expander cycle, 24−26, 24f
　hot bleed cycle, 21−22, 22f
Nucleons. *See* Neutrons; Protons
Nuclides, scattering parameters for, 84t
Nusselt number, 122−123, 125

O

Obninsk Physical Energy Institute (PEI), 8
Occupational Safety and Health Administration (OSHA), 231
One group, one region neutron diffusion equation, 100−105, 101f, 105t
One group, two region neutron diffusion equation, 106−110, 106f, 108f−109f
　core, 106−110
　reflector, 107−110
$1/V$ region, 64
Open cycle gas core rocket, 273−289, 273f
　core temperature distribution, 277−280, 278f−279f
　neutronics, 273−277
　uranium loss rare calculations, 284−289, 285f, 288f
　wall temperature calculation, 280−284, 281f, 284f
Orbital energy characteristics, 36t
Orbital parameters, 35f
　for planetary arrivals, 41f
　for planetary departures, 38f
Orion, 259−272, 260f, 265f−267f, 269f, 272f

P

Parabolic trajectory, 36−37
Particle Bed Reactor (PBR), 2, 5−7, 117, 237
　friction factor coefficients for, 199, 199t
　frits and fuel particles, 7f
　fuel elements, 7f
　　thermal fluid stability of, 200−201, 200f
Patched conic approximation, 31
Patched conic equations, 37−43, 38f, 41f
Peewee reactor, 3
Perigee, 46
Phase shift, 191−192, 193b−195b
Phoebus, 3
Phoebus 2A, 3, 5f
Pions, 56−57
Pit, 260−261
　isentropic compression of, 265
　shock compression of, 265−266
Planck's law, 277−279, 278f
Planck opacity, 280
　mean opacity, 284f
Planetary arrivals, orbital parameters for, 41f
Planetary departures, orbital parameters for, 38f
Point attenuation kernel, 223, 226
Point kinetics equations
　derivation of, 165−168, 168f, 168t, 183−185
　solution of, 169−173, 172f
Point of incipient cavitation, 152
Poisson's equation, 119−120, 140
Prandtl number, 122−123
Pressure drop, 133, 134f

Production rate
 fission, 90
 scattering, 90–91
Prompt fission gamma attenuation, 220–227, 220f–222f, 223t, 225f
Prompt jump, 173
Prompt neutrons, 165
 lifetime of, 166, 173–175, 177
Propellant, 1. *See also* Fuel
Protons, 55–56, 58–59
Pseudo-nuclide, 166
^{239}Pu, 235–236
Pulsed nuclear rocket (Orion), 259–272, 260f, 265f–267f, 269f, 272f
Pump(s)
 axial flow, 151
 cavitation effects, 153f
 cavitation susceptibility of, 154f
 centrifugal flow, 151
 characteristics, 151–156, 155f
 configurations, 152f
 turbopump, 149–151, 150f
Putt-Putts, Orion flight testing using, 259, 260f

Q

Quark–Gluon interactions, 57f
Quarks, 55–56
 flavor properties, 56t

R

Radial flow geometry, nuclear reactor fuel element temperature distributions in, 138–143, 139f, 143f
Radial turbines, 157f
 performance characteristics, 160f
 stress factors, 159f
Radiation attenuation, in multilayer shield, 227, 228f
Radiation protection and health physics, 228–233, 230t–231t, 232f, 233t
Radiation shielding, for nuclear rockets, 217–234
 formula derivation, 217–227
  neutron attenuation, 219–220
  prompt fission gamma attenuation, 220–227, 220f–222f, 223t, 225f
  radiation attenuation in multilayer shield, 227, 228f
 multilayer shield configuration, 219f
 radiation protection and health physics, 228–233, 230t–231t, 232f, 233t
Radiators, 21, 143–148, 144f, 146f–147f
Radioactive fission. *See* Nuclear fission
RD-410, 2, 8, 9f
Reactions rates, 67–69
Reactivity temperature coefficient, 185, 189–190

Reactor kinetics transfer function, 185, 186f
Reactor stability model, 185–195, 186f, 192f, 192b, 195f
Recuperation, 28
Reflector, 107–110
 critical length, 108f
 savings, 108–109
Relative biological effectiveness (RBE), 230, 230t
Relative roughness, 124–125
Research Institute of Heat Releasing Elements (RIHRE), 8
Research Institute of Thermal Processes (RITP), 8
Resonance region, 64
 unresolved, 64–67, 65f–66f
Reynolds number, 122–125, 149–150, 154–156, 196, 199–201
Rocket(s)
 engine fundamentals, 11–20
 engine testing, 249–257
 equation, 13
 gas core nuclear, 273–289, 273f, 278f–279f, 281f, 284f–285f, 288f
 materials for, 235–247
 nozzle characteristics, 12f
 nuclear thermal, 1–2, 8
 pulsed nuclear, 259–272, 260f, 265f–267f, 269f, 272f
 radiation shielding for, 217–234
 Russian nuclear, 8
 stability, 183–201
 thermal fluid aspects of, 117–148
Roentgen Equivalent Man (REM), 230–232
Rosseland opacity, 279, 279f
ROVER, 3, 8
Russian nuclear rockets, 8
Rutherford, Ernest, 55

S

Samarium 149 (^{135}Sm), 203, 204f
 depletion chain of, 211f
 poisoning, 211–213
 during reactor operation and shutdown, transient behavior of, 213f
Scattering loss rate, 91
Scattering production rate, 90–91
Schwarzschild's equation, 280
Second law of thermodynamics, 14
Serber, Robert, 264
Serber equation, 264
Shadow shields, 217–218
Shock compression, 265–266
Slowing-down range, neutron-scattering interactions in, 84–86, 85f
Solar proton events (SPEs), 228–229
Solid-core NTR engines, 1–2

Space Nuclear Propulsion Office (SNPO), 3
Space shuttle main engine (SSME), 19–20
Specific diameter, of turbopumps, 149–150, 154–156
Specific impulse, 13–15, 18f, 19b, 31
Specific speed, of turbopumps, 149–150, 154–156
Spherical harmonics method, 92
Square lattice reactor geometry, 98f
Stanton number, 122
Static tests, 250
Steady-state neutron balance equation, 91
Stefan–Boltzmann equation, 144
Stirling cycle, 28–30, 29f
Strategic Defense Initiative Organization, 5
Structural materials, 243–246, 243t–244t, 245f–246f
Sublett, Carey, 264
Subsonic flow, 15
Subsurface active filtering of exhaust (SAFE), 254, 255f
Suction-specific speed, 154
Super prompt critical state, 173
Supersonic flow, 15

T

T6061 aluminum, properties of, 244t
Tamper, 260
Taylor, Ted, 259
Temperature distribution, 117–118, 121f
Thermal energy range, neutron-scattering interactions in, 86–87, 87f
Thermal feedback, 185–195
Thermal fluid aspects, of nuclear rockets, 117–148
   axial flow geometry, nuclear reactor temperature and pressure distributions in, 128–138, 129f, 132f, 134f, 137f
   convection processes in nuclear reactor fuel elements, 122–128, 123f, 126f
   heat conduction in nuclear reactor fuel elements, 117–122, 118f–119f, 121f
   radial flow geometry, nuclear reactor fuel element temperature distributions in, 138–143, 139f, 143f
   radiators, 143–148, 144f, 146f–147f
Thermal fluid instabilities, 195–201, 196t, 198f, 199t, 200f
Thermal reactors, 98
Thermodynamics
   first law of, 13, 15
   nozzle, 15–20, 17f–18f, 18t
   second law of, 14
Thoma cavitation parameter, 153–154
316L stainless steel, properties of, 243t
Thrust, 11–12, 11f
Timberwind, 5–6, 21, 23
TNT, 266
Total exhaust containment, nuclear rocket engine testing with, 254–256, 256f
Total mission velocity, 46–48
Transmutation, 203–207
Transpiration heat transfer correlation, 125–127
Transport theory, 92–94
TRISO (TRIiSOtopic) particles, 237, 237f
Turbine characteristics, 157–163, 157f–160f
Turbomachinery, 149–163
   pump characteristics, 151–156, 152f–155f
   turbine characteristics, 157–163, 157f–160f
   turbopump. *See* Turbopump
Turbopump, 21, 23–24, 149–151, 150f
Two group, two region neutron diffusion equation, 110–116, 115f

U

Universal gas law, 133
Unresolved resonance region, 64–67, 65f–66f
Uranium
   ^{235}U, 235–236
      compression, 265–266, 266f–267f
      loss rare calculations, 284–289, 285f, 288f
      pulse unit explosive yields for, 265f
      Rosseland opacity, 279, 279f
   ^{238}U, transmutation chain for, 204–205, 205f, 208f
Uranium dioxide (UO_2), 235–236

V

Volute, 151

W

Wood's correlation, 125

X

Xenon 135 (^{135}Xe), 203, 204f
   depletion chain of, 208f
   poisoning, 207–210
   during reactor operation and shutdown, transient behavior of, 208f

Y

Yield stress, 244–245
Yukawa potential, 57–59, 58f

Z

Zero-power criticals, 250
ZrC coating, 237–238
Zirconium hydride, 241
Zweig, George, 55

Edwards Brothers Malloy
Ann Arbor MI. USA
October 27, 2016